Engineering: Modeling and Simulation Technology

Volume I

Edited by **Tommy Haynes**

*C*LANRYE
*I*NTERNATIONAL

New Jersey

Published by Clanrye International,
55 Van Reypen Street,
Jersey City, NJ 07306, USA
www.clanryeinternational.com

Engineering: Modeling and Simulation Technology
Volume I
Edited by Tommy Haynes

© 2015 Clanrye International

International Standard Book Number: 978-1-63240-212-7 (Hardback)

Printed in the United States of America.

Contents

Preface

The field of modeling and simulation in engineering focuses on providing an opportunity for the discussion of methodologies, simulation tools, and formalisms that intended to support the new, much broader understanding of engineering. Modeling and simulation can be called a discipline on its own. In such a context the conventional interpretation of the engineering field related to actual assembly needs to be expanded to incorporate the integration of outsourced components and the consideration of human, economical and logistic factors in the designing of engineering services and products. The numerous application domains of modeling and simulation often leads to the assumption and misconception that M&S is a pure application discipline, though this is not the case as it is now being recognized by engineers around the world. It is being ensured that the results of such modeling and simulations are applicable in the real world and the engineers work towards understanding the conceptualizations, assumptions and implementation restrictions of this emerging field. Modeling and simulation in engineering reports on and draws from leading-edge scientific contributions from computer science, mathematics, numerous branches of engineering, management as well as psychology and cross-cultural communication, all of which are mostly human-centered engineering systems.

This book is an attempt to compile and collate all available research on modeling and simulation in engineering under one aegis. I am grateful to those who put their hard work, effort and expertise into these researches as well as those who were supportive in this endeavor. I also wish to acknowledge the efforts of the publishing team in the completion of this book. Lastly, I wish to thank my family for their efforts and support.

Editor

Simulation Modeling for Analysis of a (Q, r) Inventory System under Supply Disruption and Customer Differentiation with Partial Backordering

Parham Azimi, Mohammad Reza Ghanbari, and Hasan Mohammadi

Department of Industrial and Mechanical Engineenng, Qazvin Branch, Islamic Azad University, Barajin, Daneshgah St., Nokhbegan Blvd., P.O. Box 34185141, Qazvin, Iran

Correspondence should be addressed to Parham Azimi, p.azimi@yahoo.com

Academic Editor: Azah Mohamed

We have modeled a new (Q, r) inventory system which involves a single product, a supplier, and a retailer with customer differentiation under continuous review inventory policy. The supplier provides the retailer with all requirements, and the retailer sells products to the customers. The supplying process is randomly subject to disruptions. Partial backordering is applied when a stock out occurs, and customer can select either to leave the system without purchasing or to backorder products. The customers are categorized into two main classes regarding to their backordering probabilities. The main contribution of this paper is including the customer differentiation in the inventory model. We used simulation technique to verify the impact of supply disruptions and customer differentiation and carried out sensitivity analysis. To test the performance of the model, we have compared our model to one from the latest related research. As the results show, the average of total annual cost of the (Q, r) inventory system is lower than that of the previously developed models such as (r, T) inventory systems.

1. Introduction

Within the last years, inventory management has received wide attention in such a way the recent researches in this field have focused on the study of inventory system in the presence of supply chain problems such as supply disruption. Various factors can disrupt a supply chain system, including an equipment breakdown, a strike, bad weather, natural disasters, political instability, traffic interruptions, terrorism, and so on [1]. For example, on March 17, 2000, lighting hit a power line in Albuquerque, New Mexico, which in turn started a fire at a local plant owned by Royal Philips Electronics, damaging millions of microchips.

The representative literature of inventory management with respect to supply disruptions includes [2–4]. In general, the previous literature divides inventory systems into two categories. One is continuous-review-based- and another is periodic-review-based. Parlar [5] considers a continuous-review stochastic inventory problem with random demand and random lead time in the situation where supply may be disrupted. Gürler and Parlar [6] make further research contribution by considering an additional randomly available supplier in the problem that Parlar [5] addresses. Arreola-Risa and DeCroix [7] studied inventory management under random supply disruptions and partial backorders, with an (s, S) policy being considered. Mohebbi [8, 9] assumes that the sales are lost when the stock is out. The references to periodic review aspect include [10, 11]. Parlar et al. [10] analyze a finite-horizon periodic-review inventory model with backlogging. Samvedi and Jain [12] studied the impact of changes in the parameter values of periodic inventory policy on supply disruption situations. The process is simulated using discrete event simulation with the inventory and backorder levels taken as the output parameters. The study shows that there is a definite connection between the costs experienced at a level in the chain and its distance from the disruption point. Due to the difficulty of handling partial backorders, the inventory literature in this area

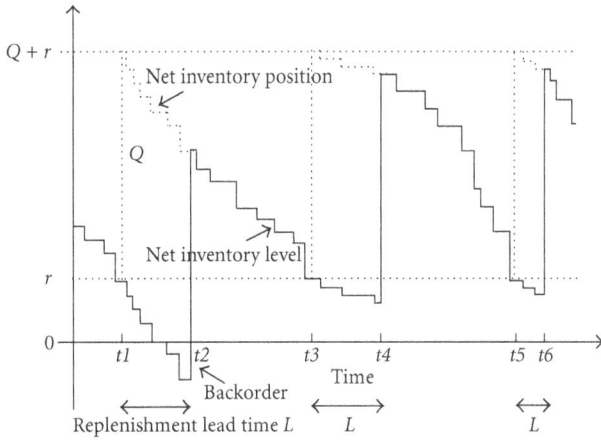

FIGURE 1: A standard (Q, r) inventory policy without supply disruption.

is limited. Moinzadeh [13] considers a base-stock level inventory system with Poisson demand, constant resupply times, and partial backorders. Other inventory models with partial backorders can be found in Montgomery et al. [14], Kim and Park [15], and Posner and Yansouni [16]. Parlar and Berkin [17] study the classic EOQ problem with supply disruptions. Parlar and Perry [18] extend this analysis to a system, where orders may be placed before the inventory level reaches zero, and where there is a fixed cost for determining the state of the supplier. Weiss and Rosenthal [19] determine the optimal inventory policy when the timing (but not the duration) of supply disruptions is known in advance. To our knowledge, the only papers dealing with supply disruptions and random demand are [2, 5, 16, 17, 20, 21].

In this study, we consider a continuous-review inventory system in the presence of supply disruptions, with the relaxation of some assumptions that are made in the previous researches. For instance, we consider backordering in the stockout situations, which means that customers can choose to backorder unfulfilled products or not. The assumption relaxation makes the problem more realistic. In addition, customer differentiation which is included in the proposed model has not been considered in the previous continuous-review inventory systems. Due to the complexity of the problem, we have used simulation modeling to develop the mentioned inventory system. Furthermore, we investigate the impacts of supply disruptions and customer differentiation on the inventory system.

In Section 2, there is a description of the inventory policy that the retailer adopts and the considered problem. In Section 3, the process of simulation modeling for the concerned inventory system is explained. In Section 4, the simulation output is examined to determine the impacts of supply disruption and customer differentiation on the inventory system, and finally in Section 5, summary of the results and future opportunities are explained.

2. Problem Description

This paper considers a continuous-review inventory system with single product of a retailer, where supply may be

disrupted. The supplier is not always available. The retailer sells products to the customers and replenishes the stock from its single supplier. When a supply disruption occurs, the supplier cannot fulfil the orders from the retailer. Only when the disruption issue is resolved can the orders be processed. We define the time period during which the supplier is available (i.e., under normal conditions) as its ON period, and the time period during which the supplier is not available (i.e., under disruption conditions) as its OFF period. ON and OFF periods represent the frequency and duration of supply disruptions, respectively. In other words, ON and OFF periods reflect the disruption severity of an unreliable supplier. The longer the ON periods, the less frequent the disruptions and the slighter the disruptions. On the contrary, the longer length of OFF periods, the longer the disruption duration and the more severe the disruptions. The standard (Q, r) policies used when the supplier is available (ON), that is, when the inventory position reaches the reorder point r, Q units are ordered to raise the inventory position. The form of the policy changes when the supplier becomes unavailable (OFF) in which case orders cannot be placed when the reorder point r is reached. However, as soon as the supplier becomes available again one orders enough to bring the inventory position up. In this paper, we use different combinations of mean values of ON and OFF periods to represent different supply disruption scenarios. In addition, in this paper replenishment lead time is deemed to be stochastic, which is consistent with the reality.

The retailer adopts a continuous-review inventory policy (Q, r), where r is reorder point (number of parts on hand when we placed an order), and Q is reorder quantity. This policy means that, continuously, the retailer reviews its product inventory position and compares it with r value and decides whether a replenishment is needed or not. When the inventory position is equal or less than r, the retailer orders Q units of product. Figure 1 shows a standard (Q, r) inventory policy that does not consider supply disruptions. Times $t1$, $t3$, and $t5$ are the points where inventory position reaches the r, and an order is placed to the supplier. $t2$, $t4$, and $t6$ are time points when the ordered products are received by the retailer. Time periods $t2$-$t1$, $t4$-$t3$, and $t6$-$t5$ are three realizations of stochastic replenishment lead time L.

Figure 2 shows an (Q, r) inventory system where supply disruptions are taken into account. As seen in this figure, the red line segments on time axes represent OFF periods of the supplier, and other line segments on the axes represent ON periods. Similar to those in Figure 1, time points $t1$, $t3$, and $t5$ are three reorder points. However, the orders placed at these time points receive different treatments. At time points $t1$ and $t3$, the supplier is available (i.e., in ON periods), and the orders are processed and shipped out immediately. The retailer receives the products at time points $t2$ and $t4$, respectively. However, at time point $t5$, the supplier is in an OFF period (i.e., under disruption status), so the order cannot be processed until the supplier restores to its normal status. Hence, the order is processed and shipped out after the OFF period ends, which occurs at time point $t6$. The retailer finally receives the products at time point $t7$. Similarly, time periods $t2$-$t1$, $t4$-$t3$, and $t7$–$t5$

Simulation Modeling for Analysis of a (Q, r) Inventory System under Supply Disruption and Customer
Differentiation with Partial Backordering

3

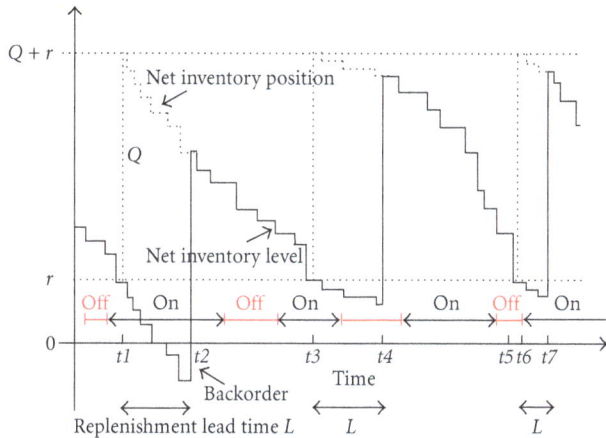

FIGURE 2: A (Q, r) inventory policy with supply disruption.

are three realizations of stochastic replenishment lead time L. Comparing Figure 2 with Figure 1, it is obvious that supply disruptions delay order replenishments.

In this paper, we study partial backordering in the stockout situations. When the retailer is out of stock, a customer may choose to backorder the products he/she needs or to abandon the purchase order and leave for other sellers (i.e., lost sale). The retailer incurs backorder cost or lost sale cost accordingly. In general, unit backorder cost per time unit is less than unit lost sale cost, since the retailer may obtain profits from selling backorders.

In addition, we allow the number of outstanding orders to be more than one. We also incorporate customer differentiation in the discussed inventory system. Customers are segmented based on their backorder probabilities in the stockout situations and are differentiated into two classes. One class has higher backorder probability, while the other class has lower backorder probability. For convenience, these two classes are denoted by classes I and II, respectively. To acknowledge class I for their higher backorder probability, the retailer provides them with high priority to receive backorders.

All the above considerations, combined with the complex nature of a continuous-review inventory system, make it very difficult to study this inventory management problem by using an analytical method. In this paper, we will utilize simulation techniques [22] to investigate the concerned inventory system. The used measure for performance is the average annual total cost of the retailer, which includes annual ordering cost, annual inventory holding cost, annual backorder cost, and annual lost sale cost. The following is the to-be-used notations [23] (see Nomenclature).

The calculation formulas of the annual ordering cost, annual holding cost, annual backorder cost, and annual lost sale cost from each customer class are as follows:

$$\text{AOC} = s * \text{number of ordering during } T$$

$$\text{AHC} = \int_0^T h * \max(\text{IL}(t), 0) dt,$$

$$\text{ABC}_i = \int_0^T b_i * \text{BoQ}_i(t) dt, \quad i \in \{\text{I}, \text{II}\},$$

$$\text{ALC}_i = l_i * \text{NLS}_i, \quad i \in \{\text{I}, \text{II}\}.$$

$$(1)$$

Therefore, the annual total cost of the retailer (i.e., the sum of the above costs) is

$$\begin{aligned}
\text{ATC} &= \text{AOC} + \text{AHC} + \text{ATBC} + \text{ATLC} \\
&= \text{AOC} + \text{AHC} + \sum_{i \in \{\text{I}, \text{II}\}} \text{ABC}_i + \sum_{i \in \{\text{I}, \text{II}\}} \text{ALC}_i.
\end{aligned} \quad (2)$$

3. The Simulation Model

The structure of the model is made up of two subsystems: customer demand subsystem and inventory replenishment subsystem, as shown in Figures 3 and 4, respectively. The two subsystems can be realized by using Enterprise Dynamics 8 (ED. 8) simulation software.

3.1. Customer Demand Subsystem. Figure 3 describes the customer demand subsystem. When a customer arrives, the retailer checks its net inventory level. There exist three situations based on net inventory level. The first situation is that the net inventory level is positive, and there is enough stock to satisfy the demand of the customer. Under this situation, the customer purchases the products with satisfaction. The retailer then updates its net inventory level and inventory position accordingly.

The second situation is that the net inventory level is positive, but there is no enough stock for the customer's demand. Under this situation, the customer takes all available products and decides whether to backorder the unfulfilled products or not. The third situation is that the net inventory level is nonpositive, and there are no products available at all. Under this situation, the customer can either backorder the unfulfilled products or leave without ordering. In the second and third situations, we differentiate customers since different customer classes have different backorder probabilities.

If a customer chooses to backorder, the net inventory level and inventory position of the retailer decrease by the demand size of the customer. Besides, the backorder quantity of the corresponding customer class is equal to minus net inventory level if it is the second situation or increases by the demand size of the customer if it is the third situation. Now consider the other case. If the customer chooses not to backorder, part of the sale is lost when it is the second situation or the entire sale is lost when it is the third situation. The resulting lost sale cost is calculated accordingly, as shown in Figure 3. Note that the number of lost sale cost is the demand size of the customer minus current net inventory level in the second situation, and it is equal to the demand size of the customer in the third situation. In addition, if it is the second situation, the inventory position decreases by the current net inventory level which is the quantity of all

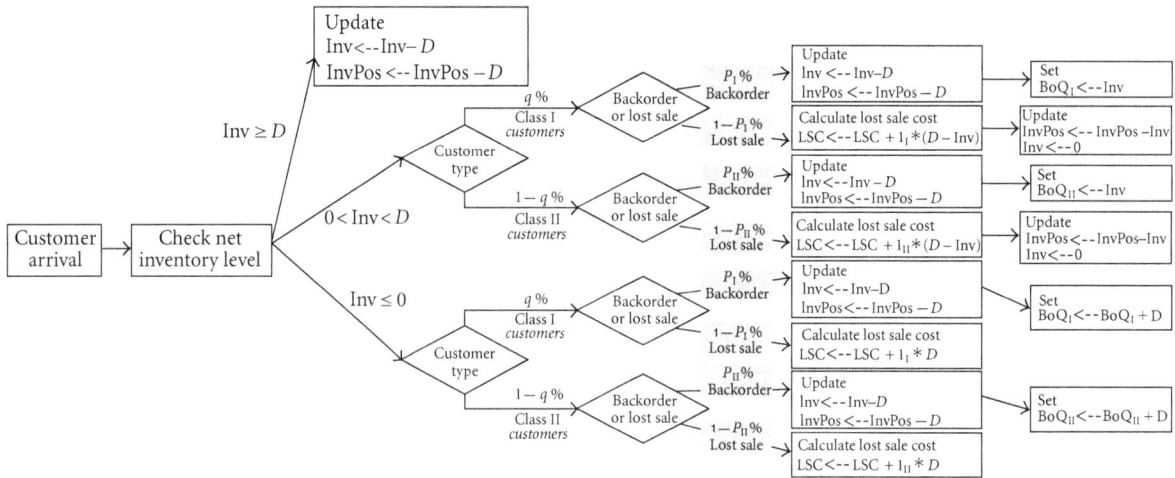

Notations:

Inv: net inventory level
InvPos: inventory position
D: demand size of a customer
LSC: lost sale cost

li: unit lost sale cost for class I customers
BoQi: backorder quantity of class I customers
a <-- b: set the value of a to be equal to b

FIGURE 3: Customer demand subsystem.

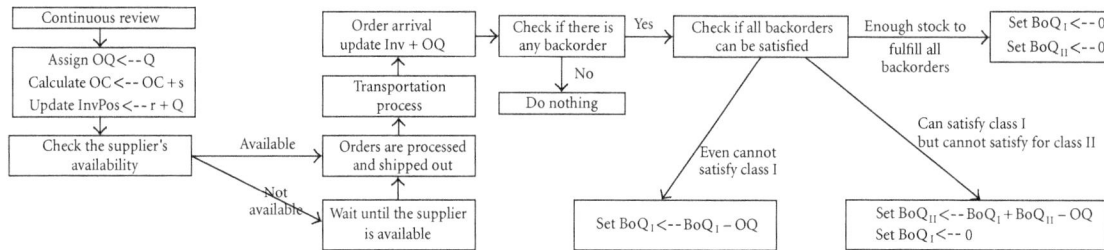

Notations:

Inv: net inventory level
InvPos: inventory position
s : setup cost for each order placement
OQ: order quantity

OC: ordering cost
BoQi: backorder quantity from customer class I
a <-- b: set the value of a to be equal to b

FIGURE 4: Inventory replenishment subsystem.

products on stock. The net inventory level is subsequently set to be zero, since the customer takes all available products [23].

3.2. Inventory Replenishment Subsystem.

Figure 4 describes the inventory replenishment subsystem. Continuously, the retailer reviews its inventory position and determines whether replenishment is needed or not. According to the adopted (Q, r) inventory policy, the inventory position needs to be increased to $r + Q$. Therefore, the needed product quantity (i.e., order quantity) is equal to Q. The corresponding ordering cost is then calculated as shown in Figure 4. When an order is placed, the availability of the supplier needs to be checked. If the supplier is in its

normal condition, it processes the order immediately. If the supplier encounters a disruption at that time, the order has to wait for being processed until the supplier restores to its normal status. After being processed, the order is shipped out. Going through the transportation process, the order arrives at the retailer, and the net inventory level is increased correspondingly. If there exist unfulfilled backorders, the retailer needs to fulfil them. Three situations need to be considered. The first situation is that the stock is enough for all backorders to be fulfilled. Mathematically, that is, $OQ \geq BoQ_I + BoQ_{II}$, where OQ and BoQ_i (i = I; II) represent order quantity and backorder quantity from customer class i, respectively. The second situation is that the backorders from customer class I can be satisfied but only part of the backorders from customer class II can be satisfied. That is,

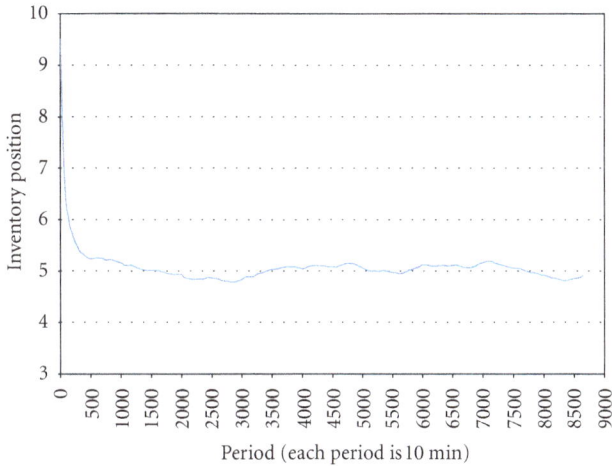

FIGURE 5: Determining warmup period.

TABLE 1: Empirical distribution of customer's demand size.

Demand size	1	2	3	4	5
Probability	0.1	0.25	0.3	0.25	0.1

TABLE 2: The design for the first experiment.

Parameter	Values (units: days)
u	20; 60; 120
v	1; 5; 10

$BoQ_I \leq OQ < BoQ_I + BoQ_{II}$. The third situation is that the arriving products even cannot satisfy the backorders of customer class I (i.e., $OQ < BoQ_I$) [23].

3.3. The Input Data of the Model. In this section, we present the input data of the model. The customer demand of the retailer is assumed to follow an empirical distribution, where customer interarrival time follows an exponential distribution with a mean of $\lambda = 0.2$ days, and each customer's demand size D has a probability distribution as shown in Table 1. Without loss of reasonability, we make the following assumptions. The setup cost for each order placement is $s = \$10$; unit product cost is $c = \$10$; unit holding cost per time unit is $h = \$2$; unit backorder costs per time unit from customer classes I and II are $b_I = \$1.8$ and $b_{II} = \$1.5$, respectively; unit lost sale costs from customer classes I and II are $l_I = \$4$ and $l_{II} = \$3$, respectively. The supplier's ON and OFF periods are supposed to follow exponential distributions, with means being u days and v days, respectively. Note that u and v are used to represent the magnitude of supply disruptions. The transportation duration of an order is assumed to follow a normal distribution with a mean of 4 days and a standard deviation of 0.5. The initial net inventory level and inventory position of the retailer are arbitrarily set to be $IL_0 = IP_0 = 10$. Such settings prevent the initial inventory status from being unrealistically "empty and idle." Later we will warmup the simulation model to remove the influences that the initial settings bring about. The inventory policy parameters r and Q are decision variables. Their values are to be determined by the experiments.

3.4. Warmup Period. In the beginning of the simulation, the model is empty without any inventory. Therefore, the

data obtained from that may not be appropriate criteria for analysis. To avoid this matter, a period of time is taken into account for the model as the warmup period. In this study we have used the Welch method [11]. The index we have used here is the average of the inventory position. By drawing the graphic diagram of the moving average of the index calculated it was determined that after period 1500 (each period is 10 minutes), the model shows a stable behaviour against the index under consideration. Therefore, in the analysis of the model we will suppose $1500 * 10$ minutes $= 15000$ minutes for the warmup period. Figure 5 determines that the system becomes stable after period 1500.

4. Experiment and Simulation Result Analysis

In this section, we simulate the inventory system according to the model described in Sections 2 and 3. We also conduct experiments for investigating the impacts of supply disruptions and customer differentiation on the inventory system. We design several scenarios for the experiments. For each scenario, the optimal inventory policy (Q, r) and the corresponding minimum average annual total cost are obtained. Because the goal of the retailer is to minimize its annual total cost, the minimum average annual total cost is taken as performance measure for each scenario. We then examine the obtained minimum average annual total costs from the experiments, expecting to discover the influences of the above two factors on the inventory system and to obtain some managerial insights for the retailer.

4.1. Experimental Design and Simulation Settings. To investigate the impact of supply disruptions on the inventory system, we conduct the following experiment. We reasonably assume that 10% of customers belong to class I, and that when a stockout occurs, 80% of customer class I and 10% of customer class II choose to backorder, that is, $q = 10\%$; $p_I = 80\%$; $p_{II} = 10\%$. We use u and v to construct different scenarios of supply disruptions, as listed in Table 2. Since u represents the frequency of supply disruptions, three values of u denote severe, moderate, and slight disruptions, respectively. Similarly, as the indicator of supply disruption duration, three values of v denote slight, moderate, and severe supply disruptions, respectively. We combine u and v and generate 9 scenarios. For each scenario, we utilize simulation techniques to obtain optimal reorder point r^* and order quantity Q^*, so that such an inventory policy can lead to the minimum average annual total cost of the retailer.

On the other hand, when looking into the impact of customer differentiation on the inventory system, we consider two cases of supply disruptions for the sake of completeness: $\{u = 60; v = 1\}$ and $\{u = 60; v = 10\}$.

TABLE 3: The design for the second experiment.

Parameter	Values
q	5%; 10%; 20%; 40%
p_{I}	60%; 90%
p_{II}	5%; 20%

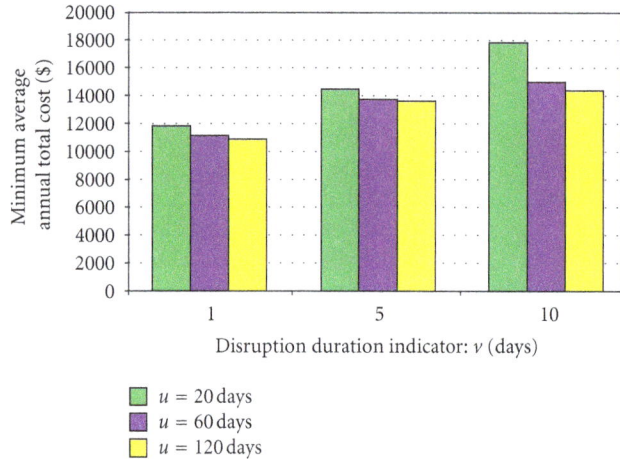

FIGURE 6: The impact of u on ATC under different scenarios of v.

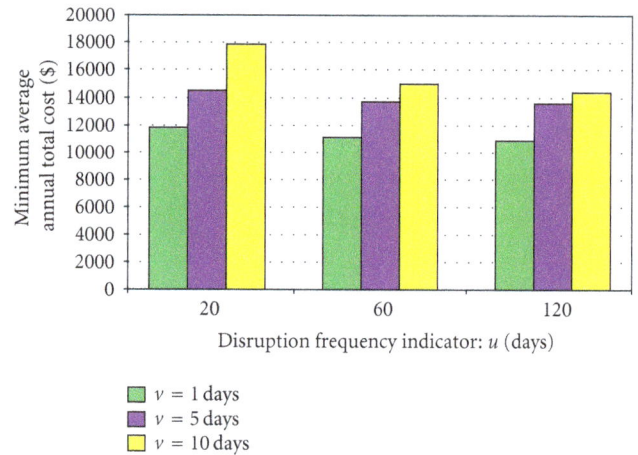

FIGURE 7: The impact of v on ATC under different scenarios of u.

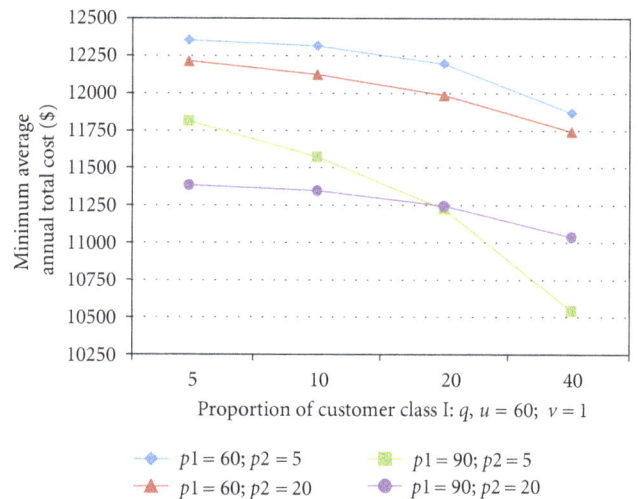

FIGURE 8: The impact of customer differentiation on ATC under the disruption scenario of $\{u = 60; v = l\}$.

Under each case, we take into account different scenarios of customer differentiation as shown in Table 3. The reason we do not consider bigger values for q (e.g., 60% and 80%) is that, in real life the customers who would choose to backorder only take a small fraction and not to mention the fraction of customer class I who have large probability to backorder products. In addition, based on the definitions, p_{I} would take bigger values, and p_{II} would take smaller values. We consider two values for p_{I} and p_{II}, respectively. Totally in this experiment, for each supply disruption case, 16 scenarios of customer differentiation are studied. As in the above experiment regarding supply disruptions' impact, for each scenario, optimal reorder point r^* and order quantity Q^* are obtained for the minimum average annual total cost of the retailer.

As we showed in Section 3.4, the warmup period is obtained to be equal to 15000 minutes. In the experiments, furthermore, for each scenario in each experiment, we utilize the optimization tool Opt-Quest in the ED simulation software to obtain the optimal inventory policy (Q^*, r^*) and the minimum average annual total cost. When conducting optimization, for each possible combination of (Q, r), we run $n = 10$ replications. For each combination of (Q, r), the average annual total cost of the retailer is derived from the data generated from 10 replications. For each scenario in each experiment, the obtained minimum average annual total cost is recorded as the result.

4.2. Results and Analysis. The experimental results are illustrated in Figures 6–9. Figure 6 shows the impact of disruption frequency indicator u on the minimum average annual total cost, based on different scenarios of disruption

duration indicator v. We can see that, for each v, the minimum average annual total cost of the retailer decreases in u. Moreover, the decrease magnitude increases in v. This figure indicates that, given a fixed mean value of disruption duration (i.e., v); less frequent disruptions (i.e., bigger u) lead to smaller minimum average annual total cost. Figure 6 also implies that, when disruption duration is short, there are no big differences regarding the impacts of different disruption frequency values on the minimum average annual total cost. The differences are more significant when disruption duration is longer.

Figure 7 illustrates the impact of disruption duration indicator v on the minimum average annual total cost, based on different scenarios of disruption frequency indicator u. It is obvious that, for each u, the minimum average annual total cost of the retailer increases in v. Moreover, the increase magnitude decreases in u. This figure reveals that, when disruption frequency is given, shorter disruption

Simulation Modeling for Analysis of a (Q, r) Inventory System under Supply Disruption and Customer
Differentiation with Partial Backordering

7

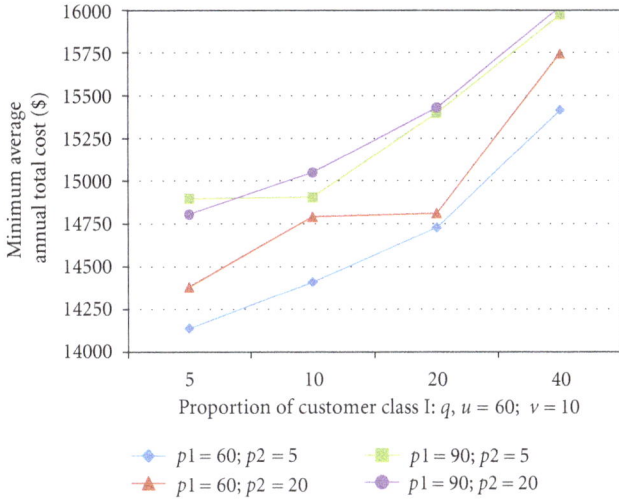

FIGURE 9: The impact of customer differentiation on ATC under the disruption scenario of $\{u = 60; v = 10\}$.

duration leads to smaller minimum average annual total cost. Figure 7 also shows that, the impacts of different disruption duration values on the minimum average annual total cost differ significantly when disruption frequency is large (i.e., small u). The differences become smaller when disruption frequency gets smaller (i.e., larger u).

Figures 6 and 7 tell the retailer that, two supply disruption magnitude indicators u and v play important roles in selecting a supplier. In the real world, a supplier is more or less subject to supply disruptions. A supplier with less frequent disruptions (i.e., bigger u) or shorter disruption duration (i.e., smaller v) is a better choice for the retailer. In addition, assume that a group of suppliers are to be selected. If these suppliers have the same disruption frequency that is little enough, then there are no big differences among the selection of these suppliers even if their disruption duration values differ dramatically. Similarly, if these suppliers have the same disruption duration that is short enough, then there are no big differences among the selection of these suppliers even if their disruption frequency values differ remarkably.

Figure 8 demonstrates the impact of customer differentiation on the minimum average annual total cost under the disruption scenario of $\{u = 60; v = 1\}$. As stated in Section 4.1, four possible q values are considered. For each q value, four scenarios of the combination of p_I and p_{II} are investigated. Figure 8 shows that, whatever scenario of the combination of p_I and p_{II}, the minimum average annual total cost decreases in q. This implies that, when the supply disruption scenario is $\{u = 60; v = 1\}$, the more customers from class I, the smaller the minimum average annual total cost of the retailer. Therefore, the retailer should attract more customers to join class I. Furthermore, from Figure 8 we can find that, when q and p_I are fixed, the minimum average annual total cost in the case of $p_{II} = 20\%$ is smaller than that in the case of $p_{II} = 5\%$, except the case for the situation of $p_I = 90\%$, $p_{II} = 5\%$ and, $q = 40\%$. Besides, when q and p_{II} are fixed, the minimum average annual total cost in the case of

$p_I = 90\%$ is smaller than that in the case of $p_I = 60\%$. Both findings imply that when the proportion of customer class I and the backorder probability of the one class are fixed, the more the customers from the other class who choose to backorder in the stockout situations, the less the minimum average annual total cost. Figure 8 also reveals that when $p_I = 60\%$, whatever p_{II} value, the decrease magnitude of the minimum average annual total cost in q is slight. However, when $p_I = 90\%$, the decrease magnitude is large. This reflects that the larger the backorder probability of class I, the more significant the decrease of the minimum average annual total cost with the increase of the proportion of class I.

Figure 9 exhibits the impact of customer differentiation on the minimum average annual total cost under the disruption scenario of $\{u = 60; v = 10\}$. Likewise, four q values and four scenarios of the combination of p_I and p_{II} are considered. Figure 9 shows that whatever scenario of the combination of p_I and p_{II}, the minimum average annual total cost increases in q. This implies that when the supply disruption scenario is $\{u = 60; v = 10\}$, the less the customers from class I and the smaller the minimum average annual total cost of the retailer. Therefore, the retailer should reduce the number of class I customers. This is totally contrary to the above situation, where the supply disruption scenario is $\{u = 60; v = 1\}$. This indicates the influence of supply disruptions on customer differentiation's impact on the inventory system. Figure 9 also shows that when q and p_I are fixed, the minimum average annual total cost in the case of $p_{II} = 5\%$ is smaller than that in the case of $p_{II} = 20\%$. This is not the case for the situation of $p_I = 90\%$ and $q = 5\%$. Besides, when q and p_{II} are fixed, the minimum average annual total cost in the case of $p_I = 60\%$ is smaller than that in the case of $p_I = 90\%$. These observations indicate that for the retailer, when the proportion of customer class I is fixed, for either value of p_{II}, smaller backorder probability of customer class I leads to smaller minimum average annual total cost. In addition, when the proportion of customer class I is fixed and $p_{II} = 5\%$, smaller backorder probability of customer class I leads to smaller minimum average annual total cost.

Now, we want to compare the results of the experiments to the one developed by Li and Chen [23]. The only difference is that they have used periodic-review model, while here we have developed a continuous-review model. First of all, the supply disruptions have been compared between the two models. In Table 4 and Figure 10, the average annual costs of the two models have been shown for different values of u and v. For each scenario, the model has been replicated 30 times, and the averages have been listed in the table.

As the results show, the continuous-review model dominates the other model in all scenarios with different values of u and v (P value is 0.0001). The main reason is that the backorders and the lost sales costs are lower in the continuous model. Now, we compare the two models based on backorder and lost sales rates. The results have been summarized in Table 5 and Figure 11.

As the results show, the average annual costs (ATC) of the continuous-review model is lower that the periodic one in all scenarios (P value is 0.0001).

TABLE 4: Compare current model with previous study under different scenarios of v and u.

Scenario number	Q	$p1$	$p2$	Disruption		Average annual total cost	
				u	v	Current model (Q, r)	Previous model (r, T)
1				120	1	10891	13900
2				60	1	11127	13900
3				20	1	11813	13950
4				120	5	13608	14150
5	10	80	10	60	5	13731	14400
6				120	10	14372	14800
7				20	5	14479	15750
8				60	10	14973	15800
9				20	10	17826	18700

TABLE 5: Compare current model with previous study under impact of customer differentiation on the inventory system.

Scenario number	U	v	q	$P1$	$P2$	Average annual total cost	
						Current model (Q, r)	Previous model (r, T)
1			40	90	5	10545	13670
2			40	90	20	11041	13050
3			20	90	5	11228	13670
4			20	90	20	11249	13270
5			10	90	20	11348	13450
6			5	90	20	11386	13600
7			10	90	5	11574	13500
8	60	1	40	60	20	11746	13370
9			5	90	5	11818	13700
10			40	60	5	11871	14050
11			20	60	20	11988	13800
12			10	60	20	12127	14050
13			20	60	5	12198	14150
14			5	60	20	12215	14200
15			10	60	5	12317	14250
16			5	60	5	12356	14300

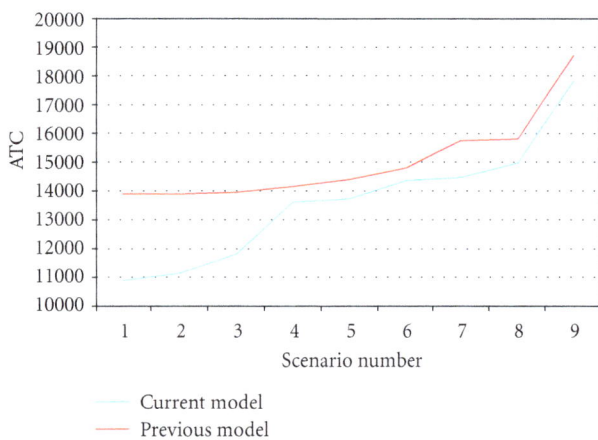

FIGURE 10: Compare current model with previous study under different scenarios of v and u.

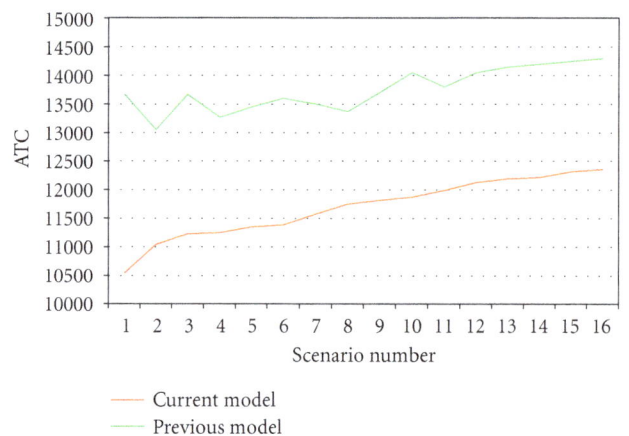

FIGURE 11: Compare current model with previous study under impact of customer differentiation on the inventory system.

Simulation Modeling for Analysis of a (Q, r) Inventory System under Supply Disruption and Customer
Differentiation with Partial Backordering

9

5. Conclusions

In the paper, a continuous-review inventory model (Q, r) with supply disruption and customer differentiation has been studied. We have considered a single supplier who may have some disruptions in its supply, one retailer who puts the orders to the supplier, and some customers for products. The first contribution of the current study is that we have considered a right for the customer to suspend the backorders or cancel them, while in majority of past studies just one of the cases has been considered, and we allowed the system to have more than one order in the pipeline, which made our model to be more realistic. For studying the inventory system, we used a simulation model consisting of two subsystems including demand subsystem and replenishment subsystem; then the effects of supply disruptions and customer differentiations on average annual costs of the inventory system. Simulation is a very powerful technique for such a complicated system where using common techniques such as mathematical programming models is very difficult. Several experiments have been carried out to optimize the system for setting the best strategies for the retailer. The results show that the both underlying factors (u and v) have significant effects on selecting the supplier. The retailer must select a supplier with low degree of frequency and duration of disruptions. If one of these factors is low, the other factor has insignificant effect on supplier selection. The other result is that the effect of customer differentiation depends on supply disruptions. When the frequency of disruptions is normal and the duration is low, the average annual costs will minimize by selecting the high-priority customers who may have the high level of probability for their backorders, and the ATC will decrease if the probability of backorders increases for both classes of customers. When the frequency of disruptions is normal and the duration increases, the situation changes. In this case, the ATC increases when the amount of high level customers increases and if the probability of backorders is low for both classes, the ATC decreases.

We have also shown that the ATC of a continuous-review model is significantly lower than a periodic one which is taken from a latest research. For future studies, we recommend to consider the price factor in the model as well. The normal fluctuation on product prices may have a great effect on the inventory system costs.

Decision Variables

Q: Order quantity
r: Reorder point.

Objective Function

ATC: Annual total cost.

Other Parameters and Notations

D: Stochastic demand size of a customer

λ: The mean interarrival time of customers
L: Replenishment lead time
T: Time horizon as one year, that is, 365 days
IL_0: The initial net inventory level of the retailer
$IL(t)$: The net inventory level of the retailer at time point t
IP_0: The initial inventory position of the retailer
$IP(t)$: The inventory position of the retailer at time point t
u: The mean duration of ON periods
v: The mean duration of OFF periods
s: Setup cost for each order placement
c: Unit product price
h: Unit holding cost per time unit
b_j: Unit backorder cost per time unit from customer class $i, i \in \{\text{I}, \text{II}\}$
l_i: Unit lost sale cost from customer class $i, i \in \{\text{I}, \text{II}\}$
$BoQ_j(t)$: The backorder quantity from customer class i at time point $t, i \in \{\text{I}, \text{II}\}$
$NILS_i$: The number of lost sales from customer class $i, i \in \{\text{I}, \text{II}\}$
P_i: The backorder probability of customer class $i, i \in \{\text{I}, \text{II}\}$
q: The proportion of customer class I
AOC: Annual ordering cost
AHC: Annual holding cost
ABC_i: Annual backorder cost from customer class $i, i \in \{\text{I}, \text{II}\}$
ATBC: Annual total backorder cost, $\text{ATBC} = \sum_{i \in \{\text{I}, \text{II}\}} \text{ABC}_i$
$AL C_i$: Annual lost sale cost from customer class $i, i \in \{\text{I}, \text{II}\}$
ATLC: Annual total lost sale cost, $\text{ATLC} = \sum_{i \in \{\text{I}, \text{II}\}} \text{ALC}_j$
r^*: The optimal reorder point
Q^*: The optimal reorder quantity
ATC^*: The minimum annual total cost.

References

[1] S. Chopra and M. S. Sodhi, "Managing risk to avoid: supply-chain breakdown," *MIT Sloan Management Review*, vol. 46, no. 1, pp. 53–87, 2004.

[2] H. P. Chao, "Inventory policy in the presence of market disruptions," *Operations Research*, vol. 35, no. 2, pp. 274–281, 1987.

[3] J. S. Song and P. H. Zipkin, "Inventory control with information about supply conditions," *Management Science*, vol. 42, no. 10, pp. 1409–1419, 1996.

[4] B. Lewis, *Inventory control with risk of major supply chain disruptions*, Ph.D. thesis, Georgia Institute of Technology, 2005.

[5] M. Parlar, "Continuous-review inventory problem with random supply interruptions," *European Journal of Operational Research*, vol. 99, no. 2, pp. 366–385, 1997.

[6] U. Gürler and M. Parlar, "An inventory problem with two randomly available suppliers," *Operations Research*, vol. 45, no.

6, pp. 904–918, 1997.

[7] A. Arreola-Risa and G. A. DeCroix, "Inventory management under random supply disruptions and partial backorders," *Naval Research Logistics*, vol. 45, no. 7, pp. 687–703, 1998.

[8] E. Mohebbi, "Supply interruptions in a lost-sales inventory system with random lead time," *Computers and Operations Research*, vol. 30, no. 3, pp. 411–426, 2003.

[9] E. Mohebbi, "A replenishment model for the supply-uncertainty problem," *International Journal of Production Economics*, vol. 87, no. 1, pp. 25–37, 2004.

[10] M. Parlar, Y. Wang, and Y. Gerchak, "A periodic review inventory model with Markovian supply availability," *International Journal of Production Economics*, vol. 42, no. 2, pp. 131–136, 1995.

[11] S. Özekici and M. Parlar, "Inventory models with unreliable suppliers in a random environment," *Annals of Operations Research*, vol. 91, pp. 123–236, 1999.

[12] A. Samvedi and V. Jain, "Studying the impact of various inventory policies on a supply chain with intermittent supply disruptions," in *Proceedings of the Winter Simulation Conference*, pp. 1641–1649, 2011.

[13] K. Moinzadeh, "Operating characteristics of the (s, S) inventory system with partial backorders and constant resupply times," *Management Science*, vol. 35, pp. 472–477, 1989.

[14] D. C. Montgomery, M. S. Bazaraa, and A. K. Keswani, "Inventory models with a mixture of backorders and lost sales," *Naval Research Logistics*, vol. 20, no. 2, pp. 255–263, 1973.

[15] D. H. Kim and K. S. Park, "(Q, r) Inventory model with a mixture of lost sales and time-weighted backorders," *Journal of the Operational Research Society*, vol. 36, no. 3, pp. 231–238, 1985.

[16] M. J. Posner and B. Yansouni, "A class of inventory models with customer impatience," *Naval Research Logistics Quarterly*, vol. 19, pp. 483–493, 1972.

[17] M. Parlar and D. Berkin, "Future supply uncertainty in EOQ models," *Naval Research Logistics*, vol. 38, pp. 50–55, 1991.

[18] M. Parlar and D. Perry, "Optimal (Q, r, T) policies in deterministic and random yield models with uncertain future supply," *European Journal of Operational Research*, vol. 84, pp. 431–443, 1993.

[19] H. J. Weiss and E. C. Rosenthal, "Optimal ordering policies when anticipating a disruption in supply or demand," *European Journal of Operational Research*, vol. 59, no. 3, pp. 370–382, 1992.

[20] H.-P. Chao, S. W. Chapel, C. E. Clark Jr., P. A. Morris, M. J. Sandling, and R. C. Grimes, "EPRI reduces fuel inventory costs in the electric utility industry," *Interfaces*, vol. 19, pp. 48–67, 1989.

[21] M. J. M. Posner and M. Berg, "Analysis of a production-inventory system with unreliable production facility," *Operations Research Letters*, vol. 8, no. 6, pp. 339–345, 1989.

[22] H. Groenevelt, L. Pintelon, and A. Seidmann, "Production lot sizing with machine breakdowns," *Management Science*, vol. 38, no. 1, pp. 104–123, 1992.

[23] X. Li and Y. Chen, "Impacts of supply disruptions and customer differentiation on a partial-backordering inventory system," *Simulation Modelling Practice and Theory*, vol. 18, no. 5, pp. 547–557, 2010.

2

CFD Analysis of the Effect of Elbow Radius on Pressure Drop in Multiphase Flow

Quamrul H. Mazumder

Mechanical Engineering, University of Michigan-Flint, Flint, MI 48502, USA

Correspondence should be addressed to Quamrul H. Mazumder, qmazumde@umflint.edu

Academic Editor: Aiguo Song

Computational fluid dynamics (CFD) analysis was performed in four different 90 degree elbows with air-water two-phase flows. The inside diameters of the elbows were 6.35 mm and 12.7 mm with radius to diameter ratios (r/D) of 1.5 to 3. The pressure drops at two different upstream and downstream locations were investigated using empirical, experimental, and computational methods. The combination of three different air velocities, ranging from 15.24 to 45.72 m/sec, and nine different water velocities, in the range of 0.1–10.0 m/s, was used in this study. CFD analysis was performed using the mixture model and a commercial code, FLUENT. The comparison of CFD predictions with experimental data and empirical model outputs showed good agreement.

1. Introduction and Background

In most industrial processes, fluids are used as a medium for material transport. A complete knowledge of the principles that rule the phenomena involving fluids transportation leads to more efficient and secure systems. However, in many industries, such as petroleum, chemical, oil, and gas industries, two-phase or multiphase flow is frequently observed [1]. Multiphase flow is defined as the simultaneous flow of several phases, with the simplest case being a two-phase flow [2]. Compared to single-phase flow, the equations associated with two-phase flow are very complex, due to the presence of different flow patterns in gas-liquid systems [3]. A detailed discussion of two-phase flow phenomenon behavior is provided by Wallis [2]. The flow patterns observed in horizontal flow are bubble, stratified, stratified wavy, slug, and annular. In vertical flows, bubble, plug, slug (or churn), annular, and wispy-annular flow patterns are present. Several investigations have been reported to determine the friction factor and pressure drops in horizontal [4] and vertical [5] two-phase and multiphase flows [6]. The presence of the two-phase flow typically produces an undesirable higher-pressure drop in the piping components. In most industrial installations, elbows are frequently used to direct the flow and provide flexibility to the system [7]. Since these fittings are also used to install instruments that monitor the main parameters of the industrial process, it is important to have a reliable way to evaluate the pressure drop in these elbows [8]. As the fluid flows through the bend, the curvature of bend causes a centrifugal force; the centrifugal force is directed toward the outer wall of the pipe from the momentary center of the curvature. The combined presence of centrifugal force and boundary layer at the wall produces the secondary flow, organized, ideally, in two identical eddies. This secondary flow is superimposed to the mainstream along the tube axis, resulting in a helical shape streamline, flowing through the bend [9].

One of the challenges with undesirable higher-pressure drop is the difficulty in determining a model for two-phase flow through pipe components. Despite unsuccessful attempts to develop an accurate model for two-phase flow through pipe components, Chisholm [10] presented an elementary model for prediction of two-phase flow in bends, based on a liquid two-phase multiplier, for different pipe diameters, r/D values, and flow rates. Detailed studies of two-phase pressure loss have largely been confined to the horizontal plane. Chenoweth and Martin [11] showed that while the two-phase pressure drop around bends was higher than in single-phase flow, it could be correlated by an adoption of the Lockhart-Martinelli [12] model, a model initially developed for straight pipe. The correlation claimed to predict loss in bends and other pipe fittings. Also, at high mass velocities, agreement was achieved with the homogeneous model. Fitzsimmons [13] presented two-phase pressure loss data for bend in terms of the equivalent length and the ratio of the bend pressure loss to the straight pipe frictional pressure gradient; the Lockhart-Martinelli multiplier referred to the single-phase gas pressure loss in the bend. The comparison against pressure drop in straight pipe gave a poor correlation. Sekoda et al. also used a two-phase multiplier, referred to as a single-phase liquid pressure loss in the bend. The two-phase bend pressure drop was found to be dependent on the r/D ratio, while being independent of pipe diameters [14]. The main focus of this paper is to investigate pressure drop for two-phase air-water mixture flow in 90 degree vertical to horizontal elbows. The first step is the prediction of the flow pattern, and then proposing an associated method of calculating the liquid holdup, which is used to determine the two-phase friction factor. Comparative studies proved that these models are inconsistently performed, as flow conditions vary. Therefore, the selection of the most appropriate flow correlation is very important in this study [15]. Reported work on the orientation of the plane of the bend has often given contrary results. Debold [16] claimed that the horizontal bend, the horizontal to vertical upbend, and the vertical down to horizontal bend all gave the same bend pressure loss. However, a horizontal to vertical downbend had a pressure drop that was 35% less. The correlation for elevation was assumed to follow the homogeneous model by Debold [16], but others, such as Alves [17], ignored head pressure differences entirely. Peshkin [18] reported that horizontal to vertical downflow had about 10% more bend pressure drop than the corresponding horizontal to vertical upflow case. Kutateladze [19], by contrast, concluded the direct opposite: that the horizontal to vertical upflow bend created the greater pressure drop. Moujaes and Aekula reported the effects of pressure drop on turning vanes in 90 degree duct elbows, using CFD models in HVAC applications area [20]. Due to the different approaches that can be used to predict pressure drop in elbows, the current study uses CFD analysis in four different elbows. The CFD analysis results were validated using two different empirical models by Azzi and Friedel [9] and Chisholm, [10] as well as experimental data.

2. The CFD Approach

Due to the advancements in computer hardware and software in recent years, the computational fluid dynamics technique has been a powerful and effective tool to understand the complex hydrodynamics of gas-liquid two-phase flows. The commercial CFD package, FLUENT, was used to model the air-water flow, in order to predict pressure drop in 90 degree elbows.

TABLE 1: Configurations of elbows used in the study.

	Pipe diameter, D (mm)	Elbow curvature radius, r (mm)	Equivalent pipe length, Le (mm)
Elbow 1	12.7	38.1	635
Elbow 2	12.7	19.05	571.5
Elbow 3	6.35	19.05	317.5
Elbow 4	6.35	9.525	285.75

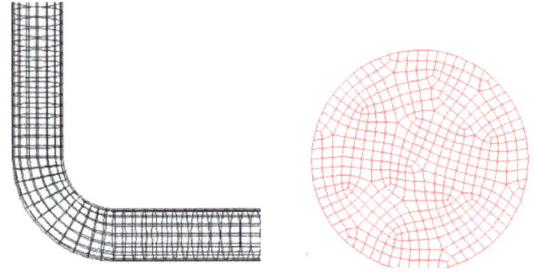

FIGURE 1: Elbow mesh with inlet domain.

3. Geometry Details

To conduct this study, four three-dimensional 90 degree vertical to horizontal elbows were created using GAMBIT. The geometries were then imported to FLUENT to simulate pressure drop. Table 1 lists the pipe diameter (D), elbow curvature radius (r), and equivalent pipe length (Le), for different elbows.

The r/D ratio of 1.5 to 3 was used to represent the standard short- and long-radius elbows. A straight pipe section of 45–50 times the pipe diameter was added at both upstream and downstream of the elbow. The Le/D ratio for single-phase flow is typically 10. However, for the two-phase flow, an Le/D of approximately 100–150 is required for fully developed flow [21]. Due to limitations of the experimental test system, the Le/D ratio used in this study was 50 for elbows 1 and 3, while the ratio was 45 for elbows 2 and 4. Structured mesh was used with the optimum number of nodes. A mesh independency study was performed to determine reasonable results that are independent of the size of the grid. Hexahedral mesh was used, due to its capabilities in providing high-quality solution, with a fewer number of cells than comparable tetrahedral mesh for simple geometry [22]. Figure 1 shows the elbow and inlet domain of three-dimensional mesh for elbow 1.

Table 2 lists the mesh details for each elbow, generated in GAMBIT, to be used in this study. CFD Analysis results at four different locations in the upstream and the downstream of the elbow were used in this study, as shown in Figure 2.

4. Multiphase Modeling

The numerical calculations of multiphase flows can be calculated through two approaches: the Euler-Euler approach and the Euler-Lagrange approach. Using the Euler-Euler approach is more efficient, because the different phases are

TABLE 2: Mesh details of all four elbows.

	Number of nodes	Number of wall faces	Number of hexahedral cells
Elbow 1	110,558	33,344	92,738
Elbow 2	97,520	29,408	81,791
Elbow 3	21,484	8,368	16,736
Elbow 4	18,942	7,376	14,752

treated mathematically, as interpenetrating continua. The concept of phasic volume fraction is used in this approach, since the volume of phase cannot be occupied by the other phases. These volume fractions are assumed to be continuous functions of space and time, and their sum is equal to one. For each phase, conservation equations are derived to obtain a set of new equations which have similar structure for all phases. By providing constitutive relations obtained from the empirical information, these sets of equations are closed. There are three available Euler-Euler multiphase models in the fluent commercial code: the Eulerian model, the mixture model, and the volume of fluid model. In this study, the mixture model is used, as it is relatively easy to understand for multiphase modeling.

5. Mixture Model Theory

The mixture model can be used to model multiphase flows, by assuming the local equilibrium over short spatial length scales, and where the phases move at different velocities. In addition, the mixture model can be used to calculate non-Newtonian viscosity. It can model a number of phases by solving the momentum, continuity, and energy equations for the mixture, the volume fraction equation for the secondary phases, and algebraic expression for the relative velocities. The mixture model is a good substitute for the full Eulerian multiphase model in several cases, as the full multiphase model may not be feasible, due to the wide distribution of a particular phase, when the interphase laws are unknown, or when the reliability of interphase laws is questioned. While solving a smaller number of variables than the full multiphase model, the mixture model can perform as well as the full multiphase model. It uses a single-fluid approach, just like the volume of fluid model but allows for the phases to be interpenetrating and for the phases to move at the different velocities, using the concept of slip velocities.

6. Mathematical Formulation

The mixture models solve the continuity equation, momentum equation, energy equation, the volume fraction equation for the secondary phases, and the algebraic expression for the relative velocities, since the two phases are moving at different velocities. The continuity, momentum, and relative velocity equations used in the mixture model are shown in the following section.

The mixture's continuity equation is given by

$$\frac{\partial}{\partial t}(\rho_m) + \nabla \cdot (\rho_m \bar{v}_m) = 0, \tag{1}$$

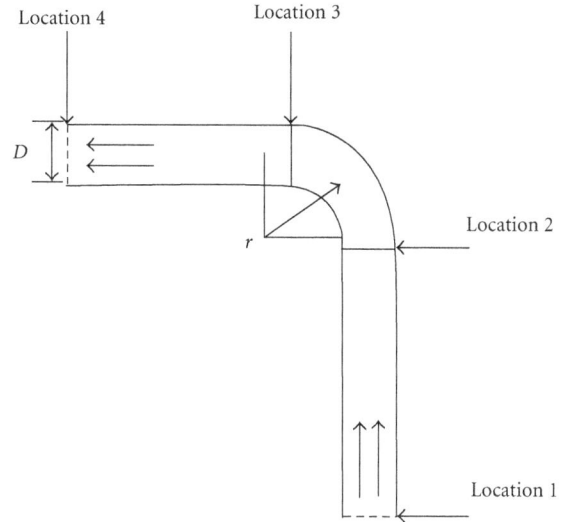

FIGURE 2: Locations of CFD and experimental data.

where the mass-averaged velocity $(\bar{v}_m)$ is

$$\bar{v}_m = \frac{\sum_{k=1}^{n} \alpha_k \rho_k \bar{v}_k}{\rho_m}, \tag{2}$$

and the mixture density (ρ_m) is given by

$$\rho_m = \sum_{k=1}^{n} \alpha_k \rho_k, \tag{3}$$

where α_k is volume fraction of phase k.

By adding the individual momentum equations for all phases, the mixture's final momentum equation can be obtained, and it is given by

$$\frac{\partial}{\partial t}(\rho_m \bar{v}_m) + \nabla \cdot (\rho_m \bar{v}_m \bar{v}_m)$$
$$= -\nabla p + \nabla \cdot \left[\mu_m \left(\nabla \bar{v}_m + \nabla \bar{v}_m^T \right) \right] \tag{4}$$
$$+ \rho_m \bar{g} + \bar{F} + \nabla \cdot \left(\sum_{k=1}^{n} \alpha_k \rho_k \bar{v}_{dr,k} \bar{v}_{dr,k} \right),$$

where $\bar{F}$ is the body force, n is the number of phases, and μ_m is the viscosity of the mixture:

$$\mu_m = \sum_{k=1}^{n} \alpha_k \mu_k, \tag{5}$$

where $\bar{v}_{dr,k}$ is the drift velocity for the secondary phase k:

$$\bar{v}_{dr,k} = \bar{v}_k - \bar{v}_m. \tag{6}$$

The relative velocity, or the slip velocity, is defined as the velocity of a secondary phase (p), relative to the velocity of the primary phase (q), and can be given by

$$\bar{v}_{pq} = \bar{v}_p - \bar{v}_q. \tag{7}$$

The mass fraction for any phase (k) is defined as

$$c_k = \frac{\alpha_k \rho_k}{\rho_m}. \tag{8}$$

The drift velocity and the relative velocity ($\bar{v}_{pq}$) are connected by the following expression:

$$\bar{v}_{dr,p} = \bar{v}_{pq} - \sum_{k=1}^{n} c_k \bar{v}_{qk}. \tag{9}$$

The algebraic slip formulation is used in the FLUENT's mixture model. Prescribing an algebraic relation for the relative velocity, a local equilibrium between phases should be reached over short spatial length scale, according to the basic assumption of the algebraic slip mixture model. The relative velocity then is given by

$$\bar{v}_{pq} = \frac{\tau_p}{f_{\text{drag}}} \frac{\left(\rho_p - \rho_m\right)}{\rho_p} \bar{a}, \tag{10}$$

where the particle relaxation time (τ_p) is given by

$$\tau_p = \frac{\rho_p d_p^2}{18 \mu_q}, \tag{11}$$

d is the diameter of the particles (or droplets or bubbles) of secondary phase, and $\bar{a}$ is the secondary-phase particle's acceleration. The default drag function f_{drag} is taken from Schiller and Naumann [19]:

$$f_{\text{drag}} = \begin{cases} 1 + 0.15\,\text{Re}^{0.687} & \text{Re} \leq 1000 \\ 0.0183\,\text{Re} & \text{Re} > 1000, \end{cases} \tag{12}$$

and the acceleration $\bar{a}$ is of the form

$$\bar{a} = \bar{g} - (\bar{v}_m \cdot \nabla)\bar{v}_m - \frac{\partial \bar{v}_m}{\partial t} a. \tag{13}$$

In the drift flux model, the acceleration of the particle is given by gravity and/or a centrifugal force, and, in order to take into account the presence of other particles, the particulate relaxation time is modified. In turbulent flows, the relative velocity should contain a diffusion term, due to the dispersion appearing in the momentum equation for the dispersed phase. FLUENT adds this dispersion to the relative velocity:

$$\bar{v}_{pq} = \frac{\left(\rho_p - \rho_m\right) d_p^2}{18 \mu_q f_{\text{drag}}} \bar{a} - \frac{v_m}{\alpha_p \sigma_D} \nabla \alpha_q, \tag{14}$$

where (v_m) is the mixture turbulent viscosity, and (σ_D) is the Prandtl dispersion coefficient.

7. Modeling Assumptions

A straight pipe section was extended at the inlet and outlet boundary to evaluate the pressure drop across each elbow. In order to ensure fully developed flow, appropriate Le/D ratio was used to calculate the length of the straight section.

The standard k-ϵ model, with wall functions, was used in this study, since it is the simplest of the "complete models" available in FLUENT. The model constants used for this analysis were $C_{1\epsilon} = 1.44$, $C_{2\epsilon} = 1.92$, $C_\mu = 0.09$, $\sigma_k = 1.0$, and $\sigma_\epsilon = 1.3$. For the mixture parameters, slip velocity was taken into consideration, as the phases had a significant difference in velocities, while the no-slip boundary conditions were assumed for the wall of tubing. Due to the complex behavior of the two-phase flow, solution strategies were followed to improve the accuracy and convergence of the solution. The mixture calculation was initialized with a low under relaxation factor of 0.2 for the slip velocity; calculations were performed by combinations of the SIMPLE pressure-velocity coupling. The first-order upwind discretization scheme was used for the momentum, volume fraction, turbulent kinetic energy, and turbulent dissipation rate. The convergence criterion was based on the residual values of the calculated variables, to ensure satisfactory accuracy, stability, and convergence. The governing equations were solved sequentially, separate from one another, requiring less memory in comparison with the coupled algorithm. SIMPLE uses the pressure-based segregated algorithm, which makes use of the relationship between the velocity and pressure corrections, to enforce mass conservation, and to obtain the pressure field.

8. CFD Analysis

CFD analysis was performed on four different elbows, at nine different water and air velocities. Thus, a total of twenty-seven different combinations of water and air velocities were used in the study, for each elbow. Each of these conditions was analyzed in CFD in order to accurately predict the effect of varying the pipe diameter and r/D ratios. Due to the limitation of the test loop system, experimental investigations could not be performed for all of these conditions. Table 3 lists all the flow conditions that were used in the pressure loss CFD analysis. Numbers 1, 2, 3, and 4 in Table 3 represent elbows 1, 2, 3, and 4, respectively.

Cross-sectional absolute pressure and radial velocity contours are presented in Figure 3. The left side of Figure 3 shows the radial velocity contours in four locations of elbow 1 at a water velocity of 0.1 m/s and air velocities of 15.24, 30.48, and 45.72 m/s, respectively. Location 1 represents the inlet of the upstream pipe, while location 2 indicates the outlet of the upstream pipe and inlet of the elbow. Similarly, location 3 indicates the outlet of elbow and inlet of the downstream pipe, and location 4 indicates the outlet of the downstream pipe. The top part of each individual contour represents the inside wall, while the bottom part of the contour represents the outer wall of the elbow. The contour maps were collected to locate, and study, any patterns that were formed. As depicted in Figure 3, the secondary flow pattern can be observed at the exit of the elbow section. The right side of Figure 3 shows the absolute pressure profiles in the four locations of elbow 1 at a water velocity of 0.1 m/s, and air velocities of 15.24, 30.48, and 45.72 m/s, respectively. The pressure distribution is dispersed without forming any regular pattern.

TABLE 3: Flow conditions used in CFD analysis.

Air velocity, (m/sec)	Water velocity, (m/sec)								
	0.1	0.5	1.0	1.3	2.0	2.5	5.0	7.5	10.0
15.24	1,2,3,4	1,2	1,2,3,4	1,2	1,2,3,4	1,2	1,2,3,4	1,2,3,4	1,2,3,4
30.48	1,2,3,4	1,2	1,2,3,4	1,2	1,2,3,4	1,2	1,2,3,4	1,2,3,4	1,2,3,4
45.72	1,2,3,4	1,2	1,2,3,4	1,2	1,2,3,4	1,2	1,2,3,4	1,2,3,4	1,2,3,4

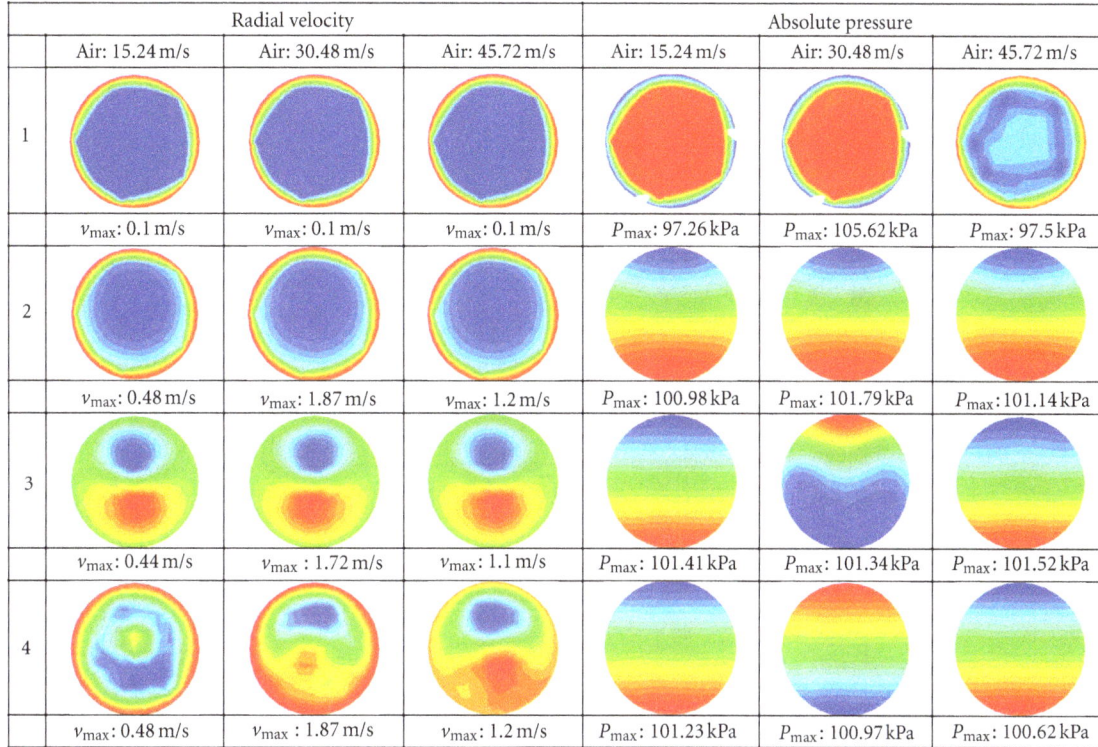

FIGURE 3: CFD-predicted velocity and pressures in elbow1 at water 0.1 m/s.

9. Pressure Drop Calculations with Empirical Models

9.1. Chisholm Model. Chisholm proposed a correlation that involves the dimensionless parameters obtained by correlating two-phase flow experimental data [10]. It also requires an equivalent pipe length Le that depends on the elbow radius to the pipe diameter ratio (r/D), as well as the angle of the elbow bend [23]. The increment of Le as a function of r/D is mainly due to friction, centrifugal force, and the secondary flow that is present in the elbows. Thus, the single-phase pressure drop in the elbow is

$$\Delta P_{1ph} = f \frac{G^2}{2\rho}\left(\frac{Le}{D}\right). \tag{15}$$

Chisholm suggested evaluating the pressure drop coefficient k_l, by assuming that the whole two-phase flow mixture flows as liquid only through the pipe fitting, as expressed by

$$k_l = f_l\left(\frac{Le}{D}\right). \tag{16}$$

Considering that only liquid phase fills up the pipe, the two-phase flow mixture pressure drop is evaluated as

$$\Delta P_{1ph,l} = \frac{k_l G_T^2}{2\rho_l}. \tag{17}$$

Therefore, pressure drop in a two-phase mixture flowing through a 90° elbow is given by an equation that already includes the mass quality, and a correlation factor, for two-phase properties:

$$\Delta P_{2ph} = \Delta P_{1ph,l}\left[1 + \{E(x(1-x)) + x^2\}\right], \tag{18}$$

where E is 90° elbow coefficient, which includes the relative radius of the elbow

$$E = 1 + \frac{2.2}{k_l(2 + (r/D))}. \tag{19}$$

9.2. Azzi-Friedel Model. According to this model, the pressure drop is based on the two-phase flow multiplier, defined as the ratio of the bend pressure loss in two-phase flow, and

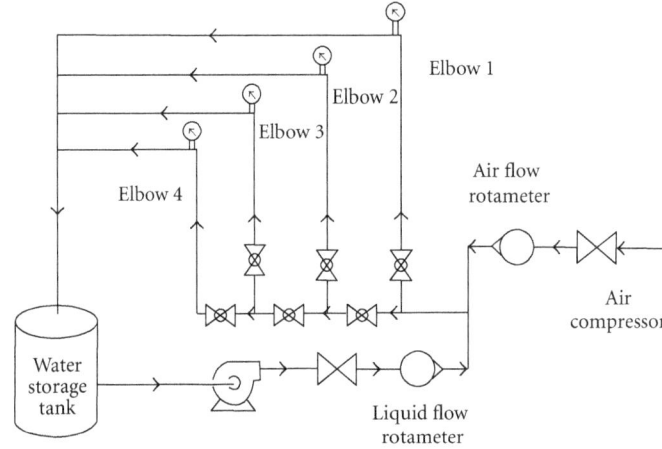

FIGURE 4: Schematic of the experimental test system.

that in the single-phase liquid flow with the same total mass flow rate as in [9]:

$$\Phi^2 = \frac{\Delta P_{2ph}}{\Delta P_{1ph,l}}, \tag{20}$$

where $\Delta P_{1ph,l}$ is the pressure drop of single-phase liquid fluid, across the same bend, defined as

$$\Delta P_{1ph,l} = k_i \frac{G_l^2}{2\rho_l}, \tag{21}$$

where

$$k_i = f_i \left(\frac{Le}{D}\right), \tag{22}$$

where Le/D is the dimensionless, single-phase equivalent length, and f_i is the single-phase flow pipe friction factor. According to Churchill [24], this factor can be calculated by using the following equation:

$$f_i = 8\left(\left(\frac{8}{Re}\right)^{12} + (A_i + B_i)^{-1.5}\right)^{1/12}, \tag{23}$$

where

$$A_i = \left(2.457 \ln\left(\left(\frac{7}{Re_i}\right)^{0.9} + 0.27\frac{\varepsilon}{D}\right)^{-1}\right)^{16},$$

$$B_i = \left(\frac{37530}{Re_i}\right)^{16}, \tag{24}$$

$$Re_i = \frac{\rho_i V_i D}{\mu_i}.$$

Subscript i in the above equations can be used for either liquid or gas.

The two-phase flow multiplier, defined by Azzi and Friedel, is given by

$$\Phi^2 = C + 7.42 Fr_l^{0.125} \frac{r}{D}^{0.502} x^{0.7}(1-x)^{0.1}$$

$$\times \left(\frac{(\rho_l - \rho_g)}{\rho_l}\right)^{0.14} \left(\frac{(\mu_l - \mu_g)}{\mu_l}\right)^{0.12}, \tag{25}$$

$$C = (1-x) + \left(\frac{\rho_l k_g}{\rho_g k_l}\right) x^2.$$

Froude number (Fr_l) is

$$Fr_l = \frac{(1-x^2)G_T^2}{\rho_l^2 rg}. \tag{26}$$

10. Experimental Investigation

To validate the CFD simulation and empirical model results, a multiphase air-water test system was designed and developed. Two different 12.7 mm and two different 6.35 mm test sections were used to conduct the pressure drop experiments. The major difference between each test section is the flow development length upstream of the test section, pipe diameter, and elbow radius to pipe diameter ratio. Figure 4 shows the schematic of the test loop, consisting of a 30-gallon water tank, a 25 HP pump, a 10 CFM air compressor, liquid and gas flow meters, four vertical to horizontal 90° elbows with two different pipe diameters and r/D ratios, inlet and outlet pressure gauges, and four differential pressure gauges. Air from the compressor enters the test loop, through a gate valve and air flow meter, that is used to control the air flow rates. Water from the tank is pumped into the test section, through a gate valve and liquid flow meter, that is used to control the water flow rates. Water is injected into the air stream through a T elbow, and the mixture then flows through a straight section of 12.7 mm pipe section to the elbows. Ball valves are used to allow the mixture to enter a specific elbow and return the water to tank. The 12.7 mm and

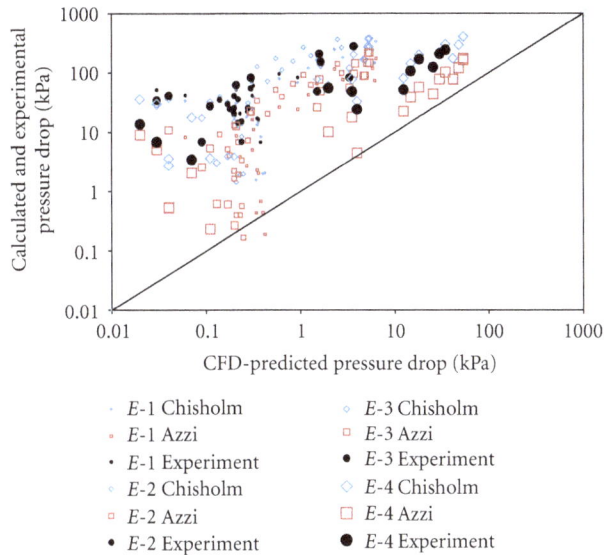

Figure 5: Comparison of CFD versus calculated and experimental pressure drop.

6.35 mm transparent sections of plexiglass pipe are used for multiphase flow pattern visualization. The pressure drop was measured across each elbow, using the differential pressure gauges hooked up across the 90 degree elbows. After the test sections, the mixture flows downstream, into the water tank, where the mixture is separated. The air is released back into the environment, while the water is recycled.

11. Comparison of CFD Results with Experimental and Empirical Results

Figure 5 shows the plot of CFD-predicted pressure drop versus two empirical and experimental pressure drop data for all elbows. The linear line shown in Figure 5 depicts a perfect line. The data above the line is overpredicted results, while the data below the line is underpredicted results. The data lying near the line shows perfect agreement. In all conditions, CFD underpredicted the pressure drop values. Azzi and Friedel, Chisholm, and the experiment overpredicted the pressure drop. However, for elbows 1 and 2, Azzi-Friedel's model underpredicted the pressure drop. On the same token, Chisholm model significantly over predicted the pressure drop for elbows 1, 2 and 3, which can be seen in Figure 5. For both empirical models, the prediction for elbows 3 and 4 was closer to the perfect agreement line.

12. Summary and Conclusion

CFD analysis of two-phase flow in a 6.35, and 12.7 mm pipe diameter with r/D ratio of 1.5 and 3 was performed using commercially available CFD code FLUENT. Analysis was performed for three different air velocities between 15.24, 30.48, and 45.72 m/s and six different water velocities, ranging from 0.1 to 10.0 m/s, in each of the four elbows. Pressure drop profiles and their respective cross-sectional

pressure contour maps were presented for characteristic flow behaviors in multiphase flows.

Nomenclature

English Letters

$1-x$:	Wetness fraction/liquid quality
A:	Coefficient 1 of pipe friction factor
a:	Acceleration
B:	Coefficient 2 of pipe friction factor
C:	Coefficient of two-phase multiplier
D:	Inner pipe diameter
E:	Chisholm coefficient
f:	Friction factor for pipe
F:	Force
Fr:	Froude number
G:	Mass flux
g:	Gravitational acceleration
K:	Friction factor for bend
Le/D:	Equivalent pipe length to pipe diameter ratio
r:	Elbow curvature radius
r/D:	Elbow curvature radius to pipe diameter ratio
Re:	Reynolds number
v:	Velocity
x:	Dryness fraction/gas quality
ΔP:	Two-phase pressure drop.

Greek Letters

μ:	Dynamic viscosity
α:	Volume fraction
ε/D:	Equivalent roughness to pipe diameter ratio
ρ:	Density
σ:	Prandtl coefficient
τ:	Relaxation time
Φ^2:	Two-phase multiplier.

Subscripts

1ph:	Single-phase flow
2ph:	Two-phase flow
d:	Dispersion
dr:	Drift
drag:	Drag force
e:	Equivalent
g:	Gas phase
i:	Subscript for either liquid or gas phase
k:	Any phase
l:	Liquid phase
m:	Mixture
p:	Secondary phase
pq:	Relative phase
q:	Primary phase
T:	Total
t:	Time.

References

[1] S. F. Sánchez, R. J. C. Luna, M. I. Carvajal, and E. Tolentino:, "Pressure drop models evaluation for two-phase flow in 90 degree horizontal elbows," *Ingenieria Mecanica Techilogia Y Desarrollo*, vol. 3, no. 4, pp. 115–122, 2010.

[2] G. B. Wallis, *One Dimensional Two-Phase Flow*, McGraw-Hill, 1969.

[3] S. Benbella, M. Al-Shannag, and Z. A. Al-Anber, "Gas-liquid pressure drop in vertical internally wavy 90° bend," *Experimental Thermal and Fluid Science*, vol. 33, no. 2, pp. 340–347, 2009.

[4] J. S. Cole, G. F. Donnelly, and P. L. Spedding, "Friction factors in two phase horizontal pipe flow," *International Communications in Heat and Mass Transfer*, vol. 31, no. 7, pp. 909–917, 2004.

[5] S. Wongwises and W. Kongkiatwanitch, "Interfacial friction factor in vertical upward gas-liquid annular two-phase flow," *International Communications in Heat and Mass Transfer*, vol. 28, no. 3, pp. 323–336, 2001.

[6] P. L. Spedding, E. Benard, and G. F. Donnelly, "Prediction of pressure drop in multiphase horizontal pipe flow," *International Communications in Heat and Mass Transfer*, vol. 33, no. 9, pp. 1053–1062, 2006.

[7] J. Hernández Ruíz, *Estudio del comportamiento de flujo de fluidos en tuberías curvas para plicaciones en metrología [Tesis de Maestría]*, IPN-ESIME, 1998.

[8] A. M. Chan, K. J. Maynard, J. Ramundi, and E. Wiklund, "Qualifying elbow meters for high pressure flow measurements in an operating nuclear power plant," in *Proceedings of the 14th International Conference on Nuclear Engineering (ICONE '06)*, Miami, Fla, USA, July 2006.

[9] A. Azzi and L. Friedel, "Two-phase upward flow 90° bend pressure loss model," *Forschung im Ingenieurwesen*, vol. 69, no. 2, pp. 120–130, 2005.

[10] D. Chisholm, *Two-Phase Flow in Pipelines and Heat Exchangers*, Godwin, 1983.

[11] J. M. Chenoweth and M. W. Martin, "Turbulent two-phase flow," *Petroleum Refiner*, vol. 34, no. 10, pp. 151–155, 1955.

[12] R. W. Lockhart and R. C. Martinelli, "Proposed correlation of data for isothermal two-phase two-component flow in pipes," *Chemical Engineering Progress*, vol. 45, no. 1, pp. 39–48, 1949.

[13] P. E. Fitzsimmons, "Two phase pressure drop in pipe components," Tech. Rep. HW-80970 Rev 1, General Electric Research, 1964.

[14] K. Sekoda, Y. Sato, and S. Kariya, "Horizontal two-phase air-water flow characteristics in the disturbed region due to a 90-degree bend," *Japan Society Mechanical Engineering*, vol. 35, no. 289, pp. 2227–2333, 1969.

[15] A. Asghar, R. Masoud, S. Jafar, and A. A. Ammar, "CFD and artificial neural network modeling of two-phase flow pressure drop," *International Communications in Heat and Mass Transfer*, vol. 36, no. 8, pp. 850–856, 2009.

[16] T. L. Deobold, "An experimental investigation of two-phase pressure losses in pipe elbows," Tech. Rep. HW-SA, 2564, MSc. University of Idaho, Chemical Engineering, 1962.

[17] G. E. Alves, "Co-current liquid-gas flow in a pipe-line contactor," *Chemical Engineering Progress*, vol. 50, no. 9, pp. 449–456, 1954.

[18] M. A. Peshkin, "About the hydraulic resistance of pipe bends to the flow of gas-liquid mixtures," *Teploenergetika*, vol. 8, no. 6, pp. 79–80, 1961.

[19] S. S. Kutateladze, *Problems of Heat Transfer and Hydraulics of Two-Phase Media*, Pergamon Press, Oxford, UK.

[20] S. F. Moujaes and S. Aekula, "CFD predictions and experimental comparisons of pressure drop effects of turning vanes in 90° duct elbows," *Journal of Energy Engineering*, vol. 135, no. 4, pp. 119–126, 2009.

[21] Q. H. Mazumder, S. A. Shirazi, and B. S. McLaury, "Prediction of solid particle erosive wear of elbows in multiphase annular flow-model development and experimental validation," *Journal of Energy Resources Technology*, vol. 130, no. 2, Article ID 023001, 10 pages, 2008.

[22] I. Fluent, *Fluent 6. 3 User Guide*, Fluent Inc., Lebanon, NH, USA, 2002.

[23] N. P. Cheremisinoff, Ed., *Encyclopedia of Fluid Mechanics: Gas Liquid Flows*, vol. 3, Gulf Publishing Company, 1986.

[24] S. W. Churchill, "Friction equation spans all fluid flow regimes," *Chemical Engineering*, vol. 84, no. 24, pp. 91–92, 1977.

Study of Swarm Behavior in Modeling and Simulation of Cluster Formation in Nanofluids

Mohammad Pirani, Hassan Basirat Tabrizi, and Ali Farshad

Department of Mechanical Engineering, Amirkabir University of Technology, P.O. BOX 15875-4413, Tehran 159163411, Iran

Correspondence should be addressed to Hassan Basirat Tabrizi; hbasirat@aut.ac.ir

Academic Editor: Azah Mohamed

Modeling the multiagents cooperative systems inspired from biological self-organized systems in the context of swarm model has been under great considerations especially in the field of the cooperation of multi robots. These models are trying to optimize the behavior of artificial multiagent systems by introducing a consensus, which is a mathematical model between the agents as an intelligence property for each member of the swarm set. The application of this novel approach in the modeling of nonintelligent multi agents systems in the field of cohesion and cluster formation of nanoparticles in nanofluids has been investigated in this study. This goal can be obtained by applying the basic swarm model for agents that are more mechanistic by considering their physical properties such as their mass, diameter, as well as the physical properties of the flow. Clustering in nanofluids is one of the major issues in the study of its effects on heat transfer. Study of the cluster formation dynamics in nanofluids using the swarm model can be useful in controlling the size and formation time of the clusters as well as designing appropriate microchannels, which the nanoparticles are plunged into.

1. Introduction

Swarming, as a novel approach in modeling the dynamics of multiagent systems, inspires from the behavior of biological self-organized and decentralized systems that cooperate to do a special task. Decentralization means that the swarm has no central leader or boss and each member does its work with a kind of imitation. First attempts to describe the behavior of such biological systems from the mathematical points of view belong to Breder [1] who developed the motion equation of schools of fish and claimed that the motion of each member is the resultant of a long-range attraction and short-range repulsion components. Gazi and Passino [2, 3] described the stability of swarm systems.

Many efforts have been investigated to introduce the behavior of multiagent systems whose members have mechanical interaction with each other. It means they may collide to one another and exchange some momentum or arrange in special configurations that can be seen in cluster formations from multiphase and granular flows to nanofluids. A particle dynamic description of solid particles in multiphase flows is one of the aspects of multiagent mechanical systems. Therefore, many attempts were dedicated to describe the behavior of particles in multiphase flows (see Hase [4], Li and Kuipers [5]).

Dorigo et al. [6] introduced an optimization method based on the dynamics of the swarm of ant colonies. The essence of the swarm model is to introduce a consensus between the agents in the form of the summation of attraction and repulsion components between the agents. There have been some efforts to introduce the interactions between the particles as an algebraic summation of the attraction and repulsion coefficients. Sadus [7] proposed an experimental correlation as a potential field for nonpolar particles comprises of attraction and repulsion terms. Zohdi [8] proposed the idea of breaking all of the interactions between particles in a multiphase flow into four major forces, for example, thermal force, drag force, inter particle contact force, and near field force. He modeled the near field force with a linear combination of an attraction and repulsion components just as what is considered in the swarm model.

Nanofluids are used in many applications because of their specific characteristics. Their capability to enhance the heat transfer is one of those most obvious characteristics. Various kinds of nanoparticles used for these applications such as Al_2O_3, Fe_3O_4, and CuO. Increasing the heat transfer depends on the shape and the size of the particles and their volume fraction. Decreasing the size of the nanoparticles causes the conducting heat transfer coefficient to increase [9–13]. Maïga et al. [14] and Chandrasekar et al. [15] also investigated the effect of the host liquid on the heat transfer by comparing water-Al_2O_3 and the Glycol-Al_2O_3 nanofluids. Brownian motion of the particles and cluster formation in nanofluids are the major issues in the studies of nanofluids according to the important effect of their size on the heat transfer rate. Cluster formation causes the viscosity of nano fluid and thermal conductivity to change [16–19]. Although there have been some efforts in modeling the behavior of nanoparticles in nanofluids, no special research has been done to describe the cohesion and cluster formations of these particles analytically, and most of these works were numerical or experimental approach to this point.

This study proposes a model for cluster formation of the nanoparticles in nanofluids and discusses about their size according to the interparticle forces. Model is based on the swarm model, which enables us to make a better control on the cluster formation phenomenon by knowing the control variables of the system. In addition, this model can be used for analyzing the cohesiveness of nanorobot probes, as nanoparticles injected into the blood vessels [20]. By knowing the cluster size of the swarm set as well as the diameter of the vessel, one can easily compute the number of nano robot probes participating in the cohesion task.

2. Swarm Model

A swarm system consists of N members, with mass m for each individual, who are placed in an n-dimensional Euclidean space. The motion equation of each member of the swarm in the general form can be written as

$$m\ddot{r}_i + c\dot{r}_i = \sum_{j=1,j\neq i}^{N} \left(-\frac{(r_i - r_j)}{|r_i - r_j|} \right) \left(f_a\left(\|r_i - r_j\|\right) - f_r\left(\|r_i - r_j\|\right) \right). \tag{1}$$

In this equation, $f_a(\|r_i - r_j\|)$ denotes the attraction between member i with member j, and $f_r(\|r_i - r_j\|)$ represents the repulsion between the pair i and j. Thus the dynamics of member i depend on the resultant of attraction and repulsion forces between ith and other $N - 1$ members in the system. Since the attraction and repulsion forces behave like a spring, they are called spring shape forces. Therefore, (1) recalls some generalized form of a mass, spring, and damper oscillator. Coefficient c can be assumed as a damper constant for such a system and is necessary for the stability of the system.

Now define a special class of such attraction and repulsion functions to convert (1) into

$$m\ddot{r}_i + c\dot{r}_i = \sum_{j=1,j\neq i}^{N} -\frac{(r_i - r_j)}{|r_i - r_j|} \left(a\|r_i - r_j\|^{-\gamma_1} - b\|r_i - r_j\|^{-\gamma_2} \right). \tag{2}$$

Here a and b are the attraction and repulsion coefficients, respectively. γ_1 and γ_2 are some positive real numbers, and their values depend on the physics of the system which will be described in the following sections. In order to apply this equation for more real mechanical agents such as nanoparticles in nano fluid flows, it is necessary to define some appropriate attraction and repulsion functions well matched with the physics of the problem.

2.1. Adding Diameter to Each Member. In order to induce dimension, or diameter, to the point-shaped members, we must conduct a strategy to inhibit particles to obtain same coordinates in space. In other words, they should not be overlap on each other. Therefore, it is necessary to define a repulsion function that prohibits the distance to be zero or even approaches to zero. Such a repulsion function is called unbounded repulsion [3]. Thus the degree of the repulsion function's denominator, or pole, should be greater than one. Such a function can be defined as

$$\lim_{\|r_i-r_j\|\to 0^+} f_r\left(\|r_i - r_j\|\right)\|r_i - r_j\| = \infty. \tag{3}$$

This condition inhibits particles to reduce their distance to zero. It can be extended to more general that we can introduce a function that inhibits members to be closer than a particular distance like C. This is obtainable with just a modification in the above limit into the following limit:

$$\lim_{\|r_i-r_j\|\to C} f_r\left(\|r_i - r_j\|\right)\|r_i - r_j\| = \infty. \tag{4}$$

C is the distance between the centers of two members or their diameter if they are spherical shaped particles. Therefore, the repulsion function is defined as follows:

$$f_r\left(\|r_i - r_j\|\right) = \frac{b}{\left(\|r_i - r_j\| - C\right)^2}. \tag{5}$$

Thus, the value of γ_2 in (2) will be equal to 2.

A long-range attraction function should be defined as well. Since the effects of long-rage attraction function are considerable through the long distances, this function should have direct ratio with the distance between each couple. On the other hand, in realistic swarm systems when two members become too far apart from one another, they lose and forget their effects. Therefore, a relative long-range attraction function should satisfy what follows:

$$\lim_{\|r_i-r_j\|\to \infty} f_a\left(\|r_i - r_j\|\right) = 0. \tag{6}$$

The infinity in above equation is a mathematical infinity. In other words, the physical infinity can be determined according to the application. For example, it can be assumed 100 times longer than the particle's diameter. By defining $l = 100d$ in which d is the agent's diameter, so

$$\lim_{\|r_i - r_j\| \to l} f_a\left(\|r_i - r_j\|\right) = 0. \tag{7}$$

For obtaining this in mind, the basic model that indicates each agent should interact with all other agents. It will be modified into the case; each agent interacts with some agents in its neighborhood, which is according to the sensitivity of the agent for receiving the long-range attraction signals.

Therefore, the attraction function can be defined as

$$f_a\left(\|r_i - r_j\|\right) = \frac{a}{\|r_i - r_j\|} \tag{8}$$

and the value of γ_1 in (2) will be equal to 1.

It is worth to mention that this attraction function is called relative attraction function for long-range relative to the repulsion function. According to (8), it not only has direct ratio with distance, but also has an inverse ratio with it. Nevertheless, this inverse ratio is in the first power. Thus, it can be concluded this attraction function is long range relative to the repulsion that has an inverse ratio with the second power of distance.

2.2. Adding Environmental Effects.

In common swarm models, the environmental effects are considered as well. These effects are modeled as a profile that can attract or repel members. In fact, this profile is considered as a simulation of nutrient that attracts member to itself or as a model of toxin, which repels members far apart. However, it can be interpreted in a more general case as a potential field exerting on each member from the environment. The motion of individuals will be through the opposite direction of the profile's gradient.

In order to add this effect into our model, we considered a potential function like $g(\cdot) : R^n \to R$ and rewrite (1) as follows:

$$\dot{r}_i = -\nabla_{r_i} g\left(r_i\right) + \sum_{j=1, j \neq i}^{N} f\left(r_i - r_j\right). \tag{9}$$

The ambient profile function can be interpreted in many kinds. In this paper, we use the plane profile as

$$g(r) = A^T r + b. \tag{10}$$

Then the equation of motion regarding the environmental effect will be

$$\ddot{r}^i + \frac{1}{m_i}\dot{r}^i$$

$$= \frac{1}{m_i}\left(\sum_{j=1, j\neq i}^{N} -\frac{\left(r_i - r_j\right)}{\left|r_i - r_j\right|}\left(\frac{a}{\|r_i - r_j\|} - \frac{b}{\left(\|r_i - r_j\| - C\right)^2}\right)\right)$$

$$- \nabla_{r_i} g\left(r_i\right). \tag{11}$$

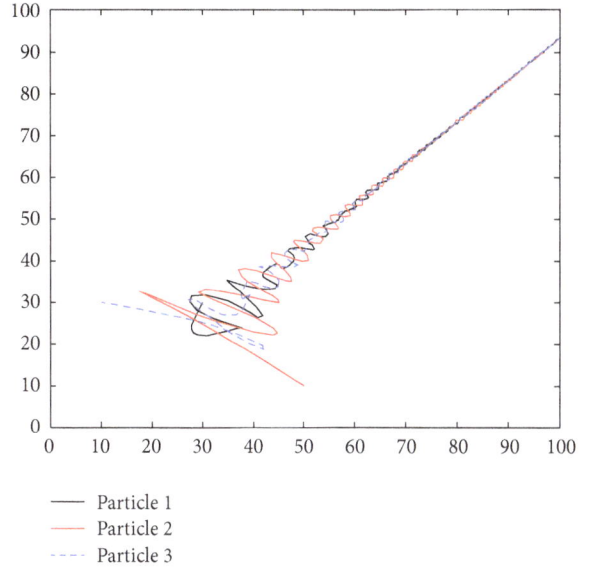

FIGURE 1: Two-dimensional simulation of (11) for three particles, $m_i = 2$, $a = 100$, $b = 10$, and $\nabla_{r_i} g(r_i) = 2$.

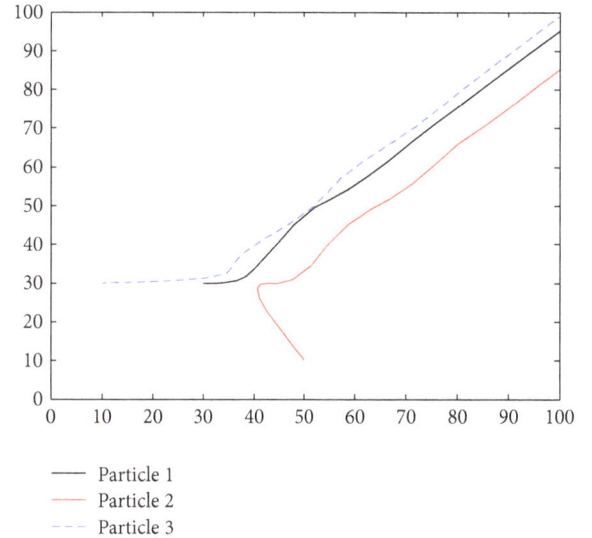

FIGURE 2: Two-dimensional simulation of (11) for three particles, $m_i = 2$, $a = 10$, $b = 100$, and $\nabla_{r_i} g(r_i) = 2$.

Two simulation examples are provided in Figures 1 and 2. Consider the effect of the attraction and the repulsion coefficients in the convergence radius of the particles. In both cases the plane profile with $\nabla g(r) = 2$ is assumed.

3. Dimensional Analysis

In order to analyze (1) and apply it for our special case, it is necessary to make a physical sense about each term in that equation. For simplicity and without losing the generality, assume the swarm system consists of just two members. Therefore, the sigma sign in (1) disappears, and we have

$$m\ddot{r} + d\dot{r} = ar^{-\gamma_1} - br^{-\gamma_2}. \tag{12}$$

Equation (12) is a nonlinear ordinary differential equation. Constants γ_1 and γ_2 can be determined according to our desire in choosing any kind of attraction and repulsion components.

It is time to apply physical properties of the system of nanoparticles in the nano fluid, which can be considered as nanorobot probes in the blood. The viscosity of blood is ten times higher than water, and we can claim that the effect of mass of each robot is negligible. To prove this claim, refer to experimental and more realistic approaches. Cavalcanti et al. [20] investigated in an experiment 10^{12} nanorobots with the total mass of just 0.2 gram in 5 lit bloods of a typical adult [21, 22].

Since, the ratio of inertia force to the viscous force is a dimensionless number called the Reynolds number, which is

$$\text{Reynolds} = \frac{d\rho\upsilon}{\mu}. \tag{13}$$

Here d, ρ, υ, and μ are the diameter of the particle, the density of the fluid, the average velocity of the flow, and the dynamic viscosity of the flow, respectively.

The velocity of the flow in a small blood vessel is assumed about 1 mm/s. The value for the density and viscosity of blood plasma is

$$\rho = 1\,\text{gr/cm}^3, \qquad \mu = 10^{-2}\,\text{gr/cm}\cdot\text{s}. \tag{14}$$

According to the above values, the Reynolds number for a nanorobot with the diameter of 1 μm is

$$\text{Reynolds} = \frac{d\rho\upsilon}{\mu} \approx 10^{-3}. \tag{15}$$

For the case of water-Al$_2$O$_3$, nanofluids the Reynold's number value is about 10^{-4}. This Reynolds number indicates that the viscosity of the host liquid is at least $O(10^3)$ higher than the mass of the particles.

Consider a dimensionless time τ as proposed in [23]

$$\tau = \frac{t}{T}. \tag{16}$$

A new derivative with respect to τ follows:

$$\dot{r} = \frac{dr}{dt} = \frac{dr}{d\tau}\frac{d\tau}{dt} = \frac{1}{T}\frac{dr}{d\tau},$$
$$\ddot{r} = \frac{d^2r}{dt^2} = \frac{d}{dt}\left(\frac{1}{T}\frac{dr}{d\tau}\right) = \frac{1}{T^2}\frac{d^2r}{d\tau^2}. \tag{17}$$

Substituting above relations into (12) gives

$$\frac{m}{T^2}\frac{d^2r}{d\tau^2} + \frac{c}{T}\frac{dr}{d\tau} = ar^{-\gamma_1} - br^{-\gamma_2}. \tag{18}$$

Dividing both sides by m yields

$$\frac{1}{T^2}\frac{d^2r}{d\tau^2} + \frac{c}{mT}\frac{dr}{d\tau} = \frac{a}{m}r^{-\gamma_1} - \frac{b}{m}r^{-\gamma_2}. \tag{19}$$

Now choose the value of T in order to satisfy the following condition:

$$\frac{c}{mT} = o(1). \tag{20}$$

An appropriate choice for T can be $T = c/m$.
So

$$\frac{1}{T^2} \ll 1 \text{ yields } \frac{m^2}{c^2} \ll 1 \implies m^2 \ll c^2. \tag{21}$$

This can be interpreted as expressing that the damping coefficient in (12), which is the viscosity of the fluid, is very strong, or the mass is very small. It exactly has the same meaning with the small Reynolds number, which was mentioned above.

In a precise sense define ε, such that

$$\varepsilon = \frac{m^2}{c^2}. \tag{22}$$

Thus, (19) becomes

$$\varepsilon\frac{d^2r}{d\tau^2} = -\frac{dr}{d\tau} + \frac{a}{m}r^{-\gamma_1} - \frac{b}{m}r^{-\gamma_2}. \tag{23}$$

The character ε is too small, so one can eliminate the right hand side as follows:

$$\frac{dr}{d\tau} = \frac{a}{m}r^{-\gamma_1} - \frac{b}{m}r^{-\gamma_2} = f(r). \tag{24}$$

The main question that naturally arises is that how much precise (24) can be and in what circumstances one can use it as a description of the system. To answer this question, we analyzed the behavior of (23) in the phase space. By defining a new parameter R, (23) will be

$$\frac{dr}{d\tau} = R,$$
$$\varepsilon\frac{dR}{d\tau} = -R + \frac{a}{m}r^{-\gamma_1} - \frac{b}{m}r^{-\gamma_2}. \tag{25}$$

By defining, $f(r) = (a/m)\,r^{-\gamma_1} - (b/m)\,r^{-\gamma_2}$ turns into

$$\frac{dr}{d\tau} = R,$$
$$\frac{dR}{d\tau} = \frac{1}{\varepsilon}(f(r) - R). \tag{26}$$

By considering that the phase plane portraits this system and curve $C : f(r) - R = 0$ in this plane. This curve demonstrates the first-order system because the second equation in (26) equals to zero. We claim that the actual second-order system will converge to this curve as time progress. For proving, consider the actual curve starts from an $O(1)$ distance bellow the curve C. According to the second equation in (26), one has $dR/d\tau = O(1/\varepsilon)$. Therefore, the curve suddenly jumps into $f(r) - R = O(\varepsilon)$, and as $\varepsilon \to 0$ this region will be indistinguishable with C.

From the above discussion, one can conclude that the system in (26) can behave like a first order system but after a time lag T. Before this time lag, we are not allowed to eliminate the second-order derivative. In other words, the first order system behaves with a desired precision at $t > T$. This is exactly what one expects from the system. Since, the desired property of this system is its convergence radius as time approaching to infinity.

Thus, according to above analysis on (26) in the phase space, we can claim that (23) can be reduced into (24) as follows:

$$\frac{dr}{d\tau} = \frac{a}{m}r^{-\gamma_1} - \frac{b}{m}r^{-\gamma_2} \qquad (27)$$

and in general form for many agents we have

$$\dot{r}_i = \sum_{j=1, j \neq i}^{N} -\frac{(r_i - r_j)}{|r_i - r_j|}\left(\frac{a}{m}\|r_i - r_j\|^{-\gamma_1} - \frac{b}{m}\|r_i - r_j\|^{-\gamma_2} \right). \qquad (28)$$

The convergence radius can be easily determined from (28). For example, for the repulsion and attraction functions which were introduced in (5) and (8), respectively, for the first order (28), Gazi and Passino [2] determined the maximum convergence radius is not greater than b/a. This bound is for the swarm set whose agents have no diameter. In addition, this bound is independent to the number of the agents. To include the volume of each agent and the number of the agents participating in the cohesion task, one need to modify this bound. To yield this, consider N swarm members, which are accumulating in 3-dimensional spaces. If all the agents stick together to form a cluster, the swarm volume size will be

$$\text{The swarm volume size} = \frac{4}{3}\pi C^3 N. \qquad (29)$$

If all these agents converge to a sphere, the minimum radius of such a sphere can be determined easily by equaling the accumulated swarm volume and the volume of the sphere as follows:

$$\frac{4}{3}\pi C^3 N < \frac{4}{3}\pi r_{\min}{}^3 \implies r_{\min} > C\sqrt[3]{N}. \qquad (30)$$

So the previous bound modifies to

$$\text{The accurate size} = \frac{b}{a}C\sqrt[3]{N}. \qquad (31)$$

4. A Design Problem: Finding the Number of Participating Nanorobot Probes

Consider a group of nanorobots in a blood vessel cooperating for a special purpose that can be finding in some cancerous cells. According to experimental notes, 10^{12} robots

are injected through the veins [20]. A typical adult has 5 lit bloods. Therefore, the density of nano robots becomes 2×10^{-4} nanorobots/μm^3.

The maximum convergence radius for a swarm system with these attraction and repulsion functions in (5) and (8) for maintaining the density constant according to (31) is

$$\varepsilon < \frac{b}{a}C\sqrt[3]{N}. \qquad (32)$$

Therefore, the density of nanorobots is

$$\text{density} = \frac{N}{(4/3)\,\pi\varepsilon^3} = \frac{N}{(4/3)\,\pi\left(C\sqrt[3]{N}b/a\right)^3} = \frac{3a^3}{4\pi b^3 C^3}. \qquad (33)$$

Equaling this formula with the experimental measured density [20] of cooperative nanorobots in blood vessels gives

$$\frac{3a^3}{4\pi b^3 C^3} = 2 \times 10^{-4} \text{ Robot}/\mu m^3. \qquad (34)$$

Considering 1 μm diameter for each agent follows

$$\left(\frac{a}{b}\right)^3 = 2 \times 10^{-4} \times 4\pi \times \frac{C^3}{3} = 8.38 \times 10^{-4}. \qquad (35)$$

Then the ratio of attraction and repulsion coefficients is

$$\frac{a}{b} = 0.094 \approx 0.1. \qquad (36)$$

One of the important aspects of maximum convergence radius in (31) is to determine the number of nanorobot agents, which can accumulate in a vessel with respect to the diameter of the vessel. For instance, suppose that the diameter of nanorobot cohesion cannot exceed a quarter of the diameter of the vein. Hence one can easily calculate the maximum number of robots participate in cohesion. Now assume the diameter of the vein is $D = 800\,\mu m$, then

$$\varepsilon_{\max} = \frac{C\sqrt[3]{N}b}{a} \implies d_{\max} = \frac{2C\sqrt[3]{N}b}{a} = \frac{D}{4}. \qquad (37)$$

As we derived a reasonable ratio of attraction and repulsion constants for nano robots in blood, it yields

$$\frac{a}{b} = 0.1 \implies \frac{b}{a} = 10, \qquad D = 800\,\mu m, \qquad C = 1\,\mu m,$$

$$d_{\max} = \frac{2C\sqrt[3]{N}b}{a} = 20C\sqrt[3]{N} = \frac{D}{4} = 200\,\mu m \implies C\sqrt[3]{N}$$

$$= 10 \implies N = 1000 \text{ Robots.} \qquad (38)$$

It means that if the robot cohesion is forced to have a diameter less than a quarter of a vein diameter, the maximum numbers of agents that can participate in the cohesion are 1000 nanorobots.

This example could be converted into a design problem of micro channels if it asks the diameter of the vessel by giving the number of participating nanorobots.

5. Conclusion

In this study we discussed about the behavior dynamics of nanoscale grains and particles moving in a host liquid. According to this model, the dynamics of each particle depend on the resultant forces between it and other particles in its neighborhood that are in the form of attraction and repulsion components. One of the most important results of this model was to derive a convergence radius for particulate clustering in nanofluids that is one of the most important phenomena in the field of the researches in micro- and nanofluids especially in the context of heat transfer.

Acknowledgments

The authors would like to thank Milad Rakhsha from Amirkabir University of Technology, Iran, for bringing some useful references in the field of nanofluidics. Also, thanks are to professor Michael Zavlanos from Duke University, NC, USA, for his helpful comments.

References

[1] C. M. Breder, "Equations descriptive of fish schools and other animal aggregations," *Ecology*, vol. 35, pp. 361–370, 1954.

[2] V. Gazi and K. M. Passino, *Swarm Stability and Optimization*, Springer, New York, NY, USA, 2011.

[3] V. Gazi and K. M. Passino, "A class of attractions/repulsion functions for stable swarm aggregations," *International Journal of Control*, vol. 77, no. 18, pp. 1567–1579, 2004.

[4] W. L. Hase, "Molecular dynamics of clusters, surfaces, liquids & interfaces," in *Advances in Classical Trajectory Methods*, vol. 4, JAI press, 1999.

[5] J. Li and J. A. M. Kuipers, "Gas-particle interactions in dense gas-fluidized beds," *Chemical Engineering Science*, vol. 58, no. 3-6, pp. 711–718, 2003.

[6] M. Dorigo, V. Maniezzo, and A. Colorni, "Ant system: optimization by a colony of cooperating agents," *IEEE Transactions on Systems, Man, and Cybernetics B*, vol. 26, no. 1, pp. 29–41, 1996.

[7] R. J. Sadus, *Molecular Simulation of Fluids: Theory, Algorithms and Object-Orientation)*,, Elsevier, New York, NY, USA, 1999.

[8] T. I. Zohdi, "Particle collision and adhesion under the influence of near-fields," *Journal of Mechanics of Materials and Structures*, vol. 2, no. 6, pp. 1011–1018, 2007.

[9] K. B. Anoop, T. Sundararajan, and S. K. Das, "Effect of particle size on the convective heat transfer in nanofluid in the developing region," *International Journal of Heat and Mass Transfer*, vol. 52, no. 9-10, pp. 2189–2195, 2009.

[10] H. Chang and Y. C. Chang, "Fabrication of Al_2O_3 nano fluid by a plasma arc nano particles synthesis system," *Journal of Materials Processing Technology*, vol. 207, no. 1–3, pp. 193–199, 2008.

[11] H. Chen, Y. Ding, and A. Lapkin, "Rheological behaviour of nanofluids containing tube / rod-like nanoparticles," *Powder Technology*, vol. 194, no. 1-2, pp. 132–141, 2009.

[12] W. Lu and Q. Fan, "Study for the particle's scale effect on some thermo physical properties of nano fluids by a simplified molecular dynamics method," *Engineering Analysis with Boundary Elements*, vol. 32, no. 4, pp. 282–289, 2008.

[13] P. K. Namburu, D. K. Das, K. M. Tanguturi, and R. S. Vajjha, "Numerical study of turbulent flow and heat transfer characteristics of nanofluids considering variable properties," *International Journal of Thermal Sciences*, vol. 48, no. 2, pp. 290–302, 2009.

[14] S. E. B. Maïga, S. J. Palm, C. T. Nguyen, G. Roy, and N. Galanis, "Heat transfer enhancement by using nanofluids in forced convection flows," *International Journal of Heat and Fluid Flow*, vol. 26, no. 4, pp. 530–546, 2005.

[15] M. Chandrasekar, S. Suresh, and A. Chandra Bose, "Experimental investigations and theoretical determination of thermal conductivity and viscosity of Al_2O_3/water nanofluid," *Experimental Thermal and Fluid Science*, vol. 34, no. 2, pp. 210–216, 2010.

[16] M. Chopkar, S. Kumar, D. R. Bhandari, P. K. Das, and I. Manna, "Development and characterization of Al_2Cu and Ag_2Al nanoparticle dispersed water and ethylene glycol based nanofluid," *Materials Science and Engineering B*, vol. 139, no. 2-3, pp. 141–148, 2007.

[17] B. Ghasemi and S. M. Aminossadati, "Brownian motion of nanoparticles in a triangular enclosure with natural convection," *International Journal of Thermal Sciences*, vol. 49, no. 6, pp. 931–940, 2010.

[18] W. Jiang, G. Ding, H. Peng, and H. Hu, "Modeling of nanoparticles' aggregation and sedimentation in nanofluid," *Current Applied Physics*, vol. 10, no. 3, pp. 934–941, 2010.

[19] N. R. Karthikeyan, J. Philip, and B. Raj, "Effect of clustering on the thermal conductivity of nano fluids," *Materials Chemistry and Physics*, vol. 109, no. 1, pp. 50–55, 2008.

[20] A. Cavalcanti, T. Hogg, B. Shirinzadeh, and H. C. Liaw, "Nanorobot communication techniques: a comprehensive tutorial," in *Proceedings of the 9th International Conference on Control, Automation, Robotics and Vision (ICARCV '06)*, Singapore, December 2006.

[21] R. A. Freitas Jr., *Basic Capabilities*, vol. 1 of *Nanomedicine*, Landes Bioscience, Georgetown, Tex, USA, 1999.

[22] R. A. Freitas Jr., *Biocompatibility*, vol. 2A of *Nanomedicine*, Landes Bioscience, Georgetown, Tex, USA, 2003.

[23] S. H. Strogatz, *Nonlinear Dynamic and Chaos: With Applications to Physics, Biology, Chemistry and Engineering*, Addison-Wesley, Reading, Mass, USA, 1994.

4

A New Hyperbolic Shear Deformation Theory for Bending Analysis of Functionally Graded Plates

Tahar Hassaine Daouadji,[1, 2] Abdelaziz Hadj Henni,[1, 2] Abdelouahed Tounsi,[2] and Adda Bedia El Abbes[2]

[1] Departement of Civil Engineering, Ibn Khaldoun University of Tiaret, BP 78 Zaaroura, 14000 Tiaret, Algeria
[2] Laboratoire des Matériaux et Hydrologie, Université de Sidi Bel Abbes, BP 89 Cité Ben M'hidi, 22000 Sidi Bel Abbes, Algeria

Correspondence should be addressed to Tahar Hassaine Daouadji, daouadjitah@yahoo.fr

Academic Editor: Guowei Wei

Theoretical formulation, Navier's solutions of rectangular plates based on a new higher order shear deformation model are presented for the static response of functionally graded plates. This theory enforces traction-free boundary conditions at plate surfaces. Shear correction factors are not required because a correct representation of transverse shearing strain is given. Unlike any other theory, the number of unknown functions involved is only four, as against five in case of other shear deformation theories. The mechanical properties of the plate are assumed to vary continuously in the thickness direction by a simple power-law distribution in terms of the volume fractions of the constituents. Numerical illustrations concern flexural behavior of FG plates with metal-ceramic composition. Parametric studies are performed for varying ceramic volume fraction, volume fractions profiles, aspect ratios, and length to thickness ratios. Results are verified with available results in the literature. It can be concluded that the proposed theory is accurate and simple in solving the static bending behavior of functionally graded plates.

1. Introduction

The concept of functionally graded materials (FGMs) was first introduced in 1984 by a group of material scientists in Japan, as ultrahigh temperature-resistant materials for aircraft, space vehicles, and other engineering applications. Functionally graded materials (FGMs) are new composite materials in which the microstructural details are spatially varied through nonuniform distribution of the reinforcement phase. This is achieved by using reinforcement with different properties, sizes, and shapes, as well as by interchanging the role of reinforcement and matrix phase in a continuous manner. The result is a microstructure that produces continuous or smooth change on thermal and mechanical properties at the macroscopic or continuum level (Koizumi, 1993 [1]; Hirai and Chen, 1999 [2]). Now, FGMs are developed for general use as structural components in extremely high-temperature environments. Therefore, it is important to study the wave propagation of functionally graded materials structures in terms of nondestructive evaluation and material characterization.

Several studies have been performed to analyze the mechanical or the thermal or the thermomechanical responses of FG plates and shells. A comprehensive review is done by Tanigawa (1995) [3]. Reddy (2000) [4] has analyzed the static behavior of functionally graded rectangular plates based on his third-order shear deformation plate theory. Cheng and Batra (2000) [5] have related the deflections of a simply supported FG polygonal plate given by the first-order shear deformation theory and third-order shear deformation theory to that of an equivalent homogeneous Kirchhoff plate. The static response of FG plate has been investigated by Zenkour (2006) [6] using a generalized shear deformation theory. In a recent study, Şimşek (2010) [7] has studied the dynamic deflections and the stresses of an FG simply supported beam subjected to a moving mass by using Euler Bernoulli, Timoshenko, and the parabolic shear deformation beam theory. Şimşek (2010) [8],

TABLE 1: Displacement models.

Model	Theory	Unknown function
CPT	Classical plate theory [12]	3
ATDSP	Analytical tree dimensional solution for plate (3D) [13]	5
SSDPT	Sinusoidal shear deformation plate theory (Zenkour) [6]	5
PSDPT	Parabolic shear deformation plate theory (Reddy) [4]	5
NHPSDT	New hyperbolic shear deformation theory (present)	4

TABLE 2: Center deflections of isotropic homogenous plates ($k = 0$, E and $a/b = 1$).

h/a	CPT [12]	3D [13] $z = 0$	SSDPT [6]	Present theory: NHPSDT	Reddy [4]
0.01	44360.9	44384.7	44383.84	44383.86	44383.87
0.03	1643.00	1650.94	1650.646	1650.652	1650.657
0.1	44.3609	46.7443	46.6548	46.65655	46.65836

Benchour et al. [9], and Abdelaziz et al. 2010 [10] studied the free vibration of FG beams having different boundary conditions using the classical, the first-order, and different higher-order shear deformation beam and plate theories. The nonlinear dynamic analysis of an FG beam with pinned-pinned supports due to a moving harmonic load has been examined by Şimşek (2010) [11] using Timoshenko beam theory.

The primary objective of this paper is to present a general formulation for functionally graded plates (FGPs) using a new higher-order shear deformation plate theory with only four unknown functions. The present theory satisfies equilibrium conditions at the top and bottom faces of the plate without using shear correction factors. The hyperbolic function in terms of thickness coordinate is used in the displacement field to account for shear deformation. Governing equations are derived from the principle of minimum total potential energy. Navier solution is used to obtain the closed-form solutions for simply supported FG plates. To illustrate the accuracy of the present theory, the obtained results are compared with three-dimensional elasticity solutions and results of the first-order, and the other higher-order theories (Table 1).

In this study, a new displacement models for an analysis of simply supported FGM plates are proposed. The plates are made of an isotropic material with material properties varying in the thickness direction only. Analytical solutions for bending deflections of FGM plates are obtained. The governing equations are derived from the principle of minimum total potential energy. Numerical examples are presented to illustrate the accuracy and efficiency of the present theory by comparing the obtained results with those computed using various other theories.

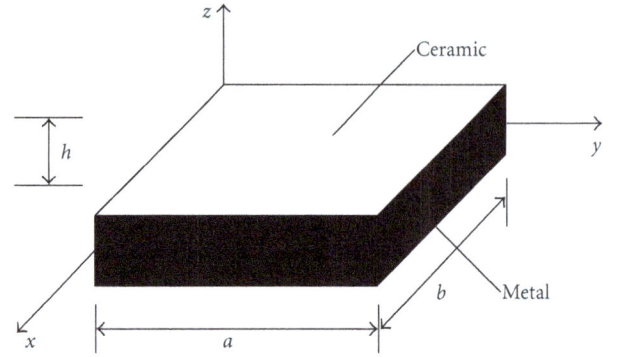

FIGURE 1: Geometry of rectangular plate composed of FGM.

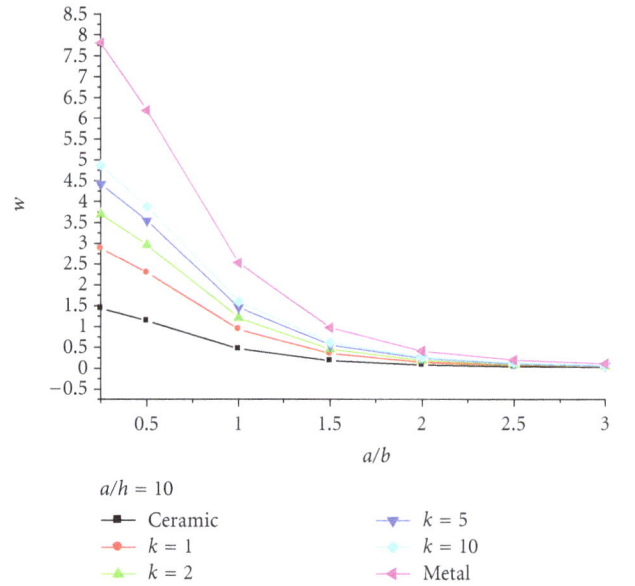

$a/h = 10$

- Ceramic
- $k = 1$
- $k = 2$
- $k = 5$
- $k = 10$
- Metal

FIGURE 2: Dimensionless center deflection (w) as function of the aspect ratio (a/b) of an FGM plate.

2. Problem Formulation

Consider a plate of total thickness h and composed of functionally graded material through the thickness. It is assumed that the material is isotropic, and grading is assumed to be only through the thickness. The xy plane is taken to be the undeformed mid plane of the plate with the z-axis positive upward from the mid plane (Figure 1).

2.1. Displacement Fields and Strains. The assumed displacement field is as follows:

$$u(x, y, z) = u_0(x, y) - z\frac{\partial w_b}{\partial x} - f(z)\frac{\partial w_s}{\partial x},$$

$$v(x, y, z) = v_0(x, y) - z\frac{\partial w_b}{\partial y} - f(z)\frac{\partial w_s}{\partial y},$$

$$w(x, y, z) = w_b(x, y) + w_s(x, y),$$

(1)

TABLE 3: Distribution of stresses across the thickness of isotropic homogenous plates (E; $a/b = 1$ and $k = 0$).

h/a	z	$\sigma_x(a/2, b/2, z)$				$\tau_{xy}(0, 0, -z)$			
		3D	SSDPT	NHPSDT	Reddy	3D	SSDPT	NHPSDT	Reddy
0.01	0.005	2873.3	2873.39	2873.422	2873.41	1949.6	1949.36	1949.086	1949.061
	0.004	2298.6	2298.57	2298.597	2298.593	1559.2	1559.04	1558.854	1558.843
	0.003	1723.9	1723.84	1723.861	1723.865	1169.1	1168.99	1168.883	1168.895
	0.002	1149.2	1149.18	1149.197	1149.205	779.3	779.18	779.127	779.151
	0.001	574.6	574.58	574.585	574.591	389.6	389.55	389.523	389.541
	0.000	0.000	0.000	0.00000	0.000	0.000	0.000	0.000	0.000
0.03	0.015	319.4	319.445	319.445	319.437	217.11	217.156	217.082	217.058
	0.012	255.41	255.415	255.416	255.413	173.26	173.282	173.255	173.244
	0.009	191.49	191.472	191.475	191.48	129.75	129.682	129.686	129.698
	0.006	127.63	127.603	127.607	127.615	86.41	86.313	86.330	86.354
	0.003	63.8	63.788	63.790	63.796	43.18	43.72	43.126	43.143
	0.000	0.000	0.000	0.0000	0.000	0.000	0.000	0.000	0.000
0.10	0.05	28.89	28.9307	28.928	28.92	19.92	20.0476	20.021	20.003
	0.04	22.998	23.0055	23.004	23.000	15.606	15.6459	15.638	15.629
	0.03	17.182	17.166	17.167	17.171	11.558	11.4859	11.494	11.504
	0.02	11.423	11.3994	11.402	11.410	7.642	7.5315	7.546	7.565
	0.01	5.702	5.6858	5.687	5.693	3.803	3.7265	3.7369	3.751
	0.00	0.000	0.000	0.000	0.000	0.000	0.000	0.000	0.000

where u_0 and v_0 are the mid-plane displacements of the plate in the x and y directions, respectively; w_b and w_s are the bending and shear components of transverse displacement, respectively, while $f(z)$ represents shape functions determining the distribution of the transverse shear strains and stresses along the thickness and is given as

$$f(z) = z\left[1 + \frac{3\pi}{2}\sec h^2\left(\frac{1}{2}\right)\right] - \frac{3\pi}{2}h\tanh\left(\frac{z}{h}\right). \quad (2)$$

It should be noted that unlike the first-order shear deformation theory, this theory does not require shear correction factors. The kinematic relations can be obtained as follows:

$$\varepsilon_x = \varepsilon_x^0 + zk_x^b + f(z)k_x^s,$$

$$\varepsilon_y = \varepsilon_y^0 + zk_y^b + f(z)k_y^s,$$

$$\gamma_{xy} = \gamma_{xy}^0 + zk_{xy}^b + f(z)k_{xy}^s,$$

$$\gamma_{yz} = g(z)\gamma_{yz}^s, \quad (3)$$

$$\gamma_{xz} = g(z)\gamma_{xz}^s,$$

$$\varepsilon_z = 0,$$

where

$$\varepsilon_x^0 = \frac{\partial u_0}{\partial x}, \quad k_x^b = -\frac{\partial^2 w_b}{\partial x^2}, \quad k_x^s = -\frac{\partial^2 w_s}{\partial x^2},$$

$$\varepsilon_y^0 = \frac{\partial v_0}{\partial y}, \quad k_y^b = -\frac{\partial^2 w_b}{\partial y^2}, \quad k_y^s = -\frac{\partial^2 w_s}{\partial y^2},$$

$$\gamma_{xy}^0 = \frac{\partial u_0}{\partial y} + \frac{\partial v_0}{\partial x}, \quad k_{xy}^b = -2\frac{\partial^2 w_b}{\partial x\partial y}, \quad k_{xy}^s = -2\frac{\partial^2 w_s}{\partial x\partial y}, \quad (4)$$

$$\gamma_{yz}^s = \frac{\partial w_s}{\partial y}, \quad \gamma_{xz}^s = \frac{\partial w_s}{\partial x}, \quad g(z) = 1 - f'(z),$$

$$f'(z) = \frac{df(z)}{dz}.$$

2.2. Constitutive Relations. In FGM, material property gradation is considered through the thickness, and the expression given below represents the profile for the volume fraction

$$P(z) = (P_t - P_b)\left(\frac{z}{h} + \frac{1}{2}\right)^k + P_b, \quad (5)$$

where P denotes a generic material property like modulus, P_t and P_b denote the property of the top and bottom faces of the plate, respectively, and k is a parameter that dictates material variation profile through the thickness. Here, it is assumed that modules E and G vary according to the equation (5), and

ν is assumed to be a constant. The linear constitutive relations are

$$\begin{Bmatrix} \sigma_x \\ \sigma_y \\ \tau_{yz} \\ \tau_{xz} \\ \tau_{xy} \end{Bmatrix} = \begin{bmatrix} Q_{11} & Q_{12} & 0 & 0 & 0 \\ Q_{12} & Q_{11} & 0 & 0 & 0 \\ 0 & 0 & Q_{44} & 0 & 0 \\ 0 & 0 & 0 & Q_{55} & 0 \\ 0 & 0 & 0 & 0 & Q_{66} \end{bmatrix} \begin{Bmatrix} \varepsilon_x \\ \varepsilon_y \\ \gamma_{yz} \\ \gamma_{xz} \\ \gamma_{xy} \end{Bmatrix}, \qquad (6)$$

where

$$Q_{11} = \frac{E(z)}{1-\nu^2}, \qquad Q_{12} = \nu Q_{11},$$

$$Q_{44} = Q_{55} = Q_{66} = \frac{E(z)}{2(1+\nu)}. \qquad (7)$$

2.3. Governing Equations. The governing equations of equilibrium can be derived by using the principle of virtual displacements. The principle of virtual work in the present case yields

$$\int_{-h/2}^{h/2} \int_\Omega \left[\sigma_x \delta\varepsilon_x + \sigma_y \delta\varepsilon_y + \tau_{xy}\delta\gamma_{xy} \right.$$

$$\left. + \tau_{yz}\delta\gamma_{yz} + \tau_{xz}\delta\gamma_{xz} \right] d\Omega\, dz - \int_\Omega q\delta w\, d\Omega = 0, \qquad (8)$$

where Ω is the top surface and $\mathbf{q}$ is the applied transverse load.

Substituting (3) and (6) into (8) and integrating through the thickness of the plate, (8) can be rewritten as

$$\int_\Omega \left[N_x \delta\varepsilon_x^0 + N_y \delta\varepsilon_y^0 + N_{xy}\delta\varepsilon_{xy}^0 + M_x^b \delta k_x^b + M_y^b \delta k_y^b \right.$$

$$+ M_{xy}^b \delta k_{xy}^b + M_x^s \delta k_x^s + M_y^s \delta k_y^s + M_{xy}^s \delta k_{xy}^s \qquad (9)$$

$$\left. + S_{yz}^s \delta\gamma_{yz}^s + S_{xz}^s \delta\gamma_{xz}^s \right] d\Omega - \int_\Omega q\delta w\, d\Omega = 0,$$

where

$$\begin{Bmatrix} N_x, N_y, N_{xy} \\ M_x^b, M_y^b, M_{xy}^b \\ M_x^s, M_y^s, M_{xy}^s \end{Bmatrix} = \int_{-h/2}^{h/2} (\sigma_x, \sigma_y, \tau_{xy}) \begin{Bmatrix} 1 \\ z \\ f(z) \end{Bmatrix} dz, \qquad (10a)$$

$$\left(S_{xz}^s, S_{yz}^s \right) = \int_{-h/2}^{h/2} \left(\tau_{xz}, \tau_{yz} \right) g(z)\, dz, \qquad (10b)$$

The governing equations of equilibrium can be derived from (9) by integrating the displacement gradients by parts and setting the coefficients δu_0, δv_0, δw_b, and δw_s zero separately.

Thus one can obtain the equilibrium equations associated with the present shear deformation theory as follows:

$$\delta u : \frac{\partial N_x}{\partial x} + \frac{\partial N_{xy}}{\partial y} = 0,$$

$$\delta v : \frac{\partial N_{xy}}{\partial x} + \frac{\partial N_y}{\partial y} = 0,$$

$$\delta w_b : \frac{\partial^2 M_x^b}{\partial x^2} + 2\frac{\partial^2 M_{xy}^b}{\partial x \partial y} + \frac{\partial^2 M_y^b}{\partial y^2} + q = 0,$$

$$\delta w_s : \frac{\partial^2 M_x^s}{\partial x^2} + 2\frac{\partial^2 M_{xy}^s}{\partial x \partial y} + \frac{\partial^2 M_y^s}{\partial y^2} + \frac{\partial S_{xz}^s}{\partial x} + \frac{\partial S_{yz}^s}{\partial y} + q = 0. \qquad (11)$$

Using (6) in (10a) and (10b) the stress resultants of a sandwich plate made up of three layers can be related to the total strains by

$$\begin{Bmatrix} N \\ M^b \\ M^s \end{Bmatrix} = \begin{bmatrix} A & B & B^s \\ A & D & D^s \\ B^s & D^s & H^s \end{bmatrix} \begin{Bmatrix} \varepsilon \\ k^b \\ k^s \end{Bmatrix}, \qquad S = A^s \gamma, \qquad (12)$$

where

$$N = \left\{ N_x, N_y, N_{xy} \right\}^t, \quad M^b = \left\{ M_x^b, M_y^b, M_{xy}^b \right\}^t,$$

$$M^s = \left\{ M_x^s, M_y^s, M_{xy}^s \right\}^t, \qquad (13a)$$

$$\varepsilon = \left\{ \varepsilon_x^0, \varepsilon_y^0, \gamma_{xy}^0 \right\}^t, \quad k^b = \left\{ k_x^b, k_y^b, k_{xy}^b \right\}^t,$$

$$k^s = \left\{ k_x^s, k_y^s, k_{xy}^s \right\}^t, \qquad (13b)$$

$$A = \begin{bmatrix} A_{11} & A_{12} & 0 \\ A_{12} & A_{22} & 0 \\ 0 & 0 & A_{66} \end{bmatrix}, \quad B = \begin{bmatrix} B_{11} & B_{12} & 0 \\ B_{12} & B_{22} & 0 \\ 0 & 0 & B_{66} \end{bmatrix},$$

$$\qquad (13c)$$

$$D = \begin{bmatrix} D_{11} & D_{12} & 0 \\ D_{12} & D_{22} & 0 \\ 0 & 0 & D_{66} \end{bmatrix},$$

$$B^s = \begin{bmatrix} B_{11}^s & B_{12}^s & 0 \\ B_{12}^s & B_{22}^s & 0 \\ 0 & 0 & B_{66}^s \end{bmatrix}, \quad D^s = \begin{bmatrix} D_{11}^s & D_{12}^s & 0 \\ D_{12}^s & D_{22}^s & 0 \\ 0 & 0 & D_{66}^s \end{bmatrix},$$

$$\qquad (13d)$$

$$H^s = \begin{bmatrix} H_{11}^s & H_{12}^s & 0 \\ H_{12}^s & H_{22}^s & 0 \\ 0 & 0 & H_{66}^s \end{bmatrix},$$

$$S = \left\{ S_{xz}^s, S_{yz}^s \right\}^t, \quad \gamma = \left\{ \gamma_{xz}, \gamma_{yz} \right\}^t, \quad A^s = \begin{bmatrix} A_{44}^s & 0 \\ 0 & A_{55}^s \end{bmatrix},$$

$$\qquad (13e)$$

where A_{ij}, B_{ij}, and so forth are the plate stiffness, defined by

$$
\begin{Bmatrix}
A_{11} & B_{11} & D_{11} & B_{11}^s & D_{11}^s & H_{11}^s \\
A_{12} & B_{12} & D_{12} & B_{12}^s & D_{12}^s & H_{12}^s \\
A_{66} & B_{66} & D_{66} & B_{66}^s & D_{66}^s & H_{66}^s
\end{Bmatrix}
$$

$$
= \int_{-h/2}^{h/2} Q_{11} \left(1, z, z^2, f(z), zf(z), f^2(z)\right) \begin{Bmatrix} 1 \\ \nu \\ \dfrac{1-\nu}{2} \end{Bmatrix} dz,
$$

$$(14a)$$

$$
(A_{22}, B_{22}, D_{22}, B_{22}^s, D_{22}^s, H_{22}^s) = (A_{11}, B_{11}, D_{11}, B_{11}^s, D_{11}^s, H_{11}^s),
$$

$$(14b)$$

$$
A_{44}^s = A_{55}^s = \int_{h_{n-1}}^{h_n} Q_{44} [g(z)]^2 dz.
$$

$$(14c)$$

Substituting from (12) into (11), we obtain the following equation:

$$
A_{11}d_{11}u_0 + A_{66}d_{22}u_0 + (A_{12} + A_{66})d_{12}v_0 - B_{11}d_{111}w_b
$$

$$
- (B_{12} + 2B_{66})d_{122}w_b - (B_{12}^s + 2B_{66}^s)d_{122}w_s \tag{15a}
$$

$$
- B_{11}^s d_{111}w_s = 0,
$$

$$
A_{22}d_{22}v_0 + A_{66}d_{11}v_0 + (A_{12} + A_{66})d_{12}u_0 - B_{22}d_{222}w_b
$$

$$
- (B_{12} + 2B_{66})d_{112}w_b - (B_{12}^s + 2B_{66}^s)d_{112}w_s \tag{15b}
$$

$$
- B_{22}^s d_{222}w_s = 0,
$$

$$
B_{11}d_{111}u_0 + (B_{12} + 2B_{66})d_{122}u_0 + (B_{12} + 2B_{66})d_{112}v_0
$$

$$
+ B_{22}d_{222}v_0 - D_{11}d_{1111}w_b - 2(D_{12} + 2D_{66})d_{1122}w_b
$$

$$
- D_{22}d_{2222}w_b - D_{11}^s d_{1111}w_s - 2(D_{12}^s + 2D_{66}^s)d_{1122}w_s \tag{15c}
$$

$$
- D_{22}^s d_{2222}w_s = q,
$$

$$
B_{11}^s d_{111}u_0 + (B_{12}^s + 2B_{66}^s)d_{122}u_0 + (B_{12}^s + 2B_{66}^s)d_{112}v_0
$$

$$
+ B_{22}^s d_{222}v_0 - D_{11}^s d_{1111}w_b - 2(D_{12}^s + 2D_{66}^s)d_{1122}w_b
$$

$$
- D_{22}^s d_{2222}w_b - H_{11}^s d_{1111}w_s - 2(H_{12}^s + 2H_{66}^s)d_{1122}w_s \tag{15d}
$$

$$
- H_{22}^s d_{2222}w_s + A_{55}^s d_{11}w_s + A_{44}^s d_{22}w_s = q,
$$

where d_{ij}, d_{ijl}, and d_{ijlm} are the following differential operators:

$$
d_{ij} = \frac{\partial^2}{\partial x_i \partial x_j}, \quad d_{ijl} = \frac{\partial^3}{\partial x_i \partial x_j \partial x_l},
$$

$$
d_{ijlm} = \frac{\partial^4}{\partial x_i \partial x_j \partial x_l \partial x_m}, \quad d_i = \frac{\partial}{\partial x_i}, \quad (i, j, l, m = 1, 2). \tag{16}
$$

2.4. Exact Solution for a Simply Supported FGM Plate. Rectangular plates are generally classified in accordance with the type of support used. We are here concerned with the exact

solution of (15a)–(15d) for a simply supported FG plate. The following boundary conditions are imposed at the side edges:

$$
v_0 = w_b = w_s = \frac{\partial w_s}{\partial y} = N_x = M_x^b = M_x^s = 0 \quad \text{at } x = -\frac{a}{2}, \frac{a}{2},
$$

$$(17a)$$

$$
u_0 = w_b = w_s = \frac{\partial w_s}{\partial x} = N_y = M_y^b = M_y^s = 0 \quad \text{at } y = -\frac{b}{2}, \frac{b}{2},
$$

$$(17b)$$

To solve this problem, Navier assumed the transverse mechanical and temperature loads, q, in the form of a double trigonometric series as

$$
q = q_0 \sin(\lambda x) \sin(\mu y), \tag{18}
$$

where $\lambda = \pi/a$, $\mu = \pi/b$, and q_0 represents the intensity of the load at the plate center.

Following the Navier solution procedure, we assume the following solution form for u_0, v_0, w_b and w_s that satisfies the following boundary conditions:

$$
\begin{Bmatrix} u_0 \\ v_0 \\ w_b \\ w_s \end{Bmatrix} = \begin{Bmatrix} U \cos(\lambda x) \sin(\mu y) \\ V \sin(\lambda x) \cos(\mu y) \\ W_b \sin(\lambda x) \sin(\mu y) \\ W_s \sin(\lambda x) \sin(\mu y) \end{Bmatrix}, \tag{19}
$$

where U, V, W_b, and W_s are arbitrary parameters to be determined subjected to the condition that the solution in (19) satisfies governing equations (15a)–(15d). One obtains the following operator equation:

$$
[C]\{\Delta\} = \{P\}, \tag{20}
$$

where $\{\Delta\} = \{U, V, W_b, W_s\}^t$ and $[C]$ is the symmetric matrix given by

$$
[C] = \begin{bmatrix}
a_{11} & a_{12} & a_{13} & a_{14} \\
a_{12} & a_{22} & a_{23} & a_{24} \\
a_{13} & a_{23} & a_{33} & a_{34} \\
a_{14} & a_{24} & a_{34} & a_{44}
\end{bmatrix}, \tag{21}
$$

in which

$$
a_{11} = A_{11}\lambda^2 + A_{66}\mu^2,
$$

$$
a_{12} = \lambda\mu(A_{12} + A_{66}),
$$

$$
a_{13} = -\lambda\left[B_{11}\lambda^2 + (B_{12} + 2B_{66})\mu^2\right],
$$

$$
a_{14} = -\lambda\left[B_{11}^s\lambda^2 + (B_{12}^s + 2B_{66}^s)\mu^2\right],
$$

$$
a_{22} = A_{66}\lambda^2 + A_{22}\mu^2,
$$

$$
a_{23} = -\mu\left[(B_{12} + 2B_{66})\lambda^2 + B_{22}\mu^2\right],
$$

$$
a_{24} = -\mu\left[(B_{12}^s + 2B_{66}^s)\lambda^2 + B_{22}^s\mu^2\right],
$$

$$
a_{33} = D_{11}\lambda^4 + 2(D_{12} + 2D_{66})\lambda^2\mu^2 + D_{22}\mu^4,
$$

$$
a_{34} = D_{11}^s\lambda^4 + 2(D_{12}^s + 2D_{66}^s)\lambda^2\mu^2 + D_{22}^s\mu^4,
$$

$$
a_{44} = H_{11}^s\lambda^4 + 2(H_{11}^s + 2H_{66}^s)\lambda^2\mu^2 + H_{22}^s\mu^4 + A_{55}^s\lambda^2 + A_{44}^s\mu^2,
$$

$$(22)$$

3. Numerical Results and Discussions

The study has been focused on the static behavior of functionally graded plate based on the present new higher-order shear deformation model. Here, some representative results of the Navier solution obtained for a simply supported rectangular plate are presented.

A functionally graded material consisting of aluminum-alumina is considered. The following material properties are used in computing the numerical values (Bouazza et al. [14]).

   (i) Metal (aluminum, Al): $E_M = 70$ GPa; $\nu = 0.3$.

   (ii) Ceramic (alumina, Al$_2$O$_3$): $E_C = 380$ GPa; $\nu = 0.3$.

Now, a functionally graded material consisting of aluminum and alumina is considered. Young's modulus for aluminum is 70 GPa while for alumina is 380 GPa. Note that, Poisson's ratio is selected constant for both and equal to 0.3. The various nondimensional parameters used are

$$\overline{w} = \frac{10h^3 E}{a^4 q_0} w\left(\frac{a}{2}, \frac{b}{2}\right), \quad \overline{u}_x = \frac{100h^3 E}{a^4 q_0} u_x\left(\frac{a}{2}, \frac{b}{2}, -\frac{h}{4}\right),$$

$$\overline{u}_y = \frac{100h^3 E}{a^4 q_0} u_y\left(\frac{a}{2}, \frac{b}{2}, -\frac{h}{6}\right), \quad \overline{\sigma}_x = \frac{h}{a q_0} \sigma_x\left(\frac{a}{2}, \frac{b}{2}, \frac{h}{2}\right),$$

$$\overline{\sigma}_y = \frac{h}{a q_0} \sigma_y\left(\frac{a}{2}, \frac{b}{2}, \frac{h}{3}\right), \quad \overline{\tau}_{xy} = \frac{h}{a q_0} \tau_{xy}\left(0, 0, -\frac{h}{3}\right),$$

$$\overline{\tau}_{yz} = \frac{h}{a q_0} \tau_{yz}\left(\frac{a}{2}, 0, \frac{h}{6}\right), \quad \overline{\tau}_{xz} = \frac{h}{a q_0} \tau_{xz}\left(0, \frac{b}{2}, 0\right).$$

$$(23)$$

It is clear that the deflection increases as the side-to-thickness ratio decreases. The same results were obtained in most literatures. In addition, the correlation between the present new higher-order shear deformation theory and different higher-order and first-order shear deformation theories is established by the author in his recent papers. It is found that this theory predicts the deflections and stresses more accurately when compared to the first- and third-order theories.

For the sake of completeness, results of the present theory are compared with those obtained using a new Navier-type three-dimensional exact solution for small deflections in bending of linear elastic isotropic homogeneous rectangular plates. The center deflection w and the distribution across the plate thickness of in-plane longitudinal stress σ_x and longitudinal tangential stress τ_{xy} are compared with the results of the 3D solution and are shown in Tables 2 and 3. The present solution is realized for a quadratic plate, with the following fixed data: $a = 1, b = 1, E_m = E_c = E = 1, q_0 = 1, \nu = 0.3$ and three values for the plate thickness: $h = 0.01, h = 0.03$, and $h = 0.1$. It is to be noted that the present results compare very well with the 3D solution. All deflections again compare well with the 3D solution and show good convergence with the average 3D solution.

In Table 4, the effect of volume fraction exponent on the dimensionless stresses and displacements of an FGM square plate ($a/h = 10$) is given. This table shows comparison between results for plates subjected to uniform or sinusoidal distributed loads, respectively. As it is well known, the uniform load distribution always overpredicts the displacements and stresses magnitude. As the plate becomes more and more metallic, the difference increases for deflection w and in-plane longitudinal stress σ_x while it decreases for in-plane normal stress σ_y. It is important to observe that the stresses for a fully ceramic plate are the same as that for a fully metal plate. This is because the plate for these two cases is fully homogeneous, and the stresses do not depend on the modulus of elasticity. Results in Table 4 should serve as benchmark results for future comparisons.

Tables 5 and 6 compare the deflections and stresses of different types of the FGM square plate ($a/b = 1, k = 0$) and FGM rectangular plate ($b = 3a, k = 2$). The deflections decrease as the aspect ratio a/b increases and this irrespective of the type of the FGM plate. All theories (SSDPT, PSDPT, and NHPSDT) give the same axial stress σ_x and σ_y for a fully ceramic plate ($k = 0$). In general, the axial stress increases with the volume fraction exponent k. The transverse shear stress for a FGM plate subjected to a distributed load. The results show that the transverse shear stresses may be indistinguishable. As the volume fraction exponent increases for FGM plates, the shear stress will increase, and the fully ceramic plates give the smallest shear stresses.

Figures 2 and 3 show the variation of the center deflection with the aspect and side-to-thickness ratios, respectively. The deflection is maximum for the metallic plate and minimum for the ceramic plate. The difference increases as the aspect ratio increases while it may be unchanged with the increase of side-to-thickness ratio. One of the main inferences from the analysis is that the response of FGM plates is intermediate to that of the ceramic and metal homogeneous plates (see also Table 4). It is to be noted that, in the case of thermal or combined loads and under certain conditions, the above response is not intermediate.

Figures 7 and 8 depict the through-the-thickness distributions of the shear stresses τ_{yz} and τ_{xz}, the in-plane longitudinal and normal stresses σ_x and σ_y, and the longitudinal tangential stress τ_{xy} in the FGM plate under the uniform load. The volume fraction exponent of the FGM plate is taken as $k = 2$ in these figures. Distinction between the curves in Figures 8 and 9 is obvious. As strain gradients increase, the inhomogeneities play a greater role in stress distribution calculations. The through-the-thickness distributions of the shear stresses τ_{yz} and τ_{xz} are not parabolic, and the stresses increase as the aspect ratio decreases. It is to be noted that the maximum value occurs at $z \cong 0.2$, not at the plate center as in the homogeneous case.

As exhibited in Figures 5 and 6, the in-plane longitudinal and normal stresses, σ_x and σ_y, are compressive throughout the plate up to $z \cong 0.155$ and then they become tensile. The maximum compressive stresses occur at a point on the bottom surface and the maximum tensile stresses occur, of course, at a point on the top surface of the FGM plate. However, the tensile and compressive values of the longitudinal tangential stress, τ_{xy} (cf. Figure 7), are maximum at a point on the bottom and top surfaces of the FGM plate, respectively. It is clear that the minimum value of zero for all

TABLE 4: Effects of volume fraction exponent and loading on the dimensionless stresses and displacements of a FGM square plate ($a/h = 10$).

k	Theory	w	σ_x	σ_y	τ_{yz}	τ_{xz}	τ_{xy}
0	NHPSDT	0.4665	2.8928	1.9104	0.4424	0.5072	1.2851
ceramic	SSDPT	0.4665	2.8932	1.9103	0.4429	0.5114	1.2850
	Reddy	0.4665	2.8920	1.9106	0.4411	0.4963	1.2855
	NHPSDT	0.9421	4.2607	2.2569	0.54404	0.50721	1.1573
1	SSDPT	0.9287	4.4745	2.1692	0.5446	0.5114	1.1143
	Reddy	0.94214	4.25982	2.25693	0.54246	0.49630	1.15725
	NHPSDT	1.2228	4.8890	2.1663	0.5719	0.4651	1.0448
2	SSDPT	1.1940	5.2296	2.0338	0.5734	0.4700	0.9907
	Reddy	1.22275	4.88814	2.16630	0.56859	0.45384	1.04486
	NHPSDT	1.3533	5.2064	1.9922	0.56078	0.4316	1.0632
3	SSDPT	1.3200	5.6108	1.8593	0.5629	0.4367	1.0047
	Reddy	1.3530	5.20552	1.99218	0.55573	0.41981	1.06319
	NHPSDT	1.4653	5.7074	1.7143	0.50075	0.4128	1.1016
5	SSDPT	1.4356	6.1504	1.6104	0.5031	0.4177	1.0451
	Reddy	1.46467	5.70653	1.71444	0.49495	0.40039	1.10162
	NHPSDT	1.6057	6.9547	1.3346	0.4215	0.4512	1.1118
10	SSDPT	1.5876	7.3689	1.2820	0.4227	0.4552	1.0694
	Reddy	1.60541	6.95396	1.33495	0.41802	0.43915	1.1119
∞	NHPSD	2.5327	2.8928	1.9104	0.4424	0.5072	1.2851
métal	SSDPT	2.5327	2.8932	1.9103	0.4429	0.5114	1.2850
	Reddy	2.5328	2.8920	1.9106	0.4411	0.4963	1.2855

TABLE 5: Comparison of normalized displacements and stresses of a FGM square plate ($a/b = 1$) and $k = 0$.

a/h	Theory	w	σ_x	σ_y	τ_{yz}	τ_{xz}	τ_{xy}
	NHPSDT	0.5866	1.1979	0.7536	0.4307	0.4937	0.4908
4	SSDPT	0.5865	1.1988	0.7534	0.4307	0.4973	0.4906
	Reddy	0.5868	1.1959	0.7541	0.4304	0.4842	0.4913
	NHPSDT	0.4665	2.8928	1.9104	0.4424	0.5072	1.2851
10	SSDPT	0.4665	2.8932	1.9103	0.4429	0.5114	1.2850
	Reddy	0.4666	2.8920	1.9106	0.4411	0.4963	1.2855
	NHPSDT	0.4438	28.7342	19.1543	0.4466	0.5119	12.9884
100	SSDPT	0.4438	28.7342	19.1543	0.4472	0.5164	13.0125
	Reddy	0.4438	28.7341	19.1543	0.4448	0.5004	12.9885

TABLE 6: Comparison of normalized displacements and stresses of a FGM rectangular plate ($b = 3a$) and $k = 2$.

a/h	Theory	w	σ_x	σ_y	τ_{yz}	τ_{xz}	τ_{xy}
	NHPSDT	4.0569	5.2804	0.6644	0.6084	0.6699	0.5900
4	SSDPT	3.99	5.3144	0.6810	0.6096	0.6796	0.5646
	Reddy	4.0529	5.2759	0.6652	0.6058	0.6545	0.5898
	NHPSDT	3.5543	12.9252	1.6938	0.61959	0.6841	1.4898
10	SSDPT	3.5235	12.9374	1.7292	0.6211	0.6910	1.4500
	Reddy	3.5537	12.9234	1.6941	0.6155	0.6672	1.4898
	NHPSDT	3.4824	25.7712	3.3971	0.6214	0.6878	2.9844
20	SSDPT	3.4567	25.7748	3.4662	0.6232	0.6947	2.9126
	Reddy	3.48225	25.7703	3.3972	0.6171	0.6704	2.9844
	NHPSDT	3.4593	128.728	17.0009	0.6220	0.6894	14.9303
100	SSDPT	3.4353	128.713	17.3437	0.6238	0.6963	14.584
	Reddy	3.45937	128.7283	17.0009	0.6177	0.67176	14.9303

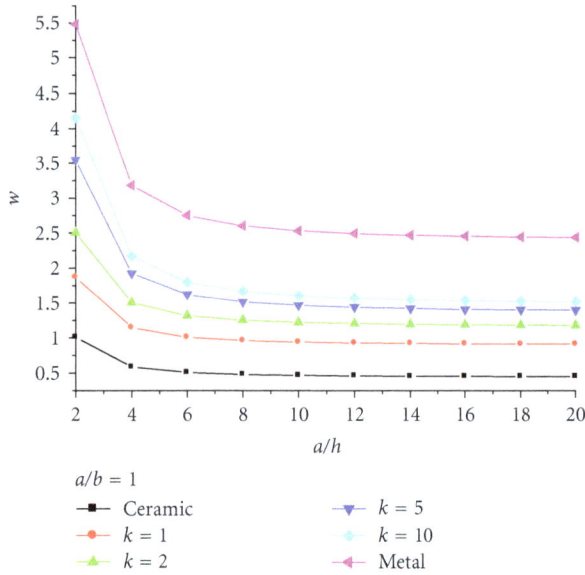

FIGURE 3: Dimensionless center deflection (w) as a function of the side-to-thickness ratio (a/h) of an FGM square plate.

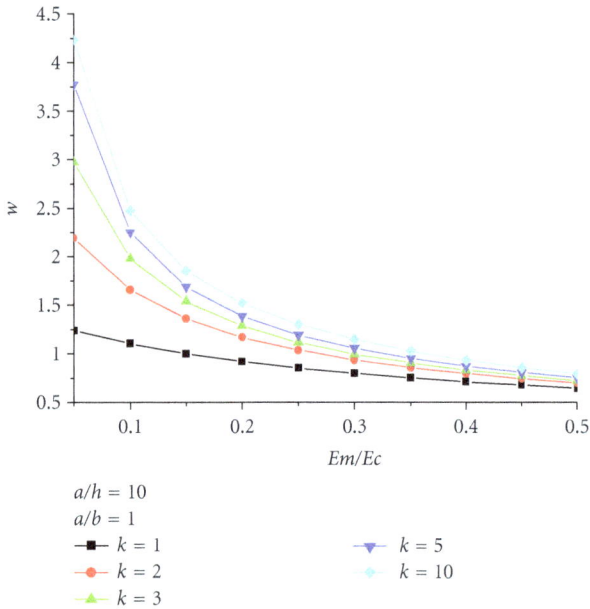

FIGURE 4: The effect of anisotropy on the dimensionless maximum deflection (w) of an FGM plate for different values of k.

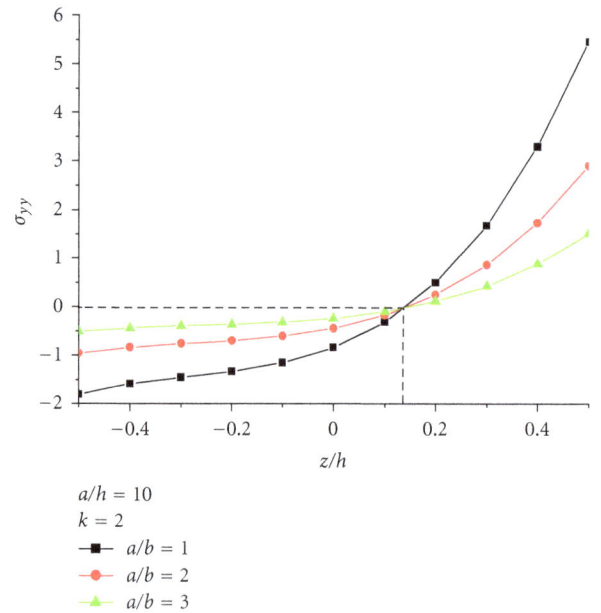

FIGURE 5: Variation of in-plane longitudinal stress (σ_{xx}) through-the thickness of an FGM plate for different values of the side-to-thickness ratio.

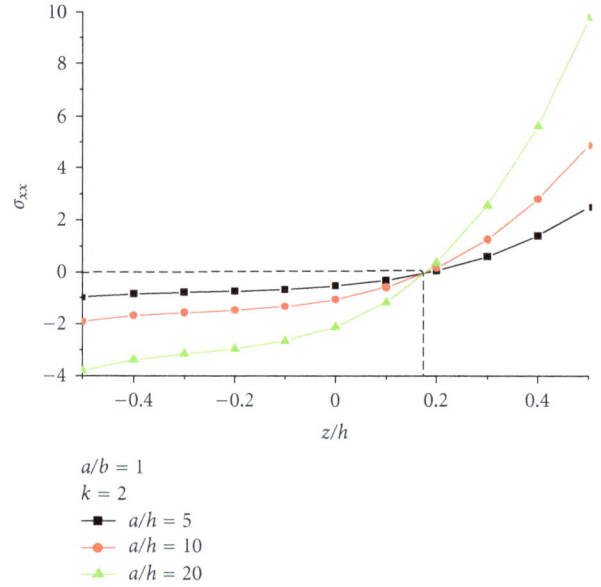

FIGURE 6: Variation of in-plane normal stress (σ_{yy}) through-the thickness of an FGM plate for different values of the aspect ratio.

in-plane stresses σ_x, σ_y and τ_{xy} occurs at $z \cong 0.153$ and this is irrespective of the aspect and side-to-thickness ratios.

Finally, the exact maximum deflections of simply supported FGM square plate are compared in Figure 4 for various ratios of module, E_m/E_c (for a given thickness, $a/h = 10$). This means that the deflections are computed for plates with different ceramic-metal mixtures. It is clear that the deflections decrease smoothly as the volume fraction exponent decreases and as the ratio of metal-to-ceramic modules increases.

4. Conclusion

In this study, a new higher-order shear deformation model is proposed to analyze the static behavior of functionally graded plates. Unlike any other theory, the theory presented gives rise to only four governing equations resulting in considerably lower computational effort when compared with the other higher-order theories reported in the literature having more number of governing equations. Bending and

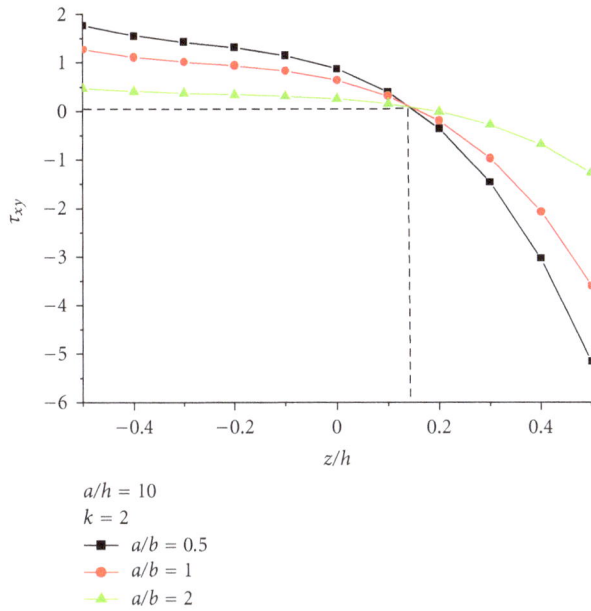

FIGURE 7: Variation of longitudinal tangential stress (τ_{xy}) through-the thickness of an FGM plate for different values of the aspect ratio.

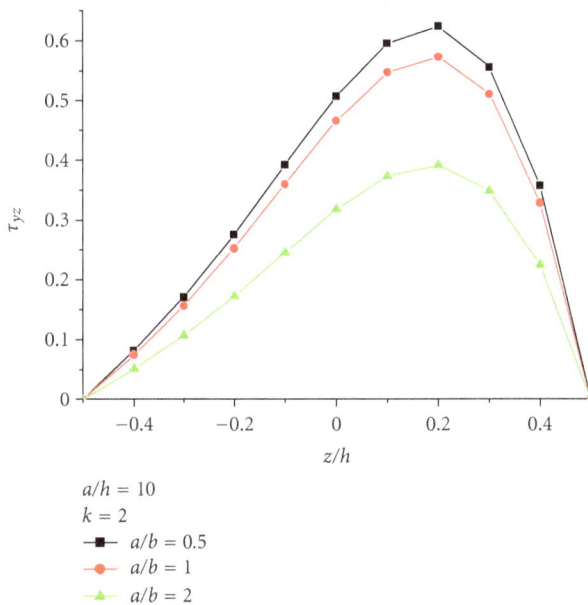

FIGURE 9: Variation of transversal shear stress (τ_{xz}) through the thickness of an FGM plate for different values of the aspect ratio.

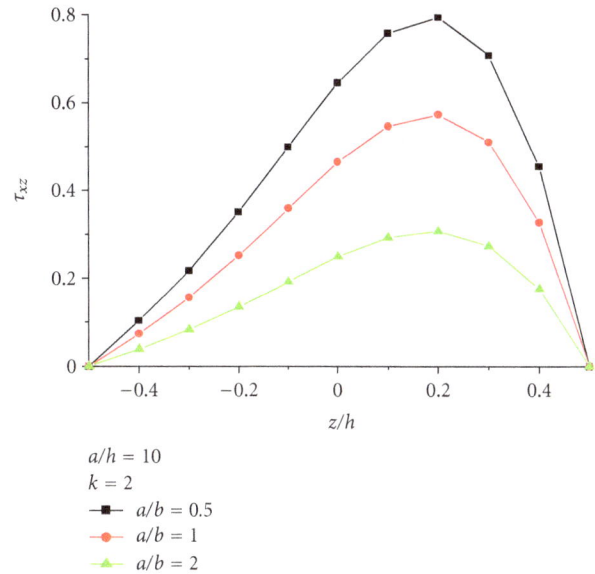

FIGURE 8: Variation of transversal shear stress (τ_{yz}) through-the thickness of an FGM plate for different values of the aspect ratio.

and stresses obtained using the present new higher-order shear deformation theories (with four unknowns) and other higher shear deformation theories such as PSDPT and SSDPT (with five unknowns) are almost identical. The extension of the present theory is also envisaged for general boundary conditions and plates of a more general shape. In conclusion, it can be said that the proposed theory NHPSDT is accurate and simple in solving the static behaviors of FGM plates.

stress analysis under transverse load were analyzed, and results were compared with previous other shear deformation theories. The developed theories give parabolic distribution of the transverse shear strains and satisfy the zero traction boundary conditions on the surfaces of the plate without using shear correction factors. The accuracy and efficiency of the present theories have been demonstrated for static behavior of functionally graded plates. All comparison studies demonstrated that the deflections

References

[1] M. Koizumi, "The concept of FGM," *Ceramic Transactions, Functionally Gradient Materials*, vol. 34, pp. 3–10, 1993.
[2] T. Hirai and L. Chen, "Recent and prospective development of functionally graded materials in Japan," *Materials Science Forum*, vol. 308–311, pp. 509–514, 1999.
[3] Y. Tanigawa, "Some basic thermoelastic problems for nonhomogeneous structural materials," *Applied Mechanics Reviews*, vol. 48, no. 6, pp. 287–300, 1995.
[4] J. N. Reddy, "Analysis of functionally graded plates," *International Journal for Numerical Methods in Engineering*, vol. 47, no. 1–3, pp. 663–684, 2000.
[5] Z. Q. Cheng and R. C. Batra, "Deflection relationships between the homogeneous Kirchhoff plate theory and different functionally graded plate theories," *Archives of Mechanics*, vol. 52, no. 1, pp. 143–158, 2000.
[6] A. M. Zenkour, "Generalized shear deformation theory for bending analysis of functionally graded plates," *Applied Mathematical Modelling*, vol. 30, no. 1, pp. 67–84, 2006.
[7] M. Şimşek, "Vibration analysis of a functionally graded beam under a moving mass by using different beam theories," *Composite Structures*, vol. 92, no. 4, pp. 904–917, 2010.
[8] M. Şimşek, "Fundamental frequency analysis of functionally graded beams by using different higher-order beam theories," *Nuclear Engineering and Design*, vol. 240, no. 4, pp. 697–705, 2010.

[9] A. Benachour, H. D. Tahar, H. A. Atmane, A. Tounsi, and M. S. Ahmed, "A four variable refined plate theory for free vibrations of functionally graded plates with arbitrary gradient," *Composites B*, vol. 42, no. 6, pp. 1386–1394, 2011.

[10] H. H. Abdelaziz, H. A. Atmane, I. Mechab, L. Boumia, A. Tounsi, and A. B. E. Abbas, "Static analysis of functionally graded sandwich plates using an efficient and simple refined theory," *Chinese Journal of Aeronautics*, vol. 24, no. 4, pp. 434–448, 2011.

[11] M. Şimşek, "Non-linear vibration analysis of a functionally graded Timoshenko beam under action of a moving harmonic load," *Composite Structures*, vol. 92, no. 10, pp. 2532–2546, 2010.

[12] S. P. Timoshenko and S. Woinowsky-Krieger, *Theory of Plates and Shells*, McGraw-Hill, New York, NY, USA, 1959.

[13] H. Werner, "A three-dimensional solution for rectangular plate bending free of transversal normal stresses," *Communications in Numerical Methods in Engineering*, vol. 15, no. 4, pp. 295–302, 1999.

[14] M. Bouazza, A. Tounsi, E. A. Adda Bedia, and M. Meguenni, "Stability analysis of fonctionnally graded plates subject to thermal load," *Advanced Structures Materials B*, vol. 15, pp. 669–680, 2011.

5

Buckling Analysis of Laminated Composite Panel with Elliptical Cutout Subject to Axial Compression

Hamidreza Allahbakhsh and Ali Dadrasi

Mechanical Department, Islamic Azad University, Shahrood Branch, Shahrood, Iran

Correspondence should be addressed to Hamidreza Allahbakhsh, allahbakhshy@gmail.com

Academic Editor: Jing-song Hong

A buckling analysis has been carried out to investigate the response of laminated composite cylindrical panel with an elliptical cutout subject to axial loading. The numerical analysis was performed using the Abaqus finite-element software. The effect of the location and size of the cutout and also the composite ply angle on the buckling load of laminated composite cylindrical panel is investigated. Finally, simple equations, in the form of a buckling load reduction factor, were presented by using the least square regression method. The results give useful information into designing a laminated composite cylindrical panel, which can be used to improve the load capacity of cylindrical panels.

1. Introduction

Laminated composite shells are widely used in many industrial structures including automotive and aviation due to their lower weights compared to metal structures [1]. Many of these shell structures have cutouts or openings that serve as doors, windows, or access ports, and these cutouts or openings often require some type of reinforcing structure to control local structural deformations and stresses near the cutout. In addition, these structures may experience compression loads during operation, and thus their buckling response characteristics must be understood and accurately predicted in order to determine effective designs and safe operating conditions for these structures.

For predicting the buckling load and buckling mode of a structure in the finite-element program, the linear (or eigenvalue) buckling analysis is an existing technique for estimation [2]. In general, the analysis of composite laminated shell is more complicated than the analysis of homogeneous isotropic ones [3].

In the literature, many published studies investigated the buckling of laminated composite plates with a cutout [4–10]. Few studies are available on buckling of composite panel. Kim and Noor [11] studied the buckling and postbuckling responses of composite panels with central circular cutouts subjected to various combinations of mechanical and thermal loads. They investigated the effect of variations in the hole diameter; the aspect ratio of the panel; the laminate stacking sequence; the fiber orientation on the stability boundary; postbuckling response and sensitivity coefficients.

Mallela and Upadhyay [12] presented some parametric studies on simply supported laminated composite panels subjected to in-plane shear loading. They analyzed many models using ANSYS, and a database was prepared for different plate and stiffener combinations. Studies are carried out by changing the panel orthotropy ratio, pitch length (number of stiffeners), stiffener depth, smeared extensional stiffness ratio of stiffener to that of the plate, and extensional stiffness to shear stiffness ratio of the shell.

Transverse central impact on thin fiber-reinforced composite cylindrical panels with emphasis on the importance of in-plane membrane effects was studied by Kistler and Waas [13]. Both small and large deformation impact responses were examined in their work. A nonlinear system of equations was derived for the impact problem, including Hertz' contact law, and solved over time using Runge-Kutta integration.

An analytical method developed for determining the interlaminar stresses at straight free boundaries was extended

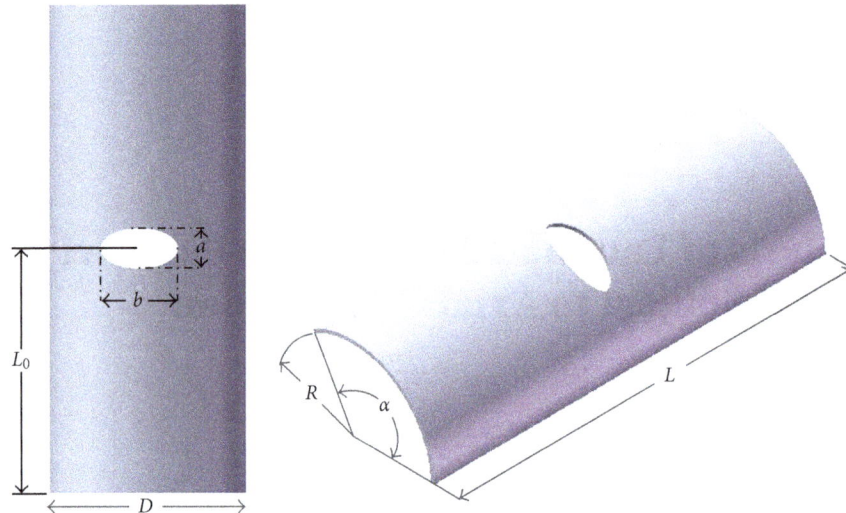

FIGURE 1: Geometry of panel.

to predict the free-edge stresses at curved boundaries of symmetric composite laminates under in plane loading by Chao Zhang et al. [14]. They described the three-dimensional (3D) stress distribution in laminates with curved boundaries on the basis of a zero-order approximation of the boundary-layer theory. The related stress functions were found by minimization of complementary energy and the variational principle and satisfy zero-order equilibrium equations, boundary conditions, and traction continuity at interfaces between plies.

Hu and Yang [15] optimized the buckling resistance of fiber-reinforced laminated cylindrical panels with a given material system and subjected to uniaxial compressive force with respect to fiber orientations by using a sequential linear programming method together with a simple move-limit strategy. The significant influences of panel thicknesses, curvatures, aspect ratios, cutouts, and end conditions on the optimal fiber orientations and the associated optimal buckling loads of laminated cylindrical panels have been shown through their investigation.

Dash et al. [16] presented vibration and stability of laminated composite curved panels with rectangular cutouts using finite-element method. The first-order shear deformation (FSDT) is used to model the curved panels, considering the effects of transverse shear deformation and rotary inertia. Dash's studies reveal that the fundamental frequencies of vibration of an angle ply flat panel decrease with introduction of small cutouts but again rise with increase in size of cutout. However, the higher frequencies of vibration continue to decrease up to a moderate size of cutout and then rise with further increase of size of cutout. The stability resistance decreases with increase in size of cutout in curved panels unlike the frequencies of vibration. Gal et al. [17] studied the buckling behavior of laminated composite panel, experimentally and numerically.

This paper studies the buckling behavior of laminated composite panel with elliptical cutout. Also, it presents parametric studies to investigate the effect of the cutout

size, cutout location, panel parameters, and ply angle on the buckling of the laminated composite panel. A set of linear analyses using the ABAQUS were carried out and were validated by comparing against solution published in literature. Finally, a set of formulas (based on the numerical results) for the computation of the buckling load reduction factor for laminated composite panel with elliptical cutouts are presented.

2. Geometry and Mechanical Properties of the Panels

The structure that is used for analyze is shown in Figure 1. The test specimen is a cylindrical panel with elliptical/circular cutout. According to this figure, parameter (a) displays the size of the cutout in longitudinal axis of the panel, and parameter (b) displays the size of the cutout along the circumferential direction of the panel. Specimens were nominated as follows: $L300$-$R250$-a-b-α. The numbers following L and R show the radius and length of the panel, respectively. The ply thickness of the composite is 0.125 mm with the laminate stacking of $[\theta/-\theta]_3$ (θ is measured from the cylinder longitudinal direction), which is antisymmetric about the middle surface, corresponding to the total thickness of $t = 0.75$ mm.

The nominal orthotropic elastic material properties are listed in Table 1 where the 1 direction is along the fibers, the 2 direction is transverse to the fibers in the surface of the lamina, and the 3 direction is normal to the lamina.

3. Numerical Analysis Using the Finite-Element Method

To obtain the buckling predictions and eigenvalue analyses with Abaqus, a "buckle" step is run. Eigenvalue analyses are performed for laminated composite cylindrical panel under axial loading that is the common type of loading studied

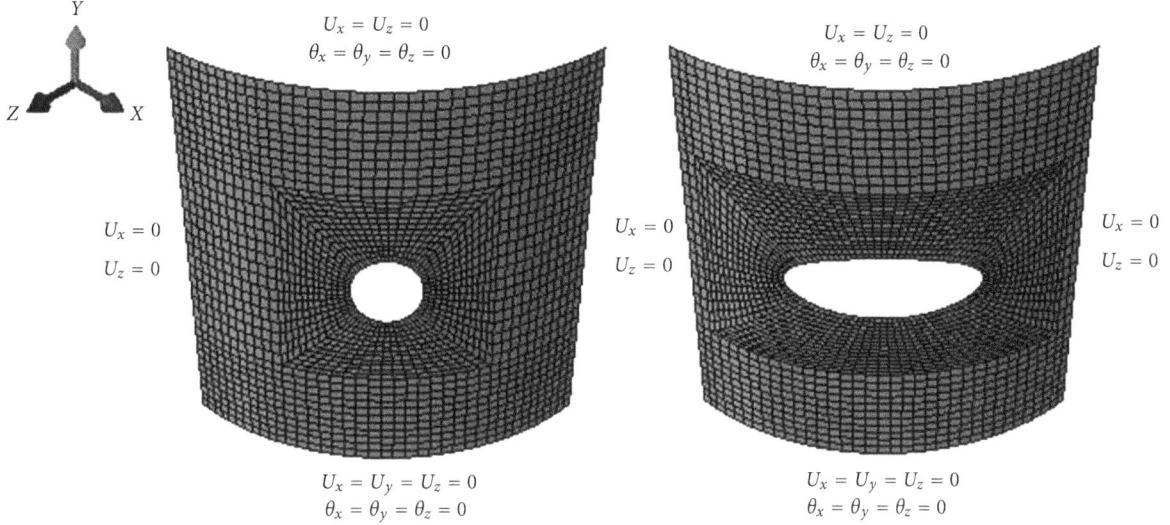

FIGURE 2: Sample mesh structure and boundary conditions.

TABLE 1: Mechanical properties of composite material.

E_{11} (kN/mm^2)	E_{22} (kN/mm^2)	G_{12}, G_{13} (kN/mm^2)	G_{23} (kN/mm^2)	ν_{12}
135	13	6.4	4.3	0.38

TABLE 2: Mesh convergence study of the cylindrical shells.

Approximate element size (mm × mm)	3	1.5	0.75	0.4
Buckling load (kN)		135	120	109	107
Difference percent with respect to previous value		11	11	9.1	1.8

for theoretical buckling studies on plates and shells, using FEM.

The panel is fully clamped on the bottom edge, clamped except for axial motion on the top edge, and simply supported along its vertical edges. The eight-node nonlinear element S8R5 which is an element with five degrees of freedom per node was used in analyses. The mesh is divided into two regions for each panel. In the region near the cutout, smaller elements are created, and also a convergence study was conducted for a composite cylindrical panel.

The results obtained from each refinement stage of the mesh were compared with previous stage and were summarized in Table 2.

In order to shun time consuming analyses, an element size equal to 3.5 mm × 3.5 mm was considered as general element size in the remaining numerical analyses. For this element size, the average aspect ratio of all elements is 1.34 which is adequate. The analyses showed that a typical element size of 0.45 mm could be used to model the area around the cutout. A typical finite-element model of a composite cylindrical panel with a cutout is shown in Figure 2.

4. Validation of FE Model for Axial Loading

To ascertain whether the FE model was sufficiently accurate, it was validated using results from existing experimental, numerical, and theoretical results. In this paper, for validation of FEA, deformation mode and buckling mode are investigated. Figures 3 and 4 show the comparison of results for the present simulations with Stanley [18] results.

5. Results of Numerical Analysis

In this section, the results of the buckling analyses of laminated cylindrical panel with elliptical/circular cutouts are presented that was done by finite-element method.

5.1. The Effect of Ply Angle on the Buckling Load. Designing an optimized composite laminate requires finding the best fiber orientation for each layer [20, 21]. Figure 5(a) shows the effect of ply angle on buckling shapes of the composite cylindrical panel in first buckling mode. Figure 5(b) displays the dependence of the first buckling load of the laminated composite cylindrical panel on the composite ply angle. For the ply angle in the range of $0 < \theta < 10$, the first buckling load is associated with buckling shape A, while increasing the ply angle causes buckling shapes B, C, D, and E to precede. For the ply angle in the range of $80 < \theta < 90$, the buckling shape A has been observed, again. The ply angle of $[70/-70]_3$ occurs to have the maximum load, having about 105% higher than that of the cylindrical panel with $[0/0]_3$ stacking. We also investigated the effect of ply angle on the first five buckling loads of the laminated composite cylindrical panels (data not shown for the sake of shortness). These results show that the sensitivity of buckling loads toward the ply angle increases a little for upper buckling modes.

5.2. The Effect of Change in Cutout Height on the Buckling Load. In this section, the effect of change in cutout height

Mode 1	Stanly	107 kN	
	FEM	109 kN	
Mode 2	Stanly	109 kN	
	FEM	112 kN	
Mode 3	Stanly	116 kN	
	FEM	118 kN	
Mode 4	Stanly	140 kN	
	FEM	145 kN	
Mode 5	Stanly	151 kN	
	FEM	162 kN	

FIGURE 3: Comparison of the numerical buckling load and mode shape with those obtained by Stanley [18].

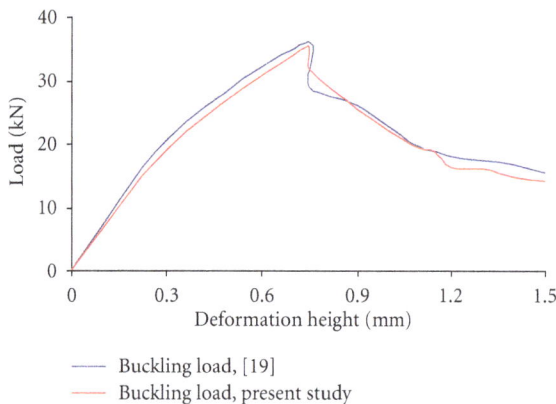

—— Buckling load, [19]
—— Buckling load, present study

FIGURE 4: Comparison of the numerical buckling curve with that obtained by Sabik and Kreja [19].

on the buckling load of laminated panel is investigated. To this investigation, cutouts with constant width (75 mm) were created in the midheight position of panels. Then, we study the change in buckling load with changing the height of the cutouts from 15 to 75 mm. The numerical results are listed in Table 4. Furthermore, Figures 6(a) and 6(b) show buckling load versus L/D and a/b ratios curves, respectively. According to Figure 6(a), it can be seen that buckling load of the laminated panel decreases slightly when the cutout height increases.

For panels with ratios $L/D = 0.7$, $L/D = 0.5$, and $L/D = 0.3$, with a radius of 400 mm, and with the increase of cutout height from 30 to 75 mm, the buckling load decreases 42, 34, and 31%, respectively. This reduction for panels with ratio $L/D = 0.75$, $L/D = 1.25$, and $L/D = 1.75$ and with a radius of 500 mm are 24, 20, and 14%, respectively. Therefore, it can be deduced that longer and slender panels are more sensitive to the change in cutout height. Also Figure 6(b) shows that shells with larger diameters and identical cutouts are more resistant to buckling.

5.3. The Effect of Change in Cutout Width on the Buckling Load. This section investigates the effect of changing the width of the cutout on the buckling load of the laminated composite cylindrical panel. So cutouts with constant height (30 mm) were created in the midheight of panels. Then, the effect of change in the width of the cutout on the buckling load was studied by changing cutouts width from 30 to 90 mm. The designation and analysis details of each model are summarized in Table 5.

Figures 7(a) and 7(b) show the buckling load versus a/b and L/D ratios curves, respectively. It can be seen that when the cutout height is fix, an increase in the width of the cutout decreases the buckling load.

In laminated cylindrical panels with a radius of 400 mm, the reduction in the buckling load with the increase of width of the cutout from 30 to 75 mm is 49, 42, and 40%, for panels with ratios $L/D = 0.7$, $L/D = 0.5$, and $L/D = 0.3$, respectively. In laminated composite panels with a radius of 1000 mm, with the increase of cutout width from 30 to 75 mm, the buckling load decreases 31, 27, and 20%, for panels with ratios $L/D = 0.75$, $L/D = 1.25$, and $L/D = 1.75$, respectively. So it is evident that longer and slender shells are more sensitive to changes in cutout width.

Comparing the results of this section with those presented in the previous section, it can be deduced that when the cutout height is fixed and cutout width increases 45 mm, the amount of decrease in the buckling load is greater than the corresponding value in the state that the cutout width is fixed and cutout height increases 45 mm. Accordingly, it is suggested that in the design of these panels, whenever possible, the bigger cutout dimension is oriented along the longitudinal axis of the panels.

5.4. Analysis of the Effect of Change in Dimensions of Fixed-Area Cutouts on the Buckling Behavior. The buckling behavior of laminated composite panels with different cutout geometries was studied in the previous sections. In

Buckling shape A

Buckling shape B

Buckling shape C

Buckling shape D

Buckling shape E

(a)

(b)

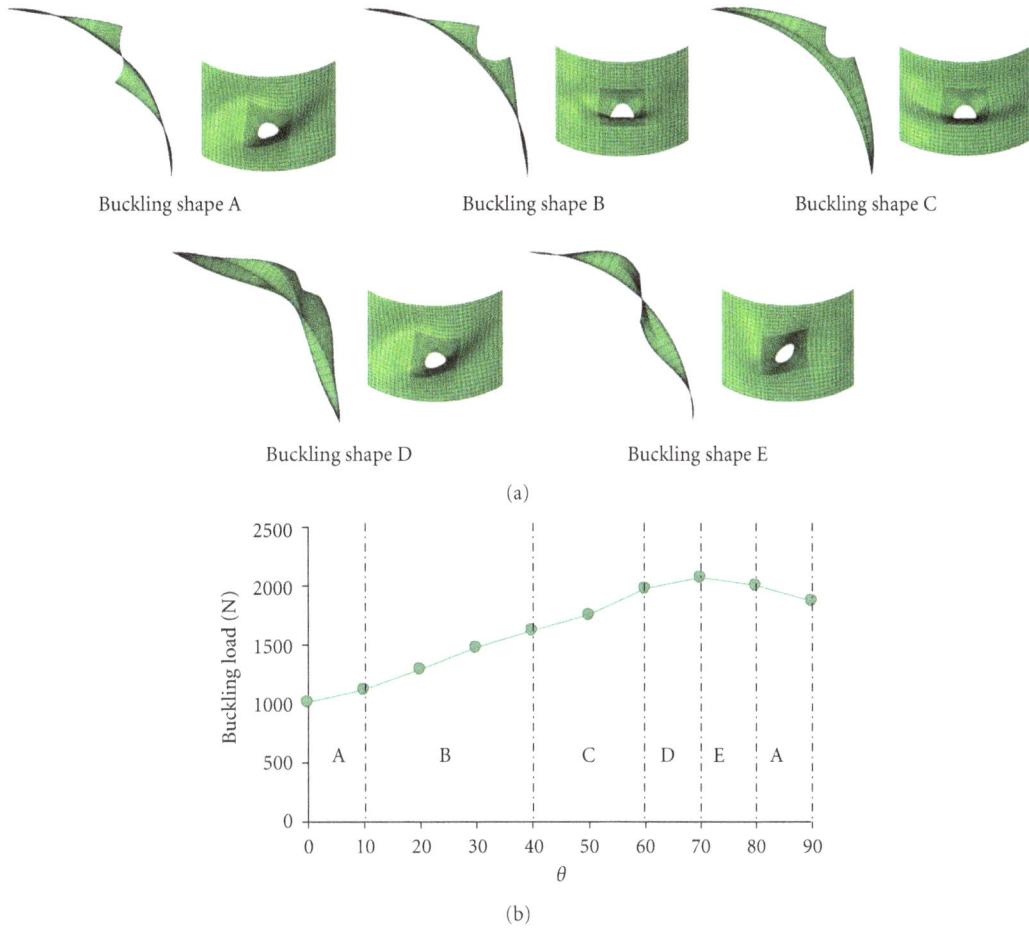

FIGURE 5: (a) Buckling shapes of a composite cylindrical panel with ply sequence of $[\theta/-\theta]_3$, which appear as the first buckling mode depending on the composite ply angle. (b) Variations of the first buckling load of composite cylindrical shell versus the composite ply angle.

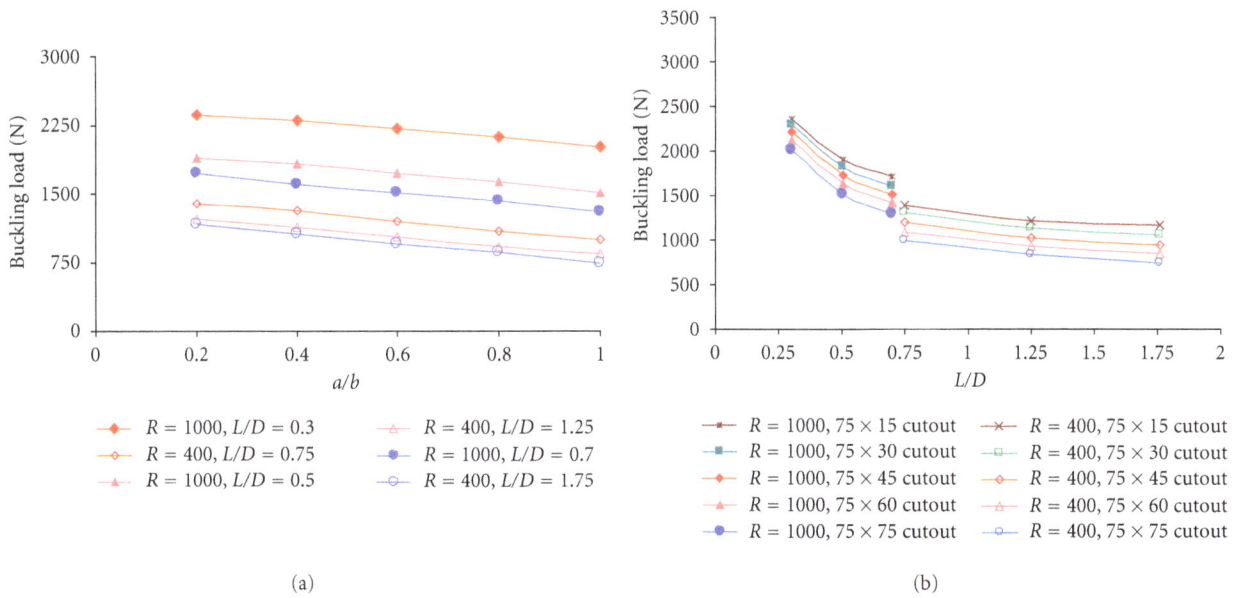

(a)

(b)

FIGURE 6: Comparison of the buckling load of laminated composite panel shells versus (a) ratios a/b and (b) L/D, for elliptical cutout with constant cutout width and various cutout heights.

(a)

(b)

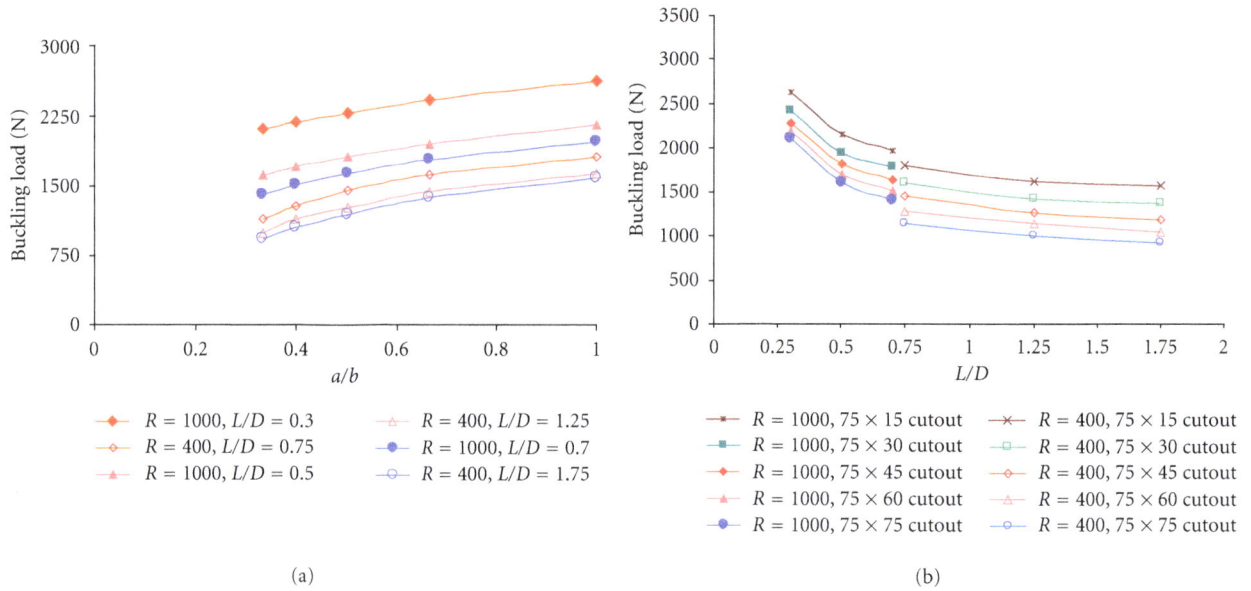

Figure 7: Comparison of the buckling load of laminated composite panel shells versus (a) ratios a/b and (b) L/D, for elliptical cutout with constant cutout heights and various cutout widths.

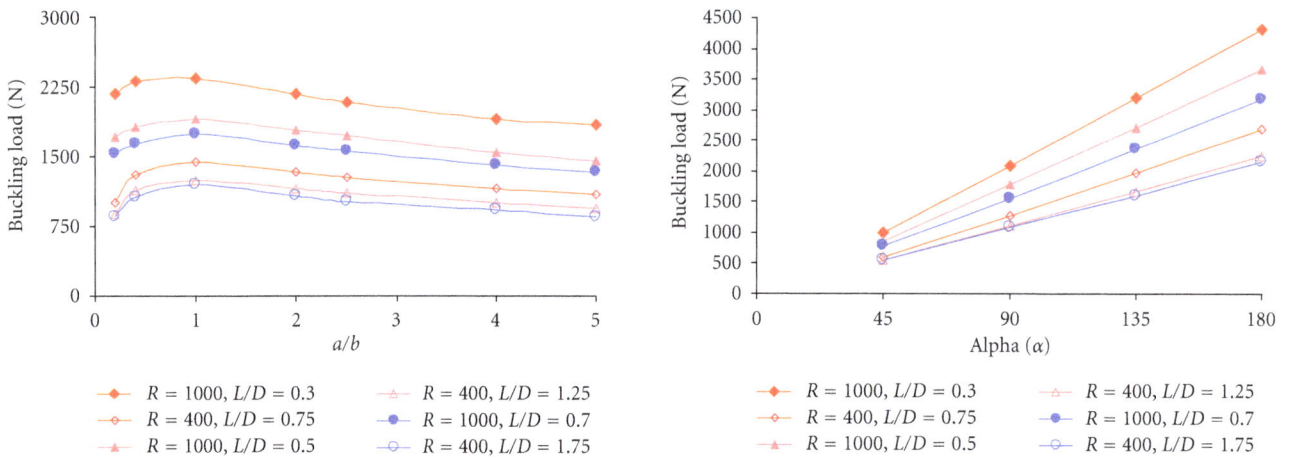

Figure 8: Plots of buckling load versus ratio a/b for cylindrical panel with an elliptical cutout with constant area.

Figure 9: Plots of buckling load versus α for cylindrical panel with an elliptical cutout.

this section, both height and width are changed, so that the product of cutout width and cutout height, which is representative of the area of the cutout, remains constant.

Therefore, cutouts with an area of $A = 8242.5 \, \text{mm}^2$ were created in the midheight of the panels. Seven different values for a/b ratio were considered. Figure 8 shows the buckling load versus the a/b ratio curves. Figure 8 clearly shows that for the a/b ratio in the range of $0 < a/b < 1$, when the cutout area is constant, an increases in a/b ratio increases the buckling load and for $a/b > 1$ with increase in a/b ratio decreases the buckling load. On the other hand, having $a/b = 1$ results in the highest load capacity.

5.5. Analysis of the Effect of Change in Panel Angle on the Buckling Behavior of Cylindrical Shells. In this section, we

investigated the relationship between the buckling load and angle of the laminated panel. For this study, we created an elliptical cutout of constant size (75×30 mm) in the midheight of the panels with various angles between $45°$ and $180°$. Figure 9 shows the buckling load versus L/D ratio. It is clear that with an increase in the panel angle, the buckling load of the panels increases. The results show that increasing the panel angle improves the shell resistance against buckling and increases the amount of the critical load. Furthermore, for short, intermediate-length, and long panels with radius of 200 mm, the buckling load increases 313, 300, and 284%, respectively. Also the buckling load for panel with radius of 500 mm increases 328, 322, and 300% for short, intermediate-length, and long panels, respectively. Therefore, slender and longer

TABLE 3: The formulas for predicting the buckling load reduction factors of laminated cylindrical panel.

Equation no.	Range	Parameters	Equations
(3)	$0.75 \leq L/D \leq 1.75$	$\left(\dfrac{L}{D}\right), \left(\dfrac{a}{D}\right)$	$K_{\text{cutout}} = 0.521 + 0.132\left(\dfrac{L}{D}\right) - 1.004\left(\dfrac{a}{D}\right) - 0.043\left(\dfrac{L}{D}\right)^2 - 0.080\dfrac{L}{D}\dfrac{a}{D} - 0.017\dfrac{a^2}{D}$
(4)	$0.30 \leq L/D \leq 0.50$	$\left(\dfrac{L}{D}\right), \left(\dfrac{a}{D}\right)$	$K_{\text{cutout}} = 0.311 + 1.662\left(\dfrac{L}{D}\right) - 0.164\left(\dfrac{a}{D}\right) - 1.327\left(\dfrac{L}{D}\right)^2 - 1.20\dfrac{L}{D}\dfrac{a}{D} - 13.015\dfrac{a^2}{D}$
(5)	$0.75 \leq L/D \leq 1.75$	$\left(\dfrac{L}{D}\right), \left(\dfrac{b}{D}\right)$	$K_{\text{cutout}} = 0.646 + 0.250\left(\dfrac{L}{D}\right) - 2.069\left(\dfrac{b}{D}\right) - 0.066\left(\dfrac{L}{D}\right)^2 - 0.421\dfrac{L}{D}\dfrac{b}{D} + 3.166\dfrac{b^2}{D}$
(6)	$0.30 \leq L/D \leq 0.50$	$\left(\dfrac{L}{D}\right), \left(\dfrac{b}{D}\right)$	$K_{\text{cutout}} = 0.376 + 1.637\left(\dfrac{L}{D}\right) - 0.777\left(\dfrac{b}{D}\right) - 1.140\left(\dfrac{L}{D}\right)^2 - 3.033\dfrac{L}{D}\dfrac{b}{D} + 0.529\dfrac{b^2}{D}$
(7)	$0.75 \leq L/D \leq 1.75$	$\left(\dfrac{L}{D}\right), \left(\dfrac{L_0}{L}\right)$	$K_{\text{cutout}} = 0.712 + 0.204\left(\dfrac{L}{D}\right) - 0.999\left(\dfrac{L_0}{L}\right) - 0.087\left(\dfrac{L}{D}\right)^2 + 0.062\dfrac{L}{D}\dfrac{L_0}{L} + 0.816\dfrac{L_0^2}{L}$
(8)	$0.30 \leq L/D \leq 0.50$	$\left(\dfrac{L}{D}\right), \left(\dfrac{L_0}{L}\right)$	$K_{\text{cutout}} = 0.350 + 1.684\left(\dfrac{L}{D}\right) - 0.399\left(\dfrac{L_0}{L}\right) - 1.595\left(\dfrac{L}{D}\right)^2 + 0.553\dfrac{L}{D}\dfrac{L_0}{L} + 0.232\dfrac{L_0^2}{L}$
(9)	$0.75 \leq L/D \leq 1.75$	$\left(\dfrac{L}{D}\right), \left(\dfrac{a}{b}\right)$	$K_{\text{cutout}} = 0.352 + 0.177\left(\dfrac{L}{D}\right) + 0.072\left(\dfrac{a}{b}\right) - 0.060\left(\dfrac{L}{D}\right)^2 - 0.005\dfrac{L}{D}\dfrac{a}{b} - 0.015\dfrac{a^2}{b}$
(10)	$0.30 \leq L/D \leq 0.50$	$\left(\dfrac{L}{D}\right), \left(\dfrac{a}{b}\right)$	$K_{\text{cutout}} = 0.292 + 1.524\left(\dfrac{L}{D}\right) + 0.014\left(\dfrac{a}{b}\right) - 1.202\left(\dfrac{L}{D}\right)^2 - 0.001\dfrac{L}{D}\dfrac{a}{b} - 0.008\dfrac{a^2}{b}$

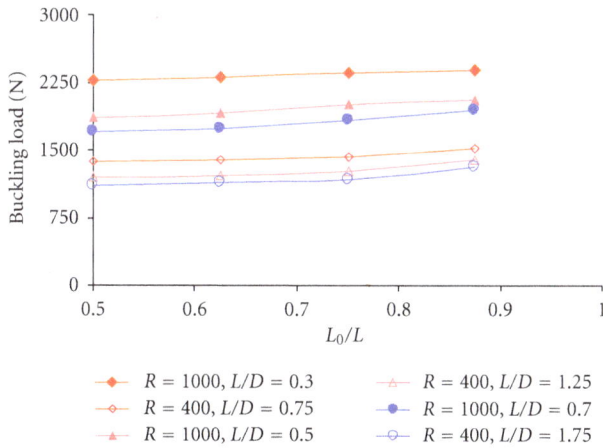

FIGURE 10: Summary of the buckling load of cylindrical panels with elliptical cutout located at various locations.

cylindrical panels are less sensitive to the changes of the panel angle.

5.6. Analysis of the Effect of Change in Cutout Position on the Buckling Load. The buckling load versus the cutout position (L_0/L) ratio curves for cylindrical laminated panel of various lengths are shown in Figure 10. This figure clearly shows that with changing the cutout position from midheight of the panels toward the edges, the buckling load slightly increases. It is clear that longer and slender panels are more sensitive to the change in the position of the cutout. For example, for panels with $L/D = 0.7$ and $R = 1000$, when the cutout is replaced from the miheight of the panels to 87.5% of its length, the buckling load increases 13.5%; while for panels with $L/D = 0.5$ and $R = 1000$, the increase in the buckling load is only 9.7%, and for panels with $L/D = 0.3$ and

$R = 1000$, the increase in the buckling load is restricted to only 4.3%. Similarly, for panels with $R = 400$, with the replace of the position of the cutout from midheight to 87.5% of panel height, the buckling load changes 17.7%, 16.1%, and 10.6% for ratios $L/D = 1.75$, 1.25, and 0.75, respectively.

6. Prediction of Buckling Load

The buckling behavior of the laminated composite cylindrical panel subjected to axial compressive loading was presented in the previous sections. Based on the numerical dimensionless buckling loads of panels, formulas are presented for the computation of the buckling load of laminated composite panels with elliptical cutouts subject to axial compression.

K_{cutout} is introduced as a buckling load reduction factor for cylindrical panels with cutout and defined according to

$$K_{\text{cutout}} = \frac{N_{\text{cutout}}}{N_{\text{Perfect}}}, \tag{1}$$

where N_{cutout} and N_{Perfect} are the buckling load for cylindrical panels with cutouts and the buckling load for cylindrical panels without cutouts, respectively.

The formulas are presented using the least-square regression method [22, 23]. Eight equations ((3)–(10)) were developed for various shell geometries, following the form:

$$\begin{aligned}
K_d(\alpha, \beta, \gamma, \eta, \lambda) &= A + B\alpha + C\beta + D\gamma + E\eta + F\lambda \\
&\quad + G\alpha^2 + H\beta^2 + I\gamma^2 + J\eta^2 + K\lambda^2 \\
&\quad + L\alpha\beta + M\alpha\gamma + N\alpha\lambda + O\alpha\lambda + P\beta\gamma \\
&\quad + Q\beta\eta + R\beta\gamma + S\gamma\eta + T\gamma\lambda + U\eta\lambda.
\end{aligned} \tag{2}$$

TABLE 4: Summary of numerical analysis for cylindrical shells including an elliptical cutout with constant width and different height.

Model designation	Shell length	Cutout size ($a \times b$)	Buckling load (N)
R400-θ90-L300-$L_0$150-α70-Perfect	300	—	2535
R400-θ90-L300-$L_0$150-α70-15 × 75	300	15 × 75	1397
R400-θ90-L300-$L_0$150-α70-30 × 75	300	30 × 75	1314
R400-θ90-L300-$L_0$150-α70-45 × 75	300	45 × 75	1201
R400-θ90-L300-$L_0$150-α70-75 × 60	300	60 × 75	1094
R400-θ90-L300-$L_0$150-α70-60 × 75	300	75 × 75	1000
R400-θ90-L500-$L_0$150-α70-Perfect	500	—	2103
R400-θ90-L500-$L_0$150-α70-15 × 75	500	15 × 75	1224
R400-θ90-L500-$L_0$150-α70-30 × 75	500	30 × 75	1139
R400-θ90-L500-$L_0$150-α70-45 × 75	500	45 × 75	1030
R400-θ90-L500-$L_0$150-α70-60 × 75	500	60 × 75	930
R400-θ90-L500-$L_0$150-α70-75 × 75	500	75 × 75	847
R400-θ90-L700-$L_0$150-α70-Perfect	700	—	2004
R400-θ90-L700-$L_0$150-α70-15 × 75	700	15 × 75	1170
R400-θ90-L700-$L_0$150-α70-30 × 75	700	30 × 75	1064
R400-θ90-L700-$L_0$150-α70-45 × 75	700	45 × 75	960
R400-θ90-L700-$L_0$150-α70-60 × 75	700	60 × 75	860
R400-θ90-L700-$L_0$150-α70-75 × 75	700	75 × 75	750
R1000-θ90-L300-$L_0$150-α70-Perfect	300	—	3489
R1000-θ90-L300-$L_0$150-α70-15 × 75	300	15 × 75	2366
R1000-θ90-L300-$L_0$150-α70-30 × 75	300	30 × 75	2300
R1000-θ90-L300-$L_0$150-α70-45 × 75	300	45 × 75	2216
R1000-θ90-L300-$L_0$150-α70-60 × 75	300	60 × 75	2125
R1000-θ90-L300-$L_0$150-α70-75 × 75	300	75 × 75	2014
R1000-θ90-L500-$L_0$150-α70-Perfect	500	—	2371
R1000-θ90-L500-$L_0$150-α70-15 × 75	500	15 × 75	1900
R1000-θ90-L500-$L_0$150-α70-30 × 75	500	30 × 75	1829
R1000-θ90-L500-$L_0$150-α70-45 × 75	500	45 × 75	1730
R1000-θ90-L500-$L_0$150-α70-60 × 75	500	60 × 75	1640
R1000-θ90-L500-$L_0$150-α70-75 × 75	500	75 × 75	1520
R1000-θ90-L700-$L_0$150-α70-Perfect	700	—	2104
R1000-θ90-L700-$L_0$150-α70-15 × 75	700	15 × 75	1720
R1000-θ90-L700-$L_0$150-α70-30 × 75	700	30 × 75	1610
R1000-θ90-L700-$L_0$150-α70-45 × 75	700	45 × 75	1520
R1000-θ90-L700-$L_0$150-α70-60 × 75	700	60 × 75	1420
R1000-θ90-L700-$L_0$150-α70-75 × 75	700	75 × 75	1300

In (2), $\alpha = a/D$, $\beta = b/D$, $\gamma = L/D$, $\eta = a/b$ and $\lambda = L_0/L$, in which a, b, D, L, and L_0 signify the cutout height, cutout width, shell diameter, shell length, and cutout location, respectively. The exact form of the resulting equations is summarized in Table 3.

Equation (3) represents the buckling load reduction factor for the cylindrical panel with various lengths ($0.75 \leq L/D \leq 1.75$), with an elliptical cutout of fixed cutout width ($b/D = 0.1875$) and various cutout heights ($0.0375 \leq a/D \leq 0.1875$) in the midheight position of the shell.

Equation (4) is the buckling load reduction factor for the cylindrical panel with various lengths ($0.3 \leq L/D \leq 0.5$), with an elliptical cutout of fixed cutout width ($b/D = 0.075$) and various cutout heights ($0.015 \leq a/D \leq 0.075$) in the midheight position of the shell.

Equation (5) represents the buckling load reduction factor for the cylindrical panel with various lengths ($0.75 \leq L/D \leq 1.75$), with an elliptical cutout of fixed cutout height ($a/D = 0.1875$) and various cutout widths ($0.0375 \leq b/D \leq 0.1875$) in the midheight position of the shell.

Equation (6) represents the buckling load reduction factor for the cylindrical panel with various lengths ($0.3 \leq L/D \leq 0.5$), with an elliptical cutout of fixed cutout height ($a/D = 0.075$) and various cutout widths ($0.015 \leq b/D \leq 0.075$) in the midheight position of the shell.

Equations (7) and (8) represent the buckling load reduction factor for the cylindrical panel with various lengths ($0.75 \leq L/D \leq 1.75$) and ($0.3 \leq L/D \leq 0.5$), with an elliptical cutout of fixed size 15 × 75 mm in different positions ($0.875 \leq L/D \leq 0.5$), respectively.

TABLE 5: Summary of numerical analysis for cylindrical shells including an elliptical cutout with constant height and different height.

Model designation	Shell length	Cutout size ($a \times b$)	Buckling load (N)
R400-θ90-L300-L$_0$150-α70-Perfect	300	—	2535
R400-θ90-L300-L$_0$150-α70-30 $\times$ 30	300	30 $\times$ 30	1806
R400-θ90-L300-L$_0$150-α70-30 $\times$ 45	300	30 $\times$ 45	1611
R400-θ90-L300-L$_0$150-α70-30 $\times$ 60	300	30 $\times$ 60	1446
R400-θ90-L300-L$_0$150-α70-30 $\times$ 75	300	30 $\times$ 75	1286
R400-θ90-L300-L$_0$150-α70-30 $\times$ 90	300	30 $\times$ 90	1140
R400-θ90-L500-L$_0$150-α70-Perfect	500	—	2103
R400-θ90-L500-L$_0$150-α70-30 $\times$ 30	500	30 $\times$ 30	1622
R400-θ90-L500-L$_0$150-α70-30 $\times$ 45	500	30 $\times$ 45	1425
R400-θ90-L500-L$_0$150-α70-30 $\times$ 60	500	30 $\times$ 60	1262
R400-θ90-L500-L$_0$150-α70-30 $\times$ 75	500	30 $\times$ 75	1141
R400-θ90-L500-L$_0$150-α70-30 $\times$ 90	500	30 $\times$ 90	995
R400-θ90-L700-L$_0$150-α70-Perfect	700	—	2004
R400-θ90-L700-L$_0$150-α70-30 $\times$ 30	700	30 $\times$ 30	1579
R400-θ90-L700-L$_0$150-α70-30 $\times$ 45	700	30 $\times$ 45	1367
R400-θ90-L700-L$_0$150-α70-30 $\times$ 60	700	30 $\times$ 60	1182
R400-θ90-L700-L$_0$150-α70-30 $\times$ 75	700	30 $\times$ 75	1056
R400-θ90-L700-L$_0$150-α70-30 $\times$ 90	700	30 $\times$ 90	926
R1000-θ90-L300-L$_0$150-α70-Perfect	300	—	3489
R1000-θ90-L300-L$_0$150-α70-30 $\times$ 30	300	30 $\times$ 30	2625
R1000-θ90-L300-L$_0$150-α70-30 $\times$ 45	300	30 $\times$ 45	2418
R1000-θ90-L300-L$_0$150-α70-30 $\times$ 60	300	30 $\times$ 60	2281
R1000-θ90-L300-L$_0$150-α70-30 $\times$ 75	300	30 $\times$ 75	2189
R1000-θ90-L300-L$_0$150-α70-30 $\times$ 90	300	30 $\times$ 90	2112
R1000-θ90-L500-L$_0$150-α70-Perfect	500	—	2371
R1000-θ90-L500-L$_0$150-α70-30 $\times$ 30	500	30 $\times$ 30	2158
R1000-θ90-L500-L$_0$150-α70-30 $\times$ 45	500	30 $\times$ 45	1950
R1000-θ90-L500-L$_0$150-α70-30 $\times$ 60	500	30 $\times$ 60	1814
R1000-θ90-L500-L$_0$150-α70-30 $\times$ 75	500	30 $\times$ 75	1699
R1000-θ90-L500-L$_0$150-α70-30 $\times$ 90	500	30 $\times$ 90	1607
R1000-θ90-L700-L$_0$150-α70-Perfect	700	—	2104
R1000-θ90-L700-L$_0$150-α70-30 $\times$ 30	700	30 $\times$ 30	1974
R1000-θ90-L700-L$_0$150-α70-30 $\times$ 45	700	30 $\times$ 45	1783
R1000-θ90-L700-L$_0$150-α70-30 $\times$ 60	700	30 $\times$ 60	1630
R1000-θ90-L700-L$_0$150-α70-30 $\times$ 75	700	30 $\times$ 75	1508
R1000-θ90-L700-L$_0$150-α70-30 $\times$ 90	700	30 $\times$ 90	1401

Equation (9) represents the buckling load reduction factor for the cylindrical panel with various lengths ($0.75 \leq L/D \leq 1.75$), with an elliptical cutout of fixed area $A = 8242.5\,mm^2$ and various dimensions ($0.2 \leq a/b \leq 0.5$) in the midheight position of the shell.

Equation (10) represents the buckling load reduction factor for the cylindrical panel with various lengths ($0.3 \leq L/D \leq 0.5$), with an elliptical cutout of fixed area $A = 8242.5\,mm^2$ and various dimensions ($0.2 \leq a/b \leq 0.5$) in the midheight position of the shell.

7. Concluding Remarks

This study investigated the effect of elliptical cutouts of various sizes in different position on the buckling load of laminated composite cylindrical panel subjected to axial load. The following results were found in this study.

(1) The laminated composite panel with the composite ply angle of $\theta = 70$ leads to maximum buckling load for the ply sequence under study, while the ply angle of $\theta = 0$ (composite fibers oriented in the longitudinal direction) exhibits the lowest load capacity.

(2) When the width of the cutout is fixed and cutout height increases, the buckling load decreases slightly. Increasing the cutout width while the height of the cutout is fixed reduces the buckling load considerably. Therefore, it is suggested that in designing the panels, the greater cutout dimension is oriented along the longitudinal axis of the panels.

(3) For the a/b ratio in the range of $0 < a/b < 1$, when the cutout area is constant an increase in a/b ratio increases the buckling load, and for $a/b > 1$ with increase in a/b ratio decrease the buckling load. On the other hand, having $a/b = 1$ results in the highest load capacity.

(4) Increasing the panel angle enhances the shell resistance against buckling and increases the amount of the critical load and slender, and longer cylindrical panels are less sensitive to the changes of the panel angle.

(5) Moving the location of the cutout from the mid-height of the laminated composite panel to their top end increases the buckling load; slender and longer panels are more sensitive to the change in cutout location.

(6) Finally, formulas were obtained for the computation of the buckling load of cylindrical panels with elliptical cutouts based on the buckling load of perfect cylindrical shells. These expressions are applicable to a wide range of cylindrical panel with elliptical cutouts.

Appendix

Tables 4 and 5 shows the effect of change in cutout height and cutout width on the buckling load.

References

[1] R. Hosseinzadeh, M. M. Shokrieh, and L. Lessard, "Damage behavior of fiber reinforced composite plates subjected to drop weight impacts," *Composites Science and Technology*, vol. 66, no. 1, pp. 61–68, 2006.

[2] M. Zor, F. Şen, and M. E. Toygar, "An investigation of square delamination effects on the buckling behavior of laminated composite plates with a square hole by using three-dimensional FEM analysis," *Journal of Reinforced Plastics and Composites*, vol. 24, no. 11, pp. 1119–1130, 2005.

[3] I. Shufrin, O. Rabinovitch, and M. Eisenberger, "Buckling of laminated plates with general boundary conditions under combined compression, tension, and shear-A semi-analytical solution," *Thin-Walled Structures*, vol. 46, no. 7-9, pp. 925–938, 2008.

[4] H. Akbulut and O. Sayman, "An investigation on buckling of laminated plates with central square hole," *Journal of Reinforced Plastics and Composites*, vol. 20, no. 13, pp. 1112–1124, 2001.

[5] F. K. Chang and L. B. Lessard, "Damage tolerance of laminated composites containing an open hole and subjected to compressive loadings—part I: analysis," *Journal of Composite Materials*, vol. 25, no. 1, pp. 2–43, 1991.

[6] P. Jain and A. Kumar, "Postbuckling response of square laminates with a central circular/elliptical cutout," *Composite Structures*, vol. 65, no. 2, pp. 179–185, 2004.

[7] M. Yazici, "Influence of cut-out variables on buckling behavior of composite plates," *Journal of Reinforced Plastics and Composites*, vol. 28, no. 19, pp. 2325–2339, 2008.

[8] S. B. Singh and D. Kumar, "Postbuckling response and failure of symmetric laminated plates with rectangular cutouts under uniaxial compression," *Structural Engineering and Mechanics*, vol. 29, no. 4, pp. 455–467, 2008.

[9] C. W. Kong, C. S. Hong, and C. G. Kim, "Postbuckling strength of composite plate with a hole," *Journal of Reinforced Plastics and Composites*, vol. 20, pp. 466–481, 2001.

[10] S. A. M. Ghannadpour, A. Najafi, and B. Mohammadi, "On the buckling behavior of cross-ply laminated composite plates due to circular/elliptical cutouts," *Composite Structures*, vol. 75, no. 1–4, pp. 3–6, 2006.

[11] Y. H. Kim and A. K. Noor, "Buckling and postbuckling of composite panels with cutouts subjected to combined loads," *Finite Elements in Analysis and Design*, vol. 22, no. 2, pp. 163–185, 1996.

[12] U. K. Mallela and A. Upadhyay, "Buckling of laminated composite stiffened panels subjected to in-plane shear: a parametric study," *Thin-Walled Structures*, vol. 44, no. 3, pp. 354–361, 2006.

[13] L. S. Kistler and A. M. Waas, "Experiment and analysis on the response of curved laminated composite panels subjected to low velocity impact," *International Journal of Impact Engineering*, vol. 21, no. 9, pp. 711–736, 1998.

[14] C. Zhang, L. B. Lessard, and J. A. Nemes, "A closed-form solution for stresses at curved free edges in composite laminates: a variational approach," *Composites Science and Technology*, vol. 57, no. 9-10, pp. 1341–1354, 1997.

[15] H. T. Hu and J. S. Yang, "Buckling optimization of laminated cylindrical panels subjected to axial compressive load," *Composite Structures*, vol. 81, no. 3, pp. 374–385, 2007.

[16] S. Dash, A. V. Asha, and S. K. Sahu, "stability of laminated composite curved panels with cutout using finite element method," in *Proceedings of the International Conference on Theoretical, Applied Computational and Experimental Mechanics*, Kharagpur, India, 2004.

[17] E. Gal, R. Levy, H. Abramovich, and P. Pavsner, "Buckling analysis of composite panels," *Composite Structures*, vol. 73, no. 2, pp. 179–185, 2006.

[18] G. M. Stanley, *Continuum-based shell elements [Ph.D. dissertation]*, Department of Mechanical Engineering, Stanford University, 1985.

[19] A. Sabik and I. Kreja, "Stability analysis of multilayered composite shells with cut-outs," *Archives of Civil and Mechanical Engineering*, vol. 11, no. 1, pp. 195–207, 2011.

[20] H. Ghiasi, D. Pasini, and L. Lessard, "Optimum stacking sequence design of composite materials—part I: constant stiffness design," *Composite Structures*, vol. 90, no. 1, pp. 1–11, 2009.

[21] H. Ghiasi, K. Fayazbakhsh, D. Pasini, and L. Lessard, "Optimum stacking sequence design of composite materials—part II: variable stiffness design," *Composite Structures*, vol. 93, no. 1, pp. 1–13, 2010.

[22] M. Shariati, H. R. Allahbakhsh, J. Saemi, and M. Sedighi, "Optimization of foam filled spot-welded column for the crashworthiness design," *Mechanika*, vol. 83, no. 3, pp. 10–16, 2010.

[23] H. R. Allahbakhsh, J. Saemi, and M. Hourali, "Design optimization of square aluminium damage columns with crashworthiness criteria," *Mechanika*, vol. 17, no. 2, pp. 187–192, 2011.

6

Supplementary High-Input Impedance Voltage-Mode Universal Biquadratic Filter Using DVCCs

Jitendra Mohan[1] and Sudhanshu Maheshwari[2]

[1]*Department of Electronics and Communications, Jaypee Institute of Information Technology,*
  Noida 201304, India
[2]*Department of Electronics Engineering, Z. H. College of Engineering and Technology, Aligarh Muslim University,*
  Aligarh 202002, India

Correspondence should be addressed to Jitendra Mohan, jitendramv2000@rediffmail.com

Academic Editor: Andrzej Dzielinski

To further extend the existing knowledge on voltage-mode universal biquadratic filter, in this paper, a new biquadratic filter circuit with single input and multiple outputs is proposed, employing three differential voltage current conveyors (DVCCs), three resistors, and two grounded capacitors. The proposed circuit realizes all the standard filter functions, that is, high-pass, band-pass, low-pass, notch, and all-pass filters simultaneously. The circuit enjoys the feature of high-input impedance, orthogonal control of resonance angular frequency (ω_o), and quality factor (Q) via grounded resistor and the use of grounded capacitors which is ideal for IC implementation.

1. Introduction

Analog filters are the basic building blocks and widely used for continuous-time signal processing. Application of analog filters employing current-mode active elements extends over a large number of fields [1]. The filter circuits may be used in phase-locked loop frequency modulation stereo demodulators, touch-tone telephone tone decoder, and cross-over networks in a three-way high fidelity loudspeaker [2, 3]. In the literature several voltage-mode biquadratic filters are presented which uses different types of current conveyors [4–28]. The voltage-mode filters with high-input impedance are of great interest because they can be easily cascaded to synthesize higher-order filters [7, 10, 13, 17, 26, 28]. In the literature there are a number of voltage-mode universal biquadratic filters with single-input multiple-outputs (SIMO) [4–12, 18] that are available. However, these reported circuits suffer from one or more of the following drawbacks:

(i) excessive use of passive components [4, 6, 7];

(ii) low-input impedance [5, 6, 8, 9, 11, 18];

(iii) lack of orthogonal control over the resonance angular frequency (ω_o) and quality factor (Q) [10, 12].

In this paper, a new voltage-mode universal biquadratic filter using three differential voltage current conveyors [29], two grounded capacitors, and three resistors is presented. The circuit realizes the entire standard filter functions, that is, high pass (HP), band pass (BP), low pass (LP), notch (NH), and all pass (AP) from the same circuit configuration. The circuit also possesses high-input impedance and provides orthogonal control of the ω_o and Q via grounded resistor.

2. Differential Voltage Current Conveyor (DVCC)

The DVCC is first introduced long back as a modified current conveyor by Pal [30] and developed by Elwan and Soliman in 1997 [29]. It is a versatile building block, especially designed to handle differential and floating input signals [30, 31]. Using standard notation, the terminal relations of a DVCC

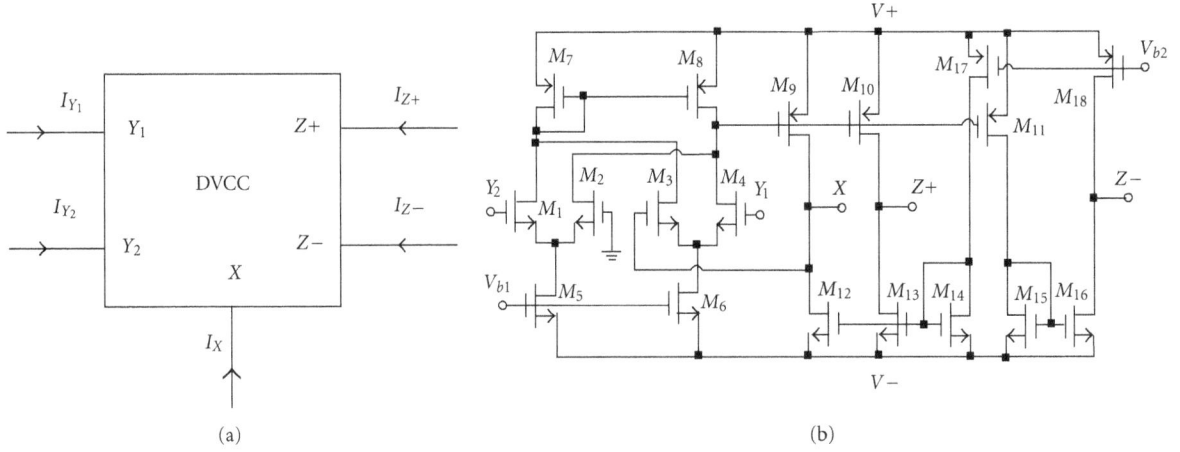

Figure 1: (a) Symbol of DVCC. (b) CMOS implementation of DVCC.

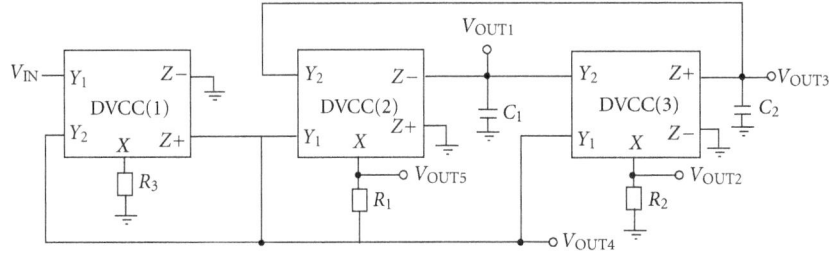

Figure 2: Voltage-mode universal biquadratic filter.

are shown in Figure 1 and described by the following matrix equation:

$$
\begin{bmatrix} I_{Y_1} \\ I_{Y_2} \\ V_X \\ I_{Z+} \\ I_{Z-} \end{bmatrix} = \begin{bmatrix} 0 & 0 & 0 & 0 & 0 \\ 0 & 0 & 0 & 0 & 0 \\ 1 & -1 & 0 & 0 & 0 \\ 0 & 0 & 1 & 0 & 0 \\ 0 & 0 & -1 & 0 & 0 \end{bmatrix} \begin{bmatrix} V_{Y_1} \\ V_{Y_2} \\ I_X \\ V_{Z+} \\ V_{Z-} \end{bmatrix}. \quad (1)
$$

The difference of the Y_1 and Y_2 terminal voltages is conveyed to the X terminal; the current input at the X terminal is conveyed to the $Z+$, with the same polarity and $Z-$ terminal with inverse polarity. The DVCC is characterized by high-input impedance at the Y_1 and Y_2 terminals, high-output impedance at the $Z+$ and $Z-$ terminals, and low impedance at the X terminal.

3. Proposed Circuit

The proposed voltage-mode universal biquadratic filter is shown in Figure 2, employing three DVCCs, three resistors, and two grounded capacitors. The routine analysis of the

proposed circuit in Figure 2 using (1) yields the filter transfer functions as

$$
\frac{V_{OUT1}}{V_{IN}} = \frac{1}{s^2 C_1 C_2 R_1 R_2 + s C_1 R_3 + 1}
$$

$$
\frac{V_{OUT2}}{V_{IN}} = \frac{s^2 C_1 C_2 R_1 R_2}{s^2 C_1 C_2 R_1 R_2 + s C_1 R_3 + 1}
$$

$$
\frac{V_{OUT3}}{V_{IN}} = \frac{s C_1 R_1}{s^2 C_1 C_2 R_1 R_2 + s C_1 R_3 + 1} \quad (2)
$$

$$
\frac{V_{OUT4}}{V_{IN}} = \frac{s^2 C_1 C_2 R_1 R_2 + 1}{s^2 C_1 C_2 R_1 R_2 + s C_1 R_3 + 1}
$$

$$
\frac{V_{OUT5}}{V_{IN}} = \frac{s^2 C_1 C_2 R_1 R_2 - s C_1 R_1 + 1}{s^2 C_1 C_2 R_1 R_2 + s C_1 R_3 + 1}.
$$

It can be seen from (2) that an LP response is obtained from V_{OUT1}; HP response is obtained from V_{OUT2}; BP response is obtained from V_{OUT3}; NH response is obtained from V_{OUT4}; if $R_3 = R_1$, AP response is obtained from V_{OUT5}. In all the cases, the parameters ω_o and Q of the filter can be found as

$$
\omega_o = \left(\frac{1}{C_1 C_2 R_1 R_2} \right)^{1/2},
$$

$$
Q = \frac{1}{R_3} \sqrt{\frac{R_1 R_2 C_2}{C_1}}. \quad (3)
$$

TABLE 1: TSMC 0.35 μm CMOS process parameters.

NMOS

LEVEL = 3 TOX = 7.9E − 9 NSUB = 1E17 GAMMA = 0.5827871 PHI = 0.7 VTO = 0.5445549 DELTA = 0 UO = 436.256147 ETA = 0 THETA = 0.1749684

KP = 2.055786E − 4 VMAX = 8.309444E4 KAPPA = 0.2574081 RSH = 0.0559398 NFS = 1E12 TPG = 1 XJ = 3E − 7 LD = 3.162278E − 11

WD = 7.04672E − 8 CGDO = 2.82E − 10 CGSO = 2.82E − 10 CGBO = 1E − 10 CJ = 1E − 3 PB = 0.9758533 MJ = 0.3448504 CJSW = 3.777852E − 10 MJSW = 0.3508721

PMOS

LEVEL = 3 TOX = 7.9E − 9 NSUB = 1E17 GAMMA = 0.4083894 PHI = 0.7 VTO = −0.7140674 DELTA = 0 UO = 212.2319801 ETA = 9.999762E − 4

THETA = 0.2020774 KP = 6.733755E − 5 VMAX = 1.181551E5 KAPPA = 1.5 RSH = 30.0712458 NFS = 1E12 TPG = −1 XJ = 2E − 7 LD = 5.000001E − 13

WD = 1.249872E − 7 CGDO = 3.09E − 10 CGSO = 3.09E − 10 CGBO = 1E − 10 CJ = 1.419508E − 3 PB = 0.8152753 MJ = 0.5 CJSW = 4.813504E − 10 MJSW = 0.5

From (3), the Q of the proposed filter can be controlled independently of ω_o by varying R_3. Since the input voltage signal is connected directly to the Y_1 port of the DVCC (1) and the input current to the Y_1 port is zero ($I_{Y_1} = 0$), the circuit has the feature of high-input impedance. Thus the proposed circuit is capable to realize all five standard filter functions simultaneously and without changing the circuit topology, unlike the previously proposed circuit in [26]. The circuit reported in [26] uses two floating resistors but the proposed circuit employs only one floating resistor. The circuit needs no component-matching conditions except for the all-pass filter realization.

Note that the proposed circuit uses two grounded capacitors at the Z-terminal and three resistors at the X-terminal of the DVCCs. The design offers the feature of a direct incorporation of the shunt parasitic capacitances at the Z-terminals and the series parasitic resistances at the X-terminal, as a part of the main capacitors (C_1 and C_2) and resistors (R_1, R_2, and R_3).

4. Nonideal Analysis

Taking the nonidealities of the DVCC into account, the relationship of the terminal voltages and currents of the DVCC can be rewritten as

$$\begin{bmatrix} I_{Y_1} \\ I_{Y_2} \\ V_X \\ I_{Z+} \\ I_{Z-} \end{bmatrix} = \begin{bmatrix} 0 & 0 & 0 & 0 & 0 \\ 0 & 0 & 0 & 0 & 0 \\ \beta_{k1} & -\beta_{k2} & 0 & 0 & 0 \\ 0 & 0 & \alpha_{k1} & 0 & 0 \\ 0 & 0 & -\alpha_{k2} & 0 & 0 \end{bmatrix} \begin{bmatrix} V_{Y_1} \\ V_{Y_2} \\ I_X \\ V_{Z+} \\ V_{Z-} \end{bmatrix}, \quad (4)$$

where $\beta_{k1}(s)$, $\beta_{k2}(s)$ represent the frequency transfers of the internal voltage followers and $\alpha_{k1}(s)$, $\alpha_{k2}(s)$ represent the frequency transfers of the internal current followers of the kth-DVCC, respectively. They can be approximated by first-order low-pass functions, which can be considered to have a unity value for frequencies [30]. If this circuit is working at frequencies much less than the corner frequencies of $\beta_{k1}(s)$, $\beta_{k2}(s)$, $\alpha_{k1}(s)$, and, $\alpha_{k2}(s)$, namely, then $\beta_{k1}(s) = \beta_{k1} = 1 - \varepsilon_{kv1}$

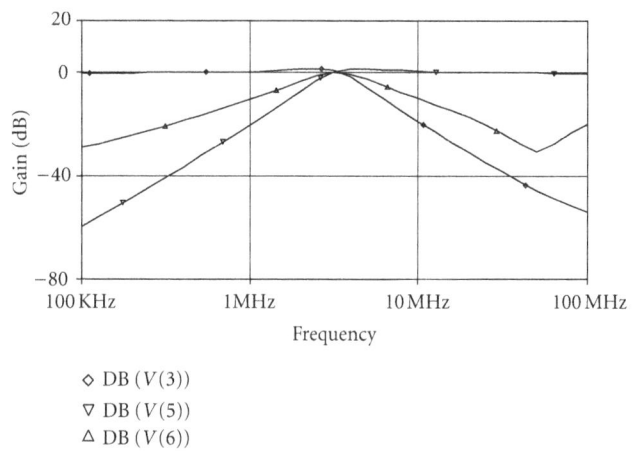

FIGURE 3: Gain plots of LP, HP, and BP filters for $Q = 1$ at 3.18 MHz.

◇ DB ($V(3)$)
▽ DB ($V(5)$)
△ DB ($V(6)$)

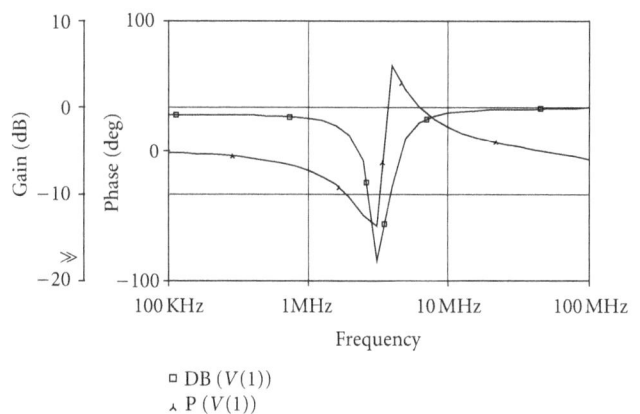

□ DB ($V(1)$)
▲ P ($V(1)$)

FIGURE 4: Gain and phase plots of NH filter for $Q = 1$ at 3.18 MHz.

and $\varepsilon_{kv1}(|\varepsilon_{kv1}| \ll 1|)$ denotes the voltage tracking error from the Y_1 terminal to the X terminal of the kth-DVCC; $\beta_{k2}(s) = \beta_{k2} = 1 - \varepsilon_{kv2}$ and $\varepsilon_{kv2}(|\varepsilon_{kv2}| \ll 1|)$ denotes the voltage tracking error from the Y_2 terminal to the X terminal of the

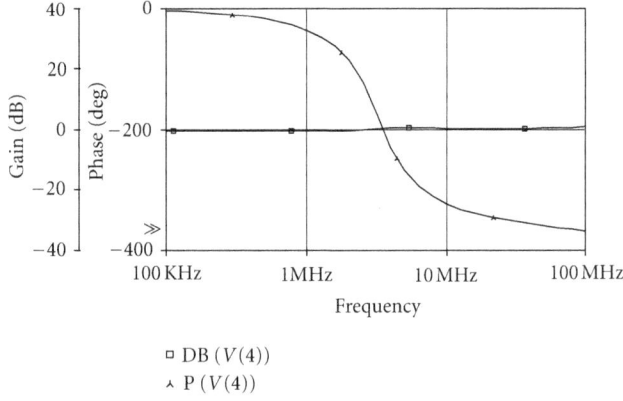

FIGURE 5: Gain and phase plots of AP filter for $Q = 1$ at 3.18 MHz.

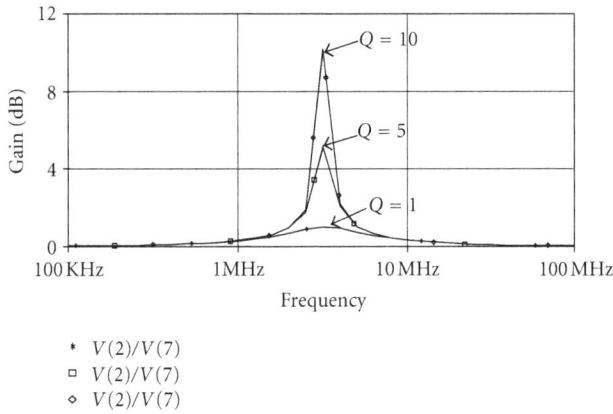

* $V(2)/V(7)$
□ $V(2)/V(7)$
◇ $V(2)/V(7)$

FIGURE 6: Q tuning ($Q = 1$, 5, and 10) of BP filter at 3.18 MHz.

kth-DVCC, $\alpha_{k1}(s) = \alpha_{k1} = 1 - \varepsilon_{ki1}$, and $\varepsilon_{ki1}(|\varepsilon_{ki1}| \ll 1|)$ denotes the current tracking error from the X terminal to the $Z+$ terminal; $\alpha_{k2}(s) = \alpha_{k2} = 1 - \varepsilon_{ki2}$ and $\varepsilon_{ki2}(|\varepsilon_{ki2}| \ll 1|)$ denotes the current tracking error from the X terminal to the $Z-$ terminal of the kth-DVCC.

Taking the tracking errors of the nonideal DVCC into account, the denominator of (2) becomes

$$
\begin{aligned}
D(s) = {} & s^2 C_1 C_2 R_2\, \alpha_{22}\left(R_3\, \alpha_{11} + R_1\, \beta_{12} - R_3\, \alpha_{11}\beta_{21}\right) \\
& + s C_1 R_3\, \beta_{22}\, \beta_{31}\, \alpha_{11}\alpha_{22} + \beta_{12}\beta_{22}\, \beta_{32}.
\end{aligned}
\tag{5}
$$

The parameters ω_o and Q from (5) can be rewritten as

$$
\omega_o = \left(\frac{\beta_{12}\beta_{22}\beta_{32}}{C_1 C_2 R_2\alpha_{22}\left(R_3\alpha_{11} + R_1\beta_{12} - R_3\alpha_{11}\beta_{21}\right)}\right)^{1/2},
$$

$$
Q = \frac{1}{R_3\beta_{31}\alpha_{11}}\left(\frac{C_2 R_2\beta_{12}\beta_{32}\left(R_3\alpha_{11} + R_1\beta_{12} - R_3\alpha_{11}\beta_{21}\right)}{C_1\beta_{22}\alpha_{22}}\right)^{1/2}.
\tag{6}
$$

The active and passive sensitivities of the proposed SIMO voltage-mode filter are derived from (6). These are as follows:

$$
S_{C_1,C_2,R_2}^{\omega_o} = -\frac{1}{2},
$$

$$
S_{R_1}^{\omega_o} = -\frac{1}{2}\frac{R_1\beta_{12}}{\left(R_3\alpha_{11} + R_1\beta_{12} - R_3\alpha_{11}\beta_{21}\right)},
$$

$$
S_{\beta_{22},\beta_{32}}^{\omega_o} = -S_{\alpha_{22}}^{\omega_o} = \frac{1}{2},
$$

$$
S_{\beta_{12}}^{\omega_o} = \frac{1}{2}\frac{\left(R_3\alpha_{11} - R_3\alpha_{11}\beta_{21}\right)}{\left(R_3\alpha_{11} + R_1\beta_{12} - R_3\alpha_{11}\beta_{21}\right)},
$$

$$
S_{R_3,\alpha_{11}}^{\omega_o} = -\frac{1}{2}\frac{R_3\alpha_{11}\left(1 - \beta_{21}\right)}{\left(R_3\alpha_{11} + R_1\beta_{12} - R_3\alpha_{11}\beta_{21}\right)},
$$

$$
S_{\beta_{21}}^{\omega_o} = \frac{1}{2}\frac{R_3\alpha_{11}\beta_{21}}{\left(R_3\alpha_{11} + R_1\beta_{12} - R_3\alpha_{11}\beta_{21}\right)},
$$

$$
S_{C_2,R_2}^{Q} = -S_{C_1}^{Q} = \frac{1}{2}, \qquad S_{\beta_{32}}^{Q} = -S_{\beta_{22},\alpha_{22}}^{Q} = \frac{1}{2},
$$

$$
S_{R_1}^{Q} = \frac{1}{2}\frac{R_1\beta_{12}}{\left(R_3\alpha_{11}\left(1 - \beta_{21}\right) + R_1\beta_{12}\right)},
$$

$$
S_{R_3,\alpha_{11}}^{Q} = \frac{\left(-2R_1\beta_{12} - R_3\alpha_{11}\left(1 - \beta_{21}\right)\right)}{2\left(R_3\alpha_{11}\left(1 - \beta_{21}\right) + R_1\beta_{12}\right)},
$$

$$
S_{\beta_{31}}^{Q} = -1,
$$

$$
S_{\beta_{12}}^{Q} = \frac{1}{2}\frac{\left(R_3\alpha_{11} + 2R_1\beta_{12} - R_3\alpha_{11}\beta_{21}\right)}{\left(R_3\alpha_{11} + R_1\beta_{12} - R_3\alpha_{11}\beta_{21}\right)},
$$

$$
S_{\beta_{21}}^{Q} = -\frac{1}{2}\frac{R_3\alpha_{11}\beta_{21}}{\left(R_3\alpha_{11}\left(1 - \beta_{21}\right) + R_1\beta_{12}\right)}.
\tag{7}
$$

From the results it is evident that the sensitivities are low and within unity in absolute value. Thus the proposed circuit can be classified as insensitive as all the active and passive sensitivities are less than or equal to unity [28].

5. Simulation Results

The performance of the proposed voltage-mode universal biquadratic filter in Figure 2 is verified using the PSPICE simulation program. The MOS transistors are simulated using 0.35 μm CMOS process parameters given in Table 1. The aspect ratios of the transistors used in the simulation are given in Table 2. The supply voltages and biasing voltage are given by $V+ = 2$ V, $V- = -1.9$ V, $V_{b1} = -1.23$ V, and $V_{b2} = -1.15$ V, respectively. The proposed circuit is designed for 3.18 MHz and $Q = 1$ by choosing $R_1 = R_2 = R_3 = 5$ kΩ and $C_1 = C_2 = 10$ pF. Figure 3 shows the simulated gain responses for the LP, HP, and BP filters and agrees well with the designed values. Figures 4 and 5 show the simulated gain and phase responses for the NH and AP filters. Next, Q tuning of BP filter at constant frequency of 3.18 MHz is shown in Figure 6. The circuit of Figure 2 is designed for Q values of 1, 5, and 10 with R_3 values as 5 kΩ, 1 kΩ, and 500 Ω, respectively, with $C_1 = C_2 = 10$ pF and $R_1 = R_2 = 5$ kΩ.

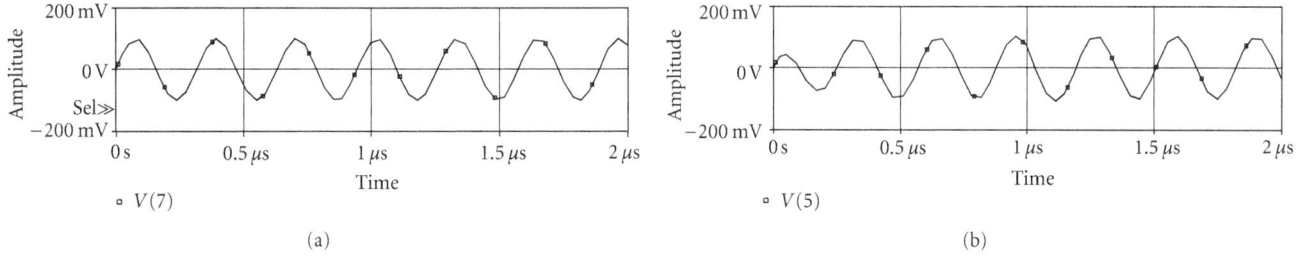

FIGURE 7: Input (a) and HP output (b) signal waveforms to demonstrate the ac dynamic range at V_{OUT2}.

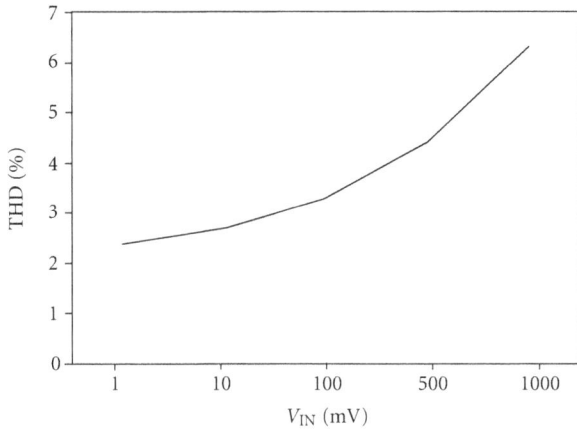

FIGURE 8: THD variation with V_{IN} for HP filter at V_{OUT2}.

TABLE 2: Aspect ratios of the MOS in Figure 1.

MOS transistors	W (μm)	L (μm)
M_1–M_4	1.2	0.8
M_7, M_8	6.6	0.8
M_9–M_{11}, M_{17}–M_{18}	18.6	0.6
M_5, M_6, M_{12}–M_{16}	24	0.8

To test the input dynamic range of the proposed filter, the simulation of the HP filter as example has been repeated for a sinusoidal input signal at 3.18 MHz. Figure 7 shows the input dynamic range of the HP filter which extends upto amplitude 200 mV (peak to peak) without significant distortion. In Figure 8, the total harmonic distortion (THD) of the V_{OUT2} output voltage is given at 3.18 MHz operation frequency. Figure 9 shows the INOISE and ONOISE simulated amplitude-frequency responses of the proposed LP filter at V_{OUT1}.

6. Conclusion

The voltage-mode universal biquadratic filter realizes all the standard filter functions by employing three DVCCs and five passive components. It offers high-input impedance, which permit the use of circuits in cascade without requiring any impedance-matching device. Moreover, the proposed circuit enjoys the following features: orthogonal control of ω_o and Q, low active and passive sensitivity, and the direct incorporation of parasitic resistance and parasitic capacitance at

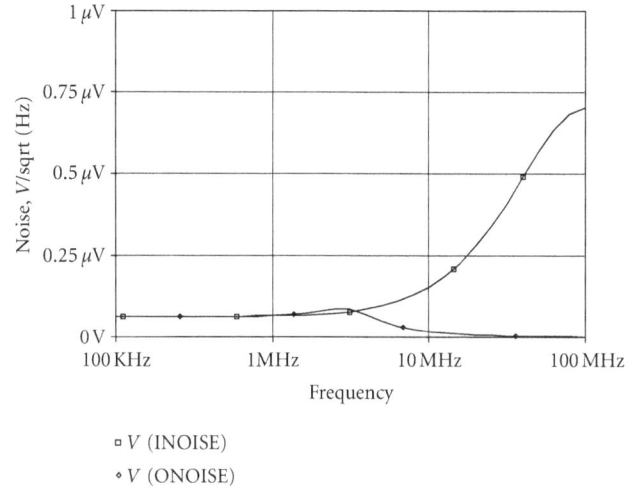

FIGURE 9: The input and output referred noise spectral densities of the proposed LP filter at V_{OUT1}.

the X-terminal and Z-terminal of the respective DVCCs, as a part of main resistor and capacitor. So, the new circuit will enhance the existing knowledge on the voltage-mode universal biquadratic filter. PSPICE simulations with TSMC 0.35 μm process confirm the theoretical prediction.

References

[1] B. Wilson, "Recent developments in current conveyors and current-mode circuits," *IEE proceedings G*, vol. 137, no. 2, pp. 63–77, 1990.

[2] A. Fabre, O. Saaid, F. Wiest, and C. Boucheron, "Low power current-mode second-order bandpass if filter," *IEEE Transactions on Circuits and Systems II*, vol. 44, no. 6, pp. 436–446, 1997.

[3] M. A. Ibrahim, S. Minaei, and H. Kuntman, "A 22.5 MHz current-mode KHN-biquad using differential voltage current conveyor and grounded passive elements," *International Journal of Electronics and Communications*, vol. 59, no. 5, pp. 311–318, 2005.

[4] M. T. Abuelma'atti and H. A. Al-Zaher, "New universal filter with one input and five outputs using current-feedback amplifiers," *Analog Integrated Circuits and Signal Processing*, vol. 16, no. 3, pp. 239–244, 1998.

[5] H. P. Chen and S. S. Shen, "A versatile universal capacitor-grounded voltage-mode filter using DVCCs," *ETRI Journal*, vol. 29, no. 4, pp. 470–476, 2007.

[6] J. W. Horng, C. L. Hou, C. M. Chang, W. Y. Chung, and H. Y. Wei, "Voltage-mode universal biquadratic filters with one input and five outputs using MOCCIIs," *Computers and Electrical Engineering*, vol. 31, no. 3, pp. 190–202, 2005.

[7] J. W. Horng, C. L. Hou, C. M. Chang, H. P. Chou, and C. T. Lin, "High input impedance voltage-mode universal biquadratic filter with one input and five outputs using current conveyors," *Circuits, Systems, and Signal Processing*, vol. 25, no. 6, pp. 767–777, 2006.

[8] H. P. Chen, "Universal voltage-mode filter using only plus-type DDCCs," *Analog Integrated Circuits and Signal Processing*, vol. 50, no. 2, pp. 137–139, 2007.

[9] H. P. Chen, "Single FDCCII-based universal voltage-mode filter," *International Journal of Electronics and Communications*, vol. 63, no. 9, pp. 713–719, 2009.

[10] H. P. Chen, "Voltage-mode FDCCII-based universal filters," *International Journal of Electronics and Communications*, vol. 62, no. 4, pp. 320–323, 2008.

[11] J. W. Horng, "Lossless inductance simulation and voltage-mode universal biquadratic filter with one input and five outputs using DVCCs," *Analog Integrated Circuits and Signal Processing*, vol. 62, no. 3, pp. 407–413, 2010.

[12] S. Minaei and E. Yuce, "All-grounded passive elements voltage-mode DVCC-based universal filters," *Circuits, Systems, and Signal Processing*, vol. 29, no. 2, pp. 295–309, 2010.

[13] M. Higashimura and Y. Fukui, "Realization of all-pass and notch filters using a single current conveyor," *International Journal of Electronics*, vol. 65, no. 4, pp. 823–828, 1988.

[14] B. Metin, S. Minaei, and O. Cicekoglu, "Enhanced dynamic range analog filter topologies with a notch/all-pass circuit example," *Analog Integrated Circuits and Signal Processing*, vol. 51, no. 3, pp. 181–189, 2007.

[15] M. A. Ibrahim, S. Minaei, and H. Kuntman, "DVCC based differential-mode all-pass and notch filters with high CMRR," *International Journal of Electronics*, vol. 93, no. 4, pp. 231–240, 2006.

[16] S. Minaei, I. C. Göknar, and O. Cicekoglu, "A new differential configuration suitable for realization of high CMRR, all-pass/notch filters," *Electrical Engineering*, vol. 88, no. 4, pp. 317–326, 2006.

[17] N. A. Shah and M. A. Malik, "High input impedance HP, BP and LP filters using FTFNs," *Indian Journal of Pure and Applied Physics*, vol. 41, no. 12, pp. 967–969, 2003.

[18] N. A. Shah, S. Z. Iqbal, and M. F. Rather, "Versatile voltage-mode CFA-based universal filter," *International Journal of Electronics and Communications*, vol. 59, no. 3, pp. 192–194, 2005.

[19] M. T. Abuelma'atti, "A novel mixed-mode current-controlled current-conveyor-based filter," *Active and Passive Electronic Components*, vol. 26, no. 3, pp. 185–191, 2003.

[20] M. T. Abuelma'atti, A. Bentrcia, and S. M. Al-Shahrani, "A novel mixed-mode current-conveyor-based filter," *International Journal of Electronics*, vol. 91, no. 3, pp. 191–197, 2004.

[21] S. Maheshwari, "High performance voltage—mode multifunction filter with minimum component count," *WSEAS Transactions on Electronics*, vol. 5, no. 6, pp. 244–249, 2008.

[22] A. M. Soliman, "Kerwin-Huelsman-Newcomb circuit using current conveyors," *Electronics Letters*, vol. 30, no. 24, pp. 2019–2020, 1994.

[23] A. M. Soliman, "Current conveyors steer universal filter," *IEEE Circuits and Devices Magazine*, vol. 11, no. 2, pp. 45–46, 1995.

[24] M. Higashimura and Y. Fukui, "Universal filter using plus-type CCIIs," *Electronics Letters*, vol. 32, no. 9, pp. 810–811, 1996.

[25] A. M. Soliman, "Generation and classification of Kerwin-Huelsman-Newcomb circuits using the DVCC," *International Journal of Circuit Theory and Applications*, vol. 37, no. 7, pp. 835–855, 2009.

[26] S. Maheshwari, J. Mohan, and D. S. Chauhan, "High input impedance voltage-mode universal filter and quadrature oscillator," *Journal of Circuits, Systems and Computers*, vol. 19, no. 7, pp. 1597–1607, 2010.

[27] S. Minaei and M. A. Ibrahim, "A mixed-mode KHN-biquad using DVCC and grounded passive elements suitable for direct cascading," *International Journal of Circuit Theory and Applications*, vol. 37, no. 7, pp. 793–810, 2009.

[28] A. Fabre, F. Dayoub, L. Duruisseau, and M. Kamoun, "High input impedance insensitive second-order filters implemented from current conveyors," *IEEE Transactions on Circuits and Systems I*, vol. 41, no. 12, pp. 918–921, 1994.

[29] H. O. Elwan and A. M. Soliman, "Novel CMOS differential voltage current conveyor and its applications," *IEE Proceedings G*, vol. 144, pp. 195–200, 1997.

[30] K. Pal, "Modified current conveyors and their applications," *Microelectronics Journal*, vol. 20, no. 4, pp. 37–40, 1989.

[31] H. Sedef and C. Acar, "New floating FDNR circuit using differential voltage current conveyors," *International Journal of Electronics and Communications*, vol. 54, no. 5, pp. 297–301, 2000.

A New Optimization via Simulation Approach for Dynamic Facility Layout Problem with Budget Constraints

7

layout problems, with one problem used separately for each period, because this approach does not consider the costs of relocating facilities from one period to the next.

Rosenblatt [8] showed the first research to develop an optimization approach based on a dynamic programming model for the DFLP. However, this approach is computationally intractable for real-life problems. The author showed that the number of layouts to be evaluated to guarantee optimality for a DFLP with N departments and T periods is $(N!)^T$. Because of the computational difficulties inherent in such a problem, several heuristics have been developed. Rosenblatt [8] proposed two heuristics that were based on dynamic programming, each of which simply considers a set of limited good layouts for a single period. Urban [9] developed a steepest-descent heuristic based on a pairwise exchange idea, which is similar to CRAFT. Lacksonen and Enscore [10] introduced and compared five heuristics to solve the DFLP, which were based on dynamic programming, a branch and bound algorithm, a cutting plane algorithm, cut trees, and CRAFT.

It should be mentioned that in addition to exact algorithms, many metaheuristic algorithms have been reported in the literature such as a genetic algorithm by [11] and a tabu search (TS) heuristic by [12]. This TS heuristic is a two-stage search process that incorporates diversification and intensification strategies. Baykasoglu and Gindy [13] developed a simulated annealing (SA) heuristic for the DFLP, in which they used the upper and lower bound of the solution of a given problem instance to determine the SA parameters. Balakrishnan et al. [14] presented a hybrid genetic algorithm. Erel et al. [15] introduced a new heuristic algorithm to solve the DFLP. They used weighted flow data from various time periods to develop viable layouts and suggested the shortest path for solving the DFLP. McKendall and Shang [16] developed three hybrid ant systems (HAS). McKendall et al. [17] introduced two (SA) heuristics. The first one (SA I) is a direct adaptation of SA for the DFLP while the second one (SA II) is the same as SA I except that it incorporates an added look-ahead/look-back strategy. A hybrid meta-heuristic algorithm based on a genetic algorithm and tabu search was introduced by Rodriguez et al. [18]. Krishnan et al. [19] used a new tool, the "Dynamic From-Between Chart," for an analysis of redesigned layouts. This tool models changes in the production rates using a continuous function. Balakrishnan and Cheng [6] investigated the performance of algorithms under fixed and rolling horizons, differing shifting costs, flow variability, and forecast uncertainty [6]. For an extensive review on the DFLP, one can refer to the studies presented by [20, 21]. The studies described above share a common assumption that all departments are of equal size. However, some studies do not make this assumption. Two recent examples of such studies are [22, 23].

It should be noticed that most previous researches did not consider the company budget for rearranging the departments. Because these rearrangements are costly activities, it is normal for a company to have a limited budget in this regard. According to the literature, there are just three studies on DFLP with the budget constraints [24–26].

The last one, which is the newest related research, used a budget constraint for each period separately. They developed a simulated annealing algorithm for the problem and showed that their algorithm is more efficient than the two previous researches.

In this paper, we first introduce the problem formulation for the DFLPB in Section 1. Then, in Section 2, this model is replaced with a similar linear model that is easier to solve. In Section 3, we introduce the idea which was proposed in [27]. They used the optimal solution of their linear model as a probability distribution in a simulation model. While they used this approach toward solving the traveling salesman problem, this empirical distribution can also be used here to determine the probability of assigning facilities to certain locations. This technique has an important role in the proposed algorithm. The number of necessary runs is also computed in this section. In Section 4, the proposed algorithm is introduced. In Section 5, computational results are summarized, and, finally, some concluding remarks are presented in Section 6.

2. DELPB Formulation

The DFLP can be modeled as a modified quadratic assignment problem, similar to the static facility layout problem (SFLP). The notations used in the model are given as follows:

$$x_{tij} = \begin{cases} 1 & \text{if department } i \text{ is assigned} \\ & \text{to location } j \text{ at period } t, \\ 0 & \text{otherwise.} \end{cases} \quad (1)$$

N is both the number of departments and the number of locations, T the number of periods in the planning horizon, C_{tijl} cost of material handling between department i in location j and department k in location l during period t, A_{tijl} cost of rearranging department i from location j to location l at the beginning of period t, LB_t leftover budget from period t to period $t+1$, B_t available budget for period t, and AB_t allocated budget for period t.

Problem 1. Problem 1 is as follows:

$$\text{Min} \cdot z = \left(\sum_{t=2}^{T}\sum_{i=1}^{N}\sum_{j=1}^{N}\sum_{l=1}^{N} A_{tijl} * x_{t-1,ij} * x_{til} + \sum_{t=2}^{T}\sum_{i=1}^{N}\sum_{j=1}^{N}\sum_{l=1}^{N} C_{tijkl} * x_{tij} * x_{tkl} \right) \quad (2)$$

s.t.

$$\sum_{j=1}^{N} x_{tij} = 1, \quad \forall i = 1,2,\ldots,N, \ \forall t = 1,2,\ldots,T, \quad (3)$$

$$\sum_{i=1}^{N} x_{tij} = 1, \quad \forall j = 1,2,\ldots,N, \ \forall t = 1,2,\ldots,T, \quad (4)$$

$$LB_t = B_t - \sum_{i=1}^{N}\sum_{j=1}^{N}\sum_{l=1}^{N} A_{tijl} * x_{t-1,ij} * x_{til}, \quad (5)$$

$$\forall t = 1,2,\ldots,T,$$

$$B_t = AB_t + LB_{t-1}, \quad \forall t = 1, 2, \ldots, T, \quad (6)$$

$$\sum_{i=1}^{N} \sum_{j=1}^{N} \sum_{l=1}^{N} A_{tijl} * x_{t-1,ij} * x_{til} \le B_t, \quad (7)$$
$$\forall t = 1, 2, \ldots, T,$$

$$x_{tij} \in \{0,1\}, \quad \forall i, j = 1, 2, \ldots, N, \ \forall t = 1, 2, \ldots, T, \quad (8)$$

$$LB_t, B_t, AB_t \ge 0, \quad \forall t = 1, 2, \ldots, T, \quad (9)$$

In Problem 1, the objective function (2) is used to minimize the sum of the rearrangement and material handling costs. Constraint set (3) restricts each location to be assigned to only one department during each period and constraint set (4) ensures that exactly one department is assigned to each location within each period. Constraint set (5) is for equating the total available budget in a period to sum of the leftover budget from previous period and the allocated budget in the current period. Finally, the constraint set (6) represents the budget constraints for each period. This zero-one programming problem has been shown to be an NP-hard model [20]. To solve the problem, a linear interpolation was used to change the objective function into a linear function. This technique makes the problem easier to solve using certain degrees of accuracy. However, we demonstrate that the computational results have sufficient accuracy in comparison to the results of the previous works. Therefore, the two nonlinear expressions $x_{t-1,ij} * x_{til}$ and $x_{tij} * x_{tkl}$ should be transformed through the linear interpolation. Assume a nonlinear function as follows:

$$f(x_i, x_j) = x_i x_j, \quad \text{where } 0 \le x_i, x_j \le 1. \quad (10)$$

By introducing two new variables:

$$x_{t-1,ij} + x_{til} = 2z_{t-1,tijl}, \quad (11)$$
$$x_{iji} + x_{tikl} = 2w_{tijkl}.$$

Replacing them in Problem 1 and relaxing the 0-1 variables, the resulting problem will form the following linear continuous model, which is the simplest model in the mathematical programming theory (complexity theory).

Problem 2. Problem 2 is as follows:

$$\text{Min} \cdot z = \left(\sum_{t=2}^{T} \sum_{i=1}^{N} \sum_{j=1}^{N} \sum_{l=1}^{N} A_{tijl} * z_{t-1,tijl} + \sum_{t=2}^{T} \sum_{i=1}^{N} \sum_{j=1}^{N} \sum_{l=1}^{N} C_{tijkl} * w_{tijkl} \right) \quad (12)$$

s.t.

$$\sum_{j=1}^{N} x_{tij} = 1, \quad \forall i = 1, 2, \ldots, N, \ \forall t = 1, 2, \ldots, T, \quad (13)$$

$$\sum_{i=1}^{N} x_{tij} = 1, \quad \forall j = 1, 2, \ldots, N, \ \forall t = 1, 2, \ldots, T, \quad (14)$$

$$x_{t-1,ij} + x_{til} = 2z_{t-1,tijl}, \quad \forall i, j, l = 1, 2, \ldots, N, \ \forall t = 2, 3, \ldots, T, \quad (15)$$

$$x_{tij} + x_{tkl} = 2w_{tijkl} \quad \forall i, j, l = 1, 2, \ldots, N, \ \forall t = 1, 2, \ldots, T, \ i \ne j, \ k \ne l, \quad (16)$$

$$LB_t = B_t - \sum_{i=1}^{N} \sum_{j=1}^{N} \sum_{l=1}^{N} A_{tijl} * z_{t-1,tijl}, \quad \forall t = 1, 2, \ldots, T, \quad (17)$$

$$B_t = AB_t + LB_{t-1}, \quad \forall t = 1, 2, \ldots, T, \quad (18)$$

$$\sum_{i=1}^{N} \sum_{j=1}^{N} \sum_{l=1}^{N} A_{tijl} * z_{t-1,tij} \le B_t, \quad \forall t = 1, 2, \ldots, T, \quad (19)$$

$$0 \le x_{tij} \le 1, \quad \forall i, j = 1, 2, \ldots, N, \ \forall t = 1, 2, \ldots, T, \quad (20)$$

$$0 \le z_{t-1,tij} \le 1, \quad \forall i, j = 1, 2, \ldots, N, \ \forall t = 2, 3, \ldots, T, \quad (21)$$

$$0 \le w_{tijkl} \le 1, \quad \forall i, j, k, l = 1, 2, \ldots, N, \ \forall t = 1, 2, \ldots, T, \ i \ne j, \ k \ne l, \quad (22)$$

$$LB_t, B_t, AB_t \ge 0, \quad \forall t = 1, 2, \ldots, T, \quad (23)$$

Problem 2 has more variables and constraints compared to Problem 1, but because this is a linear model, the computational time will be much lower than Problem 1, according to the computational results.

3. Input Values for Simulation Model

Before describing the simulation model for the DFLPB, an idea that was developed by [27] is presented for the traveling salesman problem. In this idea, because all variables in Problem 2 are between zero and one and according to constraints (13) and/or (14), their summations are equal to 1, the definition of x_{tij} can be interpreted as the probability of assigning department i to location j during period t.

As mentioned before, this concept plays a key role in the simulation model, where it must use a probability distribution for randomly assigning each facility to each location during a certain period. In fact, the optimal values that come from Problem 2 are empirical distributions that will be used by the simulation model. Because Problem 2 is an estimation of the real problem, these empirical distributions help us to reduce the simulation runs in order to find the best solution faster. Because the algorithm uses a linear interpolation, the optimal solution of Problem 2 will be an estimator for the real values. Now, suppose that the optimal solution of Problem 1 is known and x_{tij}^* is the optimal solution of Problem 2. Define A_t as the probability of finding the optimum solution of Problem 1 at period t as follows:

$$A_t = \prod_i \prod_j x_{tij}^*. \tag{24}$$

To find the optimal assignment values for a maximum n_t runs in the simulation model, with %$(1 - \alpha)$ as the significance level:

$$A_t + (1 - A_t)A_t + (1 - A_t)^2 A_t + \cdots + (1 - A_t)^{n_t - 1}, \\ A_t \geq 1 - \alpha. \tag{25}$$

Then,

$$n_t \log(1 - A_t) \leq \log(\alpha). \tag{26}$$

Therefore,

$$n_t \geq \frac{\log(\alpha)}{\log(1 - A_t)}. \tag{27}$$

Because n_t is the number of minimum needed runs to obtain optimal solution at period t, NR can be defined as the number of minimum needed runs; then we have

$$NR = \text{Max} \cdot_t \{n_t\}. \tag{28}$$

For calculating A_t and NR, a heuristic algorithm has been developed as follows.

Step 1. Define $K \subset N$ as a subset of facilities that have been assigned to the locations and set $K = \{\}$. Denote P as the set of correspondent probabilities of the elements in K.

Step 2. Solve Problem 2 and find the optimal solutions as x_{tij}^*.

Step 3. Select a facility that has the maximum value for the assignment, and, if there is a tie, then select a facility randomly. Assume that facility m has the maximum assignment value among all x_{tij}^*, which is denoted by x_{tmj}^*. Now, add facility m to set K and x_{tmj}^* to P.

Step 4. If $K = N$, then go to Step 5; otherwise, go back to Problem 2. For each facility in set K, set the correspondent assignment variables to 1, and then add these new constraints to Problem 2. Go to Step 2.

Step 5. Now, A_t can be estimated using the product of all elements in set P according to (24).

And the value for NR can be computed based on (28).

As it will be shown in the computational results.

(i) The value of NR is relatively low in comparison to other developed metaheuristics, which results in quick run times. This is because, the optimal values obtained from Problem 2 are near to the optimal solutions, and their values are greater than 0.76 in most cases. This fact will cause NR to be sufficiently low in the computations based on (27).

(ii) Because the simulation model does not depend on simulation clock, all runs will be completed very quickly (less than 0.01 min in most cases). Therefore, this effective criterion will help the simulation find the near optimum solutions as quickly as possible.

The simulation model has been designed according to Problem 1. All constraints have been coded using Enterprise Dynamics 8.1 software with 4D Script language. All input parameters such as the rearrangements costs, material handling costs, and the optimal solutions of Problem 2 are stored in tables within the software, and when the software assigns a facility to a location at each period, the related costs will be stored in another table that calculates the objective function of Problem 1 at the end of each run. This simulation technique is a good tool for such a difficult problem because it produces feasible solutions based on Problem 1, that is, all solutions produced by the simulation model are integers and satisfy all mentioned constraints. All necessary runs are conducted within the experimentation wizard in the simulation software. When a run is completed, the resulting feasible solutions will be stored in a table. Therefore, when an experiment ends after NR runs, one can access the best solutions based on the data set stored in the table. Because, the simulation model does not depend on a time process, it does not need to calculate a warm-up period.

4. Proposed Algorithm

Based on the previous explanations, the heuristic algorithm can be defined as follows.

Step 1. Initialize: assume that we have a DFLPB with N facilities, N locations at T periods, in which all other parameters such as C_{tijkl}, A_{tijl},... are given. Formulate Problem 2 as previously described and code the simulation model according to Problem 1. During this step, assume that the significance level is %$(1 - \alpha)$.

Step 2. Calculate the assignment probabilities: solve Problem 2. The optimal solutions of this problem will be used as an empirical distribution for assigning a facility to a location during each period.

Step 3. Calculate the minimum number of needed runs: according to the results of Step 2, calculate the minimum number of needed runs (NR) based on (28).

Step 4. Run the simulation model: run the simulation model with NR replications. At each replication r, store the objective function, Z_r and the corresponding assignments for each period.

Step 5. Find the best solution: the best solution is determined as

$$Z^* = \text{Min} \cdot_r \{Z_r\}. \qquad (29)$$

5. Computational Results

As mentioned before, Enterprise Dynamics 8.1 software has been used for the simulation, Lingo 8.0 for finding the optimal solution of Problem 2 and Microsoft Visual Basic 2007 as the coordinator between the simulation software and mathematical programming software. All computations were run on a PC with a 4.8 GHz CPU and 4 GB of RAM. All parameters for the DFLPBs were taken from a data set provided by [11]. For comparison, the results have been compared with those reported in [26]. Their report has been selected because they had compared their results with two previous papers and showed that their proposed heuristic algorithm was the best one among all others. The computational results are summarized in Table 1 over a wide range of test problems (48 problems). These problems contain cases using 6, 15, and 30 facilities ($N = 6, 15, 30$) each with 5 and 10 periods. The second column in this table is the problem number denoted by [26]. For each problem without a budget constraint, [25] shows three problems with different budget constraints. First, a total budget constraint is obtained by solving an unconstrained problem with respect to the budget and by setting the total rearrangement cost of this solution as the total budget constraint. Then the allocation of this total budget is carried out in three ways: (1) divide the total budget by the number of periods:1 (number of transactions), and allocate equally to the periods. (2) The level of the budget for each period is found by taking the half of the rearrangement cost for the same period in solution of the unconstrained problem. (3) The level of the budget for each period is found by adding 10% more to the rearrangement costs for the same period in the solution of the unconstrained problem. Through this process three sets of problems are obtained, which were denoted by 1, 2, and 3 in the third column. For the exact parameter values of the obtained problems please see [25]. The forth column, labeled "Average Probabilities," lists the average optimal solutions obtained from Problem 2. The number of needed runs is listed in the fifth column. The optimal solutions under the proposed algorithm are listed in the sixth column. The seventh column lists the optimal solutions reported by [26]. The percentage of deviation, denoted by "%Dev." of the best solution obtained from the proposed algorithm, which is lower than the best solution obtained from [26], is given in the eighth column, for each test problem. In the last column, the average run times are given in minutes.

Another important factor regarding the proposed algorithm is the "average CPU time," which is sufficiently fast for use in these applications. As previously explained, the simulation time depends on many factors such as NR, N, and T. However, [26] did not report any running time in their experiments. As listed in the table, the average percentage of deviation is -1.89% for the solved problems and the average run time is 0.33 min. The average probabilities in the forth column is 0.8348 which shows the effectiveness of mathematical programming for estimating the optimal solutions. In Tables 2 and 3, the same structures and calculations are given for $N = 15$ and $N = 30$ respectively.

As listed in Table 2, in certain cases both algorithms have the same best solutions, but the average is -1.70%, that is, the proposed algorithm is better than that of [26] when $N = 15$. In addition, the running time of the algorithm is reasonable in both cases.

In Table 3, again in some cases both algorithms have the same best solutions, but the average is -1.25%, that is, the algorithm has better solutions than that of [26] when $N = 30$. Meanwhile, the speed of the algorithm is around 1 h. This fact is very important, because according to [26], for such a large size problems, their algorithm took several hours.

To sum up, the proposed algorithm provides good solution quality in comparison to the algorithms developed in previous researches. It was able to improve the optimal solution for a known data-set by 1.72% on average. Regarding the run time, the algorithm has reasonable run time in comparison to previous researches.

6. Conclusions

In this research, a new heuristic algorithm has been developed to address the dynamic facility layout problem with budget constraints using optimization via simulation technique. The proposed heuristic algorithm integrates mathematical programming and simulation methods. The optimization via simulation approach was selected and used to show the efficiency of simulation technique to solve even such a large scale combinatorial problem. However, the simulation technique has a vast range of benefits in real world applications, but we tried to show that, simulation is a powerful technique among its so-called competitors such as genetic algorithm, ant colony Optimization, and other evolutionary algorithms. The first contribution of the current study is that it defines the optimal solution of the linear programming model in terms of empirical distributions for a simulation model. This idea can decrease the number of replications required in the simulation model, leading to better speed. The performance of the proposed algorithm was tested over a wide range of test problems taken from the literature. The proposed algorithm improved the objective function of the problem by 1.72% on average, whereas the time required for the largest problem with $N = 30$ and $T = 10$ was around 1 hour. This is the second contribution of the current research.

The proposed algorithm not only avoids uncommon issues in metaheuristic algorithms such as premature events, parameter tuning, and trapping in local optimums but also uses a simulation technique that produces feasible solutions without the use of any specific nonrealistic assumptions.

Table 1: The computational results with significant level 99% with $N = 6$.

T	Problem number	Budget	Average probabilities	NR	Best solution	Best solution by Sahin et al. [26]	% Dev.	Average time
	P01	1	0.7608	21	106419	106419	0	0.53
		2	0.7984	15	106419	106419	0	0.38
		3	0.8432	10	106419	106419	0	0.26
	P02	1	0.7745	19	105731	105731	0	0.47
		2	0.8432	10	105731	105731	0	0.26
		3	0.8004	15	103429	104834	−1.34	0.38
	P03	1	0.8346	11	103541	106011	−2.33	0.28
		2	0.8343	11	106049	107609	−1.45	0.28
		3	0.7954	16	102092	105762	−3.47	0.39
	P04	1	0.7922	16	106547	106583	−0.03	0.41
		2	0.7732	19	107984	107984	0	0.48
		3	0.851	10	106906	106906	0	0.24
5	P05	1	0.8848	7	104786	106328	−1.45	0.18
		2	0.8679	8	107870	107870	0	0.21
		3	0.821	13	106285	106.328	−0.04	0.31
	P06	1	0.8847	7	104315	104315	0	0.18
		2	0.8771	8	107698	107698	0	0.19
		3	0.8213	13	104001	104262	−0.25	0.31
	P07	1	0.9235	5	103582	107406	−3.56	0.12
		2	0.9133	5	104752	108114	−3.11	0.13
		3	0.9011	6	106173	106439	−0.25	0.15
	P08	1	0.8261	12	107248	107248	0	0.3
		2	0.8022	15	107248	107248	0	0.37
		3	0.8045	15	107248	107248	0	0.36
	P09	1	0.7892	17	220301	220367	−0.03	0.52
		2	0.7954	16	220776	220776	0	0.49
		3	0.8124	14	217251	217251	0	0.42
	P10	1	0.8103	14	216607	217106	−0.23	0.43
		2	0.8092	14	216767	217201	−0.2	0.43
		3	0.8674	8	211837	212134	−0.14	0.26
	P11	1	0.8464	10	211951	214960	−1.4	0.31
		2	0.8522	10	206178	215622	−4.38	0.3
		3	0.8775	8	215393	215393	0	0.23
	P12	1	0.8923	7	216828	216828	0	0.2
		2	0.8955	6	216828	216828	0	0.2
10		3	0.798	15	216828	216828	0	0.48
	P13	1	0.8563	9	205695	211620	−2.8	0.28
		2	0.8543	9	210958	213304	−1.1	0.29
		3	0.8439	10	205060	211620	−3.1	0.32
	P14	1	0.7845	17	211916	212341	−0.2	0.54
		2	0.798	15	207966	213430	−2.56	0.48
		3	0.8238	12	205335	213424	−3.79	0.38
	P15	1	0.8842	7	217221	217460	−0.11	0.22
		2	0.8906	7	218291	218794	−0.23	0.21
		3	0.8578	9	214136	214823	−0.32	0.28
	P16	1	0.77	20	171712	220144	−22	0.61
		2	0.7903	16	189324	220144	−14	0.51
		3	0.8431	10	181917	219177	−17	0.32
	Average		0.8348	12	157616	161343	−1.89	0.33

TABLE 2: The computational results with significant level 99% with $N = 15$.

T	Problem number	Budget	Average probabilities	NR	Best solution	Best solution by Sahin et al. [26]	% Dev.	Average time
		1	0.9306	11	481675	481675	0	0.48
	P17	2	0.88	29	480208	481682	−0.31	1.25
		3	0.7845	173	494401	480453	2.9	7.45
		1	0.8758	31	468932	484799	−3.27	1.35
	P18	2	0.7849	172	483921	490290	−1.3	7.39
		3	0.782	182	478213	486726	−1.75	7.82
		1	0.8221	85	474661	489583	−3.05	3.64
	P19	2	0.8762	31	492274	493018	−0.15.	1.34
		3	0.9911	2	489450	489450	0	0.1
		1	0.9317	11	477414	484876	−1.54	0.47
	P20	2	0.9618	6	484856	489912	−1.03	0.24
		3	0.8198	88	470294	484954	−3.02	3.8
5		1	0.8261	79	475885	488262	−2.54	3.38
	P21	2	0.811	104	476112	487935	−2.42	4.49
		3	0.8148	97	469153	487822	−3.83	4.18
		1	0.9153	15	473,148	486493	−2.74	0.64
	P22	2	0.8581	43	473392	488199	−3.03	1.87
		3	0.9523	7	485532	487360	−0.37	0.3
		1	0.9302	11	458388	478000	−4.1	0.48
	P23	2	0.8334	69	466110	487007	−4.29	2.95
		3	0.8044	118	467295	486801	−4.01	5.08
		1	0.8628	40	480468	491080	−2.16	1.71
	P24	2	0.9481	8	489292	494369	−1.03	0.33
		3	0.9867	3	476618	491237	−2.98	0.12
		1	0.9476	8	939786	981531	−4.25	0.38
	P25	2	0.8936	23	985031	985031	0	1.1
		3	0.9821	3	979638	979638	0	0.16
		1	0.906	18	979655	979655	0	0.87
	P26	2	0.7792	192	955783	981478	−2.62	9.41
		3	0.954	7	952918	977462	−2.51	0.33
		1	0.9215	13	955190	984103	−2.94	0.65
	P27	2	0.9272	12	972096	993049	−2.11	0.58
		3	0.8726	33	960196	983112	−2.33	1.63
		1	0.9512	7	950604	971759	−2.18	0.35
	P28	2	0.8484	52	974385	974385	0	2.54
10		3	0.9101	17	973223	974792	−0.16	0.81
		1	0.7854	170	936480	978456	−4.29	8.34
	P29	2	0.9871	3	980346	980346	0	0.13
		3	0.7638	260	947673	978748	−3.18	12.73
		1	0.782	182	949566	970024	−2.11	8.91
	P30	2	0.793	147	972765	972765	0	7.2
		3	0.7929	147	969998	970435	−0.04	7.22
		1	0.7887	160	962403	978549	−1.65	7.83
	P31	2	0.8457	55	990976	990976	0	2.67
		3	0.7747	210	979339	979339	0	10.27
		1	0.8746	32	971053	985001	−1.42	1.57
	P32	2	0.8432	57	958486	986493	−2.84	2.8
		3	0.9894	2	977270	985817	−0.87	0.12
	Average		0.8729	67	721720	733644	−1.7	3.19

TABLE 3: The computational results with significant level 99% with $N = 30$.

T	Problem number	Budget	Average probabilities	NR	Best solution	Best solution by Sahin et al. [26]	% Dev.	Average time
5	P33	1	0.887	166	576451	577086	−0.11	10.27
		2	0.9611	13	579704	579704	0	0.79
		3	0.9256	44	577493	577493	0	2.76
	P34	1	0.9588	14	551951	571846	−3.48	0.86
		2	0.995	2	559139	572396	−2.32	0.15
		3	0.9009	103	556359	570,537	−2.49	6.39
	P35	1	0.8387	899	566291	579113	−2.21	55.76
		2	0.8735	264	556438	579406	−3.96	16.37
		3	0.887	166	566301	574225	−1.38	10.28
	P36	1	0.9824	5	557872	572964	−2.63	0.32
		2	0.9731	8	554936	578631	−4.09	0.49
		3	0.9262	44	545506	569880	−4.28	2.7
	P37	1	0.9224	50	552347	559934	−1.35	3.08
		2	0.9568	15	551905	559078	−1.28	0.92
		3	0.9379	29	555069	559506	−0.79	1.81
	P38	1	0.9888	4	544879	569457	−4.32	0.23
		2	0.941	26	559640	567166	−1.33	1.62
		3	0.8689	310	546839	567749	−3.68	19.2
	P39	1	0.8817	199	569470	569470	0	12.33
		2	0.7843	6740	570521	570521	0	417.85
		3	0.9374	30	563648	569382	−1.01	1.84
	P40	1	0.992	3	556582	579411	−3.94	0.19
		2	0.927	42	565906	586310	−3.48	2.63
		3	0.9413	26	560792	577719	−2.93	1.61
10	P41	1	0.9291	39	1133743	1171634	−3.23	2.88
		2	0.986	4	1155647	1172520	−1.44	0.32
		3	0.9198	54	1129068	1171500	−3.62	3.96
	P42	1	0.9475	21	1166613	1174896	−0.71	1.52
		2	0.966	11	1137578	1175998	−3.27	0.77
		3	0.9457	22	1162838	1177009	−1.2	1.62
	P43	1	0.9223	50	1169208	1169208	0	3.63
		2	0.8739	260	1179660	1179660	0	19.01
		3	0.8867	167	1134677	1164129	−2.53	12.23
	P44	1	0.9085	80	1140598	1151468	−0.94	5.81
		2	0.7854	6462	1152874	1152874	0	471.72
		3	0.9123	70	1122006	1147234	−2.2	5.11
	P45	1	0.9703	9	1114861	1127044	−1.08	0.65
		2	0.858	453	1141881	1141881	0	33.09
		3	0.8781	225	1128472	1129703	−0.11	16.44
	P46	1	0.8762	240	1132099	1146000	−1.21	17.55
		2	0.7867	6149	1154691	1154691	0	448.88
		3	0.8583	449	1145044	1145858	−0.07	32.75
	P47	1	0.778	8585	1210573	1210573	0	626.71
		2	0.9444	23	1210573	1210573	0	1.7
		3	0.8937	132	1210573	1210573	0	9.62
	P48	1	0.7786	8389	1199048	1189154	0.83	612.37
		2	0.872	278	1152896	1201885	−4.08	20.3
		3	0.9003	105	1181360	1181360	0	7.68
	Average		0.9076	864	858596	870759	−1.58	63.02

Regarding the constraints, inherent in this kind of research, we think that if we use the new version of Lingo software and run the algorithm on a faster computer (in particular, one with a faster CPU) the results will be further improved. Finally, for the future works, we strongly suggest concentrating on a cost sensitivity process (including the rearrangement and material handling costs), which will occur in future periods and have a great influence on the optimal solution. As a suggestion, fuzzy costs may be useful under uncertainty conditions, or at least the time value of the monetary investment must be considered.

References

[1] A. Drira, H. Pierreval, and S. Hajri-Gabouj, "Facility layout problems: a survey," *Annual Reviews in Control*, vol. 31, no. 2, pp. 255–267, 2007.

[2] J. Tompkins, J. White, Y. Bozer, and J. Tanchoco, *Facilities Planning*, John Wiley & Sons, Hoboken, NJ, USA, 3rd edition, 2003.

[3] M. Dong, C. Wu, and F. Hou, "Shortest path based simulated annealing algorithm for dynamic facility layout problem under dynamic business environment," *Expert Systems with Applications*, vol. 36, no. 8, pp. 11221–11232, 2009.

[4] Y. Gary, K. J. Chen, and A. Rogers, "Multi-objective evaluation of dynamic facility layout using ant colony optimization," in *Proceedings of the Industrial Engineering Research Conference*, 2009.

[5] A. L. Page, "New product development survey: performance and best practices," in *Proceedings of the Product Development & Management Association Conference*, 1991.

[6] J. Balakrishnan and C. H. Cheng, "The dynamic plant layout problem: incorporating rolling horizons and forecast uncertainty," *Omega*, vol. 37, no. 1, pp. 165–177, 2009.

[7] P. Kouvelis, A. A. Kurawarwala, and G. J. Gutiérrez, "Algorithms for robust single and multiple period layout planning for manufacturing systems," *European Journal of Operational Research*, vol. 63, no. 2, pp. 287–303, 1992.

[8] M. J. Rosenblatt, "The dynamics of plant layout," *Management Science*, vol. 32, no. 1, pp. 76–86, 1986.

[9] T. L. Urban, "Solution procedures for the dynamic facility layout problem," *Annals of Operations Research*, vol. 76, pp. 323–342, 1998.

[10] T. A. Lacksonen and E. E. Enscore, "Quadratic assignment algorithms for the dynamic layout problems," *International Journal of Production Research*, vol. 31, no. 3, pp. 503–517, 1993.

[11] J. Balakrishnan and C. H. Cheng, "Genetic search and the dynamic layout problem," *Computers and Operations Research*, vol. 27, no. 6, pp. 587–593, 2000.

[12] B. K. Kaku and J. B. Mazzola, "A tabu-search heuristic for the dynamic plant layout problem," *INFORMS Journal on Computing*, vol. 9, no. 4, pp. 374–384, 1997.

[13] A. Baykasoglu and N. N. Z. Gindy, "A simulated annealing algorithm for dynamic facility layout problem," *Computers and Operations Research*, vol. 28, no. 14, pp. 1403–1426, 2001.

[14] J. Balakrishnan, C. H. Cheng, D. G. Conway, and C. M. Lau, "A hybrid genetic algorithm for the dynamic plant layout problem," *International Journal of Production Economics*, vol. 86, no. 2, pp. 107–120, 2003.

[15] E. Erel, J. B. Ghosh, and J. T. Simon, "New heuristic for the dynamic layout problem," *Journal of the Operational Research Society*, vol. 56, no. 8, p. 1001, 2005.

[16] A. R. McKendall and J. Shang, "Hybrid ant systemsfor the dynamic facility layout problem," *Computers & Operations Research*, vol. 33, no. 3, pp. 790–803, 2006.

[17] A. R. McKendall, J. Shang, and S. Kuppusamy, "Simulated annealing heuristics for the dynamic facility layout problem," *Computers and Operations Research*, vol. 33, no. 8, pp. 2431–2444, 2006.

[18] J. M. Rodriguez, F. C. MacPhee, D. J. Bonham, and V. C. Bhavsar, "Solving the dynamic plant layout problem using a new hybrid meta-heuristic algorithm," *International Journal of High Performance Computing and Networking*, vol. 4, no. 5-6, pp. 286–294, 2006.

[19] K. K. Krishnan, S. H. Cheraghi, and C. N. Nayak, "Dynamic from-between charts: a new tool for solving dynamic facility layout problems," *International Journal of Industrial and Systems Engineering*, vol. 11, no. 1-2, pp. 182–200, 2006.

[20] J. Balakrishnan and C. H. Cheng, "Dynamic layout algorithms: a state-of-the-art survey," *Omega*, vol. 26, no. 4, pp. 507–521, 1998.

[21] S. Kulturel-Konak, "Approaches to uncertainties in facility layout problems: perspectives at the beginning of the 21st Century," *Journal of Intelligent Manufacturing*, vol. 18, no. 2, pp. 273–284, 2007.

[22] T. Dunker, G. Radons, and E. Westkämper, "Combining evolutionary computation and dynamic programming for solving a dynamic facility layout problem," *European Journal of Operational Research*, vol. 165, no. 1, pp. 55–69, 2005.

[23] A. R. McKendall and A. Hakobyan, "Heuristics for the dynamic facility layout problem with unequal-area departments," *European Journal of Operational Research*, vol. 201, no. 1, pp. 171–182, 2010.

[24] J. Balakrishnan, F. R. Jacobs, and M. A. Venkataramanan, "Solutions for the constrained dynamic facility layout problem," *European Journal of Operational Research*, vol. 57, no. 2, pp. 280–286, 1992.

[25] A. Baykasoglu, T. Dereli, and I. Sabuncu, "An ant colony algorithm for solving budget constrained and unconstrained dynamic facility layout problems," *Omega*, vol. 34, no. 4, pp. 385–396, 2006.

[26] R. Sahin, K. Ertogral, and O. Turkbey, "A simulated annealing heuristic for the dynamic facility layout problem with budget constraint," *Computers & Industrial Engineering*, vol. 59, no. 2, pp. 308–313, 2010.

[27] P. Azimi and P. Daneshvar, "An efficient heuristic algorithm for the traveling salesman problem," in *Proceedings of the 8th International Heinz Nixdorf Symposium on Changing Paradigms (IHNS '10)*, Advanced Manufacturing and Sustainable Logistics, pp. 384–395, Paderborn, Germany, 2010.

8

DC Motor Parameter Identification Using Speed Step Responses

Wei Wu

Flight Control and Navigation Group, Rockwell Collins, Warrenton, VA 20187, USA

Correspondence should be addressed to Wei Wu, wu_esi@yahoo.com

Academic Editor: F. Gao

Based on the DC motor speed response measurement under a step voltage input, important motor parameters such as the electrical time constant, the mechanical time constant, and the friction can be estimated. A power series expansion of the motor speed response is presented, whose coefficients are related to the motor parameters. These coefficients can be easily computed using existing curve fitting methods. Experimental results are presented to demonstrate the application of this approach. In these experiments, the approach was readily implemented and gave more accurate estimates than conventional methods.

1. Introduction

DC motors have wide applications in industrial control systems because they are easy to control and model. For analytical control system design and optimization, sometimes a precise model of the DC motor used in a control system may be needed. In this case, the values for reference of the motor parameters given in the motor specifications, usually provided by the motor manufacturer, may not be considered adequate, especially for cheaper DC motors which tend to have relatively large tolerances in their electrical and mechanical parameters. General system identification methods [1–4] can be applied to DC motor model identification. In particular, various methods have been applied to DC motor parameter identification; that is, [5, 6] used the algebraic identification method, [7] used the recursive least square method, [8] applied the inverse theory, [9] used the least square method, and [10] applied the moments method. Identified DC motor models are often subsequently used for controller design and/or optimization, for example, [6, 9, 11].

Without expensive testing apparatus and a long testing cycle, a quick and effective system identification approach based on the motor input and output is desirable and valuable, especially for the field applications and quick controller prototyping. In this paper, a DC motor parameter identification approach based on the Taylor series expansion of the motor speed response under a constant voltage input is presented. The relationships between the motor parameters and the coefficients of the Taylor series are established. In the implementation, the motor speed response under a constant voltage is sampled, then fit the samples to obtain the coefficients of power terms in the Taylor series. Then, the DC motor mechanical and electrical time constants, back-EMF, and the friction can be computed using these coefficients. With the knowledge of these parameters, a precise motor model is obtained for the subsequent controller design.

For application point of view, this approach requires only a speed/position sensor, such as an optical encoder, and a voltage power supply, no current measurement is needed and the motor is run in open loop; thus it is practical and cost effective. The curve fitting can be performed using many existing methods, such as the least square method, and these optimization methods are widely available in commercial computing packages such as Matlab and LabVIEW.

2. Main Results

Consider the following DC motor governing equations:

$$L\frac{di}{dt} + iR + k_b\omega = V,$$
$$J\frac{d\omega}{dt} = k_t i + T_d, \tag{1}$$

where ω is the motor speed, V is the motor terminal voltage, i is the winding current, k_b is the back-EMF constant of the motor, k_t is the torque constant, R is the terminal resistance,

L is the terminal inductance, J is the motor and load inertia, and T_d is the disturbance torque. T_d is a combination of the cogging torque, T_{cog}, the kinetic friction, T_f, and the viscous friction (viscous damping force):

$$T_d = T_{\text{cog}} + T_f + c\dot{\omega}, \qquad (2)$$

where c is the damping coefficient. According to (1), the velocity response in the Laplace domain is

$$\omega(s) = \frac{1/k_b}{t_m t_e s^2 + t_m s + 1} V(s) + \frac{(1/J)t_m(t_e s + 1)}{t_m t_e s^2 + t_m s + 1} T_d(s), \quad (3)$$

where $t_e = L/R$ is the electrical time constant, $t_m = RJ/k_t k_b$ is the mechanical time constant, and s is the Laplace variable.

Based on these equations, we would like to know t_m, t_e, T_d, J, and so forth, by measuring the velocity response under a known, controlled voltage input. In this paper, we consider two application situations: the first situation is that the disturbance torque is negligible, while in the second one, the disturbance needs to be considered.

2.1. Estimation without the Disturbance Torque.

When the voltage speed response dominates; for example, the input voltage is large, we can ignore the disturbance torque in the speed response see (3). In this case, we can consider the following DC motor model:

$$\frac{\omega(s)}{V(s)} = \frac{1/k_b}{t_m t_e s^2 + t_m s + 1}. \qquad (4)$$

The transfer function can be factorized into

$$\frac{\omega(s)}{V(s)} = \frac{1/k_b}{t_m t_e (s + a)(s + b)}, \qquad (5)$$

where

$$a, b = \frac{1 \mp \sqrt{1 - 4t_e/t_m}}{2t_e}. \qquad (6)$$

Assumption. It is assumed here that there are two distinct real poles; that is, $t_m > 4t_e$.

For a constant voltage input $V(s) = V_0/s$, the speed response is

$$\omega(s) = \frac{V_0/k_b}{t_m t_e s(s + a)(s + b)} = \frac{\alpha_1}{s} + \frac{\alpha_2}{s + a} + \frac{\alpha_3}{s + b}, \qquad (7)$$

where

$$\alpha_1 = \frac{V_0}{k_b}, \qquad \alpha_2 = \frac{V_0}{k_b}\frac{b}{a - b}, \qquad \alpha_3 = \frac{V_0}{k_b}\frac{a}{b - a}. \qquad (8)$$

Consider the three terms in the step response one at a time. α_1/s is a step function in the time domain; both $\alpha_2/(s + a)$ and $\alpha_1/(s + b)$ are exponential functions in the time domain and can be expanded using the Taylor series. Expanding the term $\alpha_2/(s + a)$, we get

$$\frac{V_0}{k_b}\frac{b}{a - b}\left(1 - at + \frac{1}{2}a^2 t^2 - \frac{1}{6}a^3 t^3 + \cdots\right). \qquad (9)$$

Expanding the term $\alpha_3/(s + b)$, we get

$$\frac{V_0}{k_b}\frac{a}{b - a}\left(1 - bt + \frac{1}{2}b^2 t^2 - \frac{1}{6}b^3 t^3 + \cdots\right). \qquad (10)$$

Combining the three terms together, we have the total speed response:

$$\omega(t) = \frac{V_0}{k_b}\left(\frac{1}{2}\beta_0 t^2 + \frac{1}{6}\beta_1 t^3 + \frac{1}{24}\beta_2 t^4 + \cdots\right), \qquad (11)$$

where $\beta_0 = ab$, $\beta_1 = -ab(a + b)$, and $\beta_2 = ab(a^2 + ab + b^2)$.

According to (6),

$$ab = \frac{1}{t_m t_e}, \qquad a + b = \frac{1}{t_e}. \qquad (12)$$

Thus, we have

$$t_m = -\frac{\beta_1}{\beta_0^2}, \qquad t_e = -\frac{\beta_0}{\beta_1}. \qquad (13)$$

The above equation allows us to calculate the mechanical and electrical time constants t_m and t_e using the coefficients of the power series in (11). These coefficients can be obtained by curve fitting the motor speed step response data using power functions.

2.2. Estimation with the Disturbance Torque.

Consider that the disturbance torque in the DC motor is not negligible. The disturbance transfer function is

$$\frac{\omega(s)}{T_d(s)} = \frac{(1/J)t_m(t_e s + 1)}{t_m t_e s^2 + t_m s + 1}. \qquad (14)$$

Disturbance torque generally consists of the cogging torque and the friction torque. The cogging torque is quite complicated and is not addressed here. Both the kinetic and viscous frictions are considered and are assumed to be constant on average under a constant motor speed.

Given a constant motor terminal voltage $V(s) = V_0/s$ and the constant disturbance (ignore the cogging torque or consider the average cogging torque effect on speed over one revolution is zero) $T_d(s) = T_0/s$, the speed response is

$$\omega(s) = \frac{1/k_b}{t_m t_e s^2 + t_m s + 1}\frac{V_0}{s} + \frac{(1/J)t_m(t_e s + 1)}{t_m t_e s^2 + t_m s + 1}\frac{T_0}{s}. \qquad (15)$$

As in the previous section, applying the partial fraction expansion of the step response in the Laplace domain, then expanding the exponential terms in the time domain using the Taylor series, we obtain the total step response in the time domain:

$$\omega(t) = \beta_0 t + \beta_1 t^2 - \beta_2 t^3 + \beta_3 t^4 - \cdots. \qquad (16)$$

Based on these coefficients, we have

$$ab = \frac{18\beta_2^2 - 24\beta_3\beta_1}{3\beta_0\beta_2 + 2\beta_1^2},$$
$$a + b = \frac{6\beta_2 - \beta_0 ab}{2\beta_1}, \qquad (17)$$

and another equation for $a + b$:

$$a + b = \frac{12\beta_3 + \beta_1 ab}{2\beta_2}. \qquad (18)$$

TABLE 1: RK370CA parameter values.

Parameter	Value	Unit
Terminal resistance	$17 \pm 15\%$	Ω
Terminal inductance	N/A	Henry
Torque constant	$18.3 \pm 18\%$	mNm/A
Mass moment of inertia	9.0	gcm^2
Counter-electromotive force	0.0233	volt/(rad/sec)

TABLE 2: RK370CA test results.

Parameter	w/o dist. 20 v	w/dist. 2 v/10 v	Spec. (meas.)	Unit
k_t	0.0238	0.0207/0.0169	$0.0183 \pm 18\%$	Nm/A
t_m	0.0407	0.0211/0.0203	0.0359	sec
t_e	0.00554	0.00122/0.00134	(0.00122)	sec
T_0/J	N/A	10.551/115.758	N/A	Nm/kgm^2

Then, we can express the motor parameters as

$$t_m = \frac{a+b}{ab},$$

$$t_e = \frac{1}{a+b},$$

$$\frac{T_0}{J} = \beta_0, \tag{19}$$

$$k_b = \frac{ab}{2\beta_1}V_0.$$

In practice, fit the measured motor speed step response using power functions according to (16); then calculate the motor parameters using (19).

Remark 1. Another relationship useful for checking the algorithm is based on the steady-state response of (15), expressed by the following equation:

$$\frac{V_0}{k_b} + \beta_0 t_m = \omega_{ss}, \tag{20}$$

where $\beta_0 = T_0/J$ and ω_{ss} is the motor steady-state angular speed.

3. Implementation and Results

The proposed approaches were first applied to a Mabuchi RK370CA motor, then a Mabuchi FC130 motor. To implement the algorithms, a LabVIEW program was created to interface a pulse width modulated (PWM) motor drive and an optical encoder with quadrature digital outputs mounted on the motor shaft. The determinism of the sample time was assured by the LabVIEW real-time module. And, a national instrument (NI) LabVIEW FPGA (field programmable gate array) card was utilized to process the digital quadrature encoder signals to obtain the motor speed and to control the motor PWM drive.

Values of the motor parameters given in the motor specifications for reference are presented in Table 1.

Note that the Back-EMF and torque constant are not equal (although it should be theoretically). Inductance value is not given and was measured as 20.25 Henry. The resistance was measured as 16.4 Ω. Thus we calculated $t_e = L/R = 0.00122$ sec.

First, apply the algorithm for no-disturbance torque. To apply this algorithm, the speed response part due to the voltage input is assumed to dominate. To meet this condition,

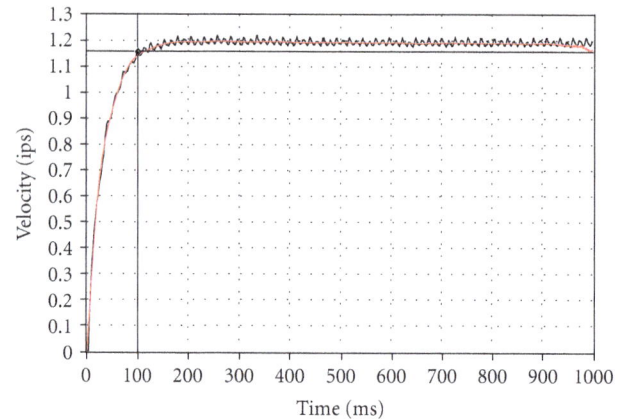

FIGURE 1: Approach w/consideration of disturbance under 2 volt input: black, measurement: red, power series fitted.

for example, the speed variation at the steady-state is small compared to the steady-state speed, we send a large voltage to the motor drive, $V = 20$ volt. Next, we apply the approach for disturbance torque. The disturbances, that is, friction, effects on the speed response are significant when the input voltage is small. To demonstrate the effectiveness of the algorithm, we sent two voltages, 2 volt and 10 volt, to the motor. Driving the motor at two different voltage levels can demonstrate that the viscous friction varies with the speed, also can allow us to calculate the viscous damping coefficient.

Usually t_e is very small compared to t_m a good estimate of both t_e and t_m at the same time is difficult. Because t_m is usually much larger than t_e, t_m and t_e were estimated separately using different data collected with different sample rates and different time durations. For estimating t_m, the motor speed in both the transient phase and the steady-state was sampled at 1 kHz for one second; for estimating t_e, the motor speed in the transient phase was sampled at 8 kHz for 200 msec. In each test, the motor was driven multiple times and parameter estimates were averaged.

Results are summarized in Table 2. Column two gives the values estimated using the algorithm for no disturbance, and column three gives the values obtained using the algorithm considering disturbance; values in the fourth column are computed using values from Table 1. Note that $R = 17\,\Omega$, $J = 9\,$gcm^2, $k_b = 0.0233$ volt/(rad/sec), and $k_t = 0.0183$ Nm/A are used to calculate t_m in the fourth column of the table. According to Table 2, the estimates of k_t, t_m, and t_e are in good agreement with those given by the motor specifications.

Time responses sampled at 1 kHz for 1 s are given in Figures 1, 2 and 3. In these figures, red curves represent the power series, $\sum_{i=1}^n x_i t^i$, resulting from curve fitting the motor

TABLE 3: FC130 test results.

Parameter	w/dist.	Spec.	Unit
k_t	0.0137	$0.0127 \pm 10\%$	Nm/A
t_m	0.0208	0.024	sec
t_e	0.251	0.214	msec

speed responses. Comparing these figures, it is obvious that the approach with disturbance consideration approximates the measurements much better, because of the existence of the linear term, $\beta_0 t$, in the power series due to the presence of the constant disturbance in the motor.

To further demonstrate the effectiveness of the proposed algorithms, we compared them to conventional identification approaches. First, we drove the motor using random voltage input (10 volts maximum) and measured the motor speed at a sampling rate of 10 kHz. Then, the motor/drive frequency response function was calculated through spectral analysis. Based on the calculated frequency response data, we used Matlab system identification toolbox to identify a second-order model. Various methods, that is, subspace approach in the system identification toolbox, were tried and compared. The best model found was

$$T(s) = \frac{9078}{s^2 + 334.6s + 18860}. \tag{21}$$

Using the model coefficients, we get

$$t_m = 0.0177 \text{ sec}, \qquad t_e = 3 \text{ msec},$$
$$k_t = 0.031 \text{ volts/(rad/sec)}. \tag{22}$$

These estimates are bad, especially the electrical time constant t_e due to the very small time scale as alluded to earlier.

Remark 2. T_0/J may be used to calculate the friction (both kinetic and viscous) if J is known. First, calculate the viscous friction coefficient $c = (T_1 - T_0)/(\omega_1 - \omega_0)$. Then, calculate the dynamic friction, $T_f = T_0 - c\omega_0$. For example, $\omega_0 = 1.21$ ips under 2 volt, $\omega_1 = 6.274$ ips under 10 volt, $J = 9.0$ gcm^2, and it renders $c = 0.0187$ mNm/ips.

Remark 3. The number of terms in the power series included for fitting the data was determined through trial and error. When disturbance was not considered, twenty-five terms were included; when disturbance was considered, including forty terms gave the best results. Since the coefficients were calculated using the polynomial curve fitting function from the math library provided inside LabVIEW, it was not difficult and time consuming to try different number of terms. Including more terms does not necessarily improve the parameter estimation accuracy.

A Mabuchi FC130 motor was tested as well. It is a smaller motor compared to RK370. Good results were obtained again this time; see Table 3. Note the very small t_e in this small motor. Algorithm considering disturbance torque was applied. In the testing, 10 volts was used as the motor drive input. For t_m estimation, the speed response was sampled at 1000 Hz for 500 samples, while for t_e estimation, it was sampled at 6000 Hz for 850 samples.

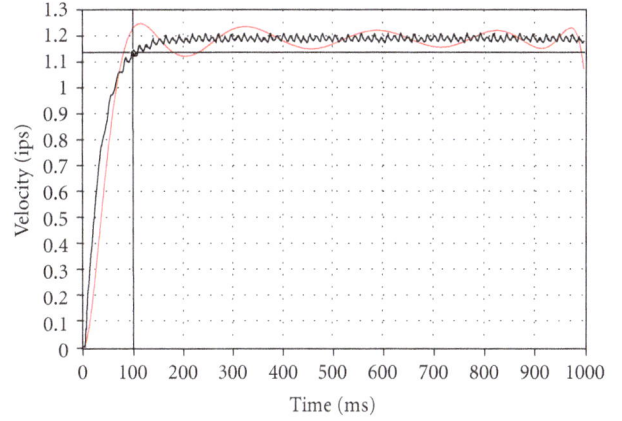

FIGURE 2: Approach *w/o* consideration of disturbance under 2 volt input: black, measurement: red, power series fitted.

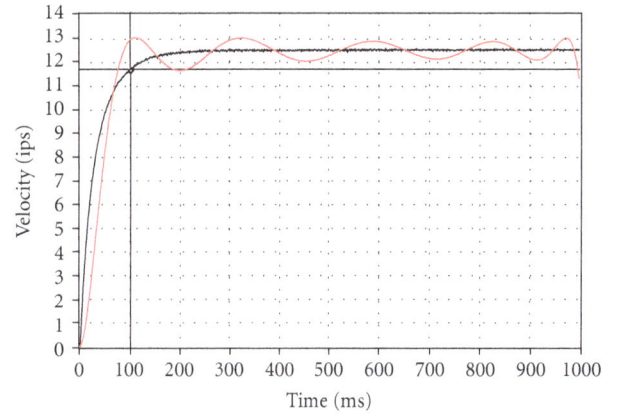

FIGURE 3: Approach *w/o* consideration of disturbance under 20 volt input:Black, measurement: Red, power series fitted.

4. Conclusions

A convenient, effective system identification approach is proposed to estimate the DC motor torque constant, mechanical time constant, electrical time constant, and friction. This approach was implemented on two Mabuchi motors, and the great test results were presented. This open-loop method requires little hardware, only a speed/position sensor and a voltage supply. The estimated motor parameters can be used to verify the DC motor performance or be used to build a model of the motor for the subsequent controller design or system optimization. This approach is especially suited to quick field applications.

Appendix

Coefficients for no disturbance case are as follows:

$$\beta_0 = ab,$$
$$\beta_1 = -ab(a + b), \tag{A.1}$$
$$\beta_2 = ab(a^2 + ab + b^2).$$

Coefficients for disturbance case are as follows:

$$\beta_0 = \frac{T_0}{J},$$

$$\beta_1 = \frac{1}{2}\frac{V_0}{k_b}ab,$$

$$\beta_2 = \frac{1}{6}\left[\frac{V_0}{k_b}ab(a+b) - \frac{T_0}{J}(a^2+ab+b^2)\right.$$
$$\left. +\frac{T_0}{J}t_m ab(a+b)\right],$$

$$\beta_3 = \frac{1}{24}\left[\frac{V_0}{k_b}ab(a^2+ab+b^2) - \frac{T_0}{J}(a^3+a^2b+ab^2+b^3)\right.$$
$$\left. +\frac{T_0}{J}t_m ab(a^2+ab+b^2)\right].$$

$$(A.2)$$

References

[1] L. Ljiung, *System Identification: Theory for the User*, Prentice Hall, 2nd edition, 1999.

[2] H. Unbehauen and G. P. Rao, "A review of identification in continuous-time systems," *Annual Reviews in Control*, vol. 22, pp. 145–171, 1998.

[3] G. F. Franklin, J. D. Powell, and M. L. Workman, *Digital Control of Dynamic Systems*, Addison Wesley, 2nd edition, 1990.

[4] J. C. Basilio and M. V. Moreira, "State-space parameter identification in a second control laboratory," *IEEE Transactions on Education*, vol. 47, no. 2, pp. 204–210, 2004.

[5] G. Mamani, J. Becedas, H. Sira-Ramirez, and V. Feliu Batlle, "Open-loop algebraic identification method for DC motors," in *Proceedings of the European Control Conference*, Kos, Greece, 2007.

[6] G. Mamani, J. Becedas, and V. Feliu-Batlle, "On-line fast algebraic parameter and state estimation for a DC motor applied to adaptive 16 control," in *Proceedings of the World Congress on Engineering*, London, UK, 2008.

[7] R. Krneta, S. Antic, and D. Stojanovic, "Recursive least square method in parameters identification of DC motors models," *Facta Universitatis*, vol. 18, no. 3, pp. 467–478, 2005.

[8] M. Hadef and M. R. Mekideche, "Parameter identification of a separately excited DC motor via inverse problem methodology," in *Proceedings of the Ecologic Vehicles andRenewable Energies*, Monaco, France, 2009.

[9] M. Ruderman, J. Krettek, F. Hoffman, and T. Betran, "Optimal state space control of DC motor," in *Proceedings of the 17th World Congress IFAC*, pp. 5796–5801, Seoul, Korea, 2008.

[10] M. Hadef, A. Bourouina, and M. R. Mekideche, "Parameter identification of a DC motor via moments method," *International Journal of Electrical and Power Engineering*, vol. 1, no. 2, pp. 210–214, 2008.

[11] A. Rubaai and R. Kotaru, "Online identification and control of a dc motor using learning adaptation of neural networks," *IEEE Transactions on Industry Applications*, vol. 36, no. 3, pp. 935–942, 2000.

9

Design of Wideband MIMO Car-to-Car Channel Models Based on the Geometrical Street Scattering Model

Nurilla Avazov and Matthias Pätzold

Faculty of Engineering and Science, University of Agder, P.O. Box 509, 4898 Grimstad, Norway

Correspondence should be addressed to Nurilla Avazov, nurilla.k.avazov@uia.no

Academic Editor: Neji Youssef

We propose a wideband multiple-input multiple-output (MIMO) car-to-car (C2C) channel model based on the geometrical street scattering model. Starting from the geometrical model, a MIMO reference channel model is derived under the assumption of single-bounce scattering in line-of-sight (LOS) and non-LOS (NLOS) propagation environments. The proposed channel model assumes an infinite number of scatterers, which are uniformly distributed in two rectangular areas located on both sides of the street. Analytical solutions are presented for the space-time-frequency cross-correlation function (STF-CCF), the two-dimensional (2D) space CCF, the time-frequency CCF (TF-CCF), the temporal autocorrelation function (ACF), and the frequency correlation function (FCF). An efficient sum-of-cisoids (SOCs) channel simulator is derived from the reference model. It is shown that the temporal ACF and the FCF of the SOC channel simulator fit very well to the corresponding correlation functions of the reference model. To validate the proposed channel model, the mean Doppler shift and the Doppler spread of the reference model have been matched to real-world measurement data. The comparison results demonstrate an excellent agreement between theory and measurements, which confirms the validity of the derived reference model. The proposed geometry-based channel simulator allows us to study the effect of nearby street scatterers on the performance of C2C communication systems.

1. Introduction

C2C communications is an emerging technology, which receives considerable attention due to new traffic telematic applications that improve the efficiency of traffic flow and reduce the number of road accidents [1]. The development of C2C communication technologies is supported in Europe by respected organizations, such as the European Road Transport Telematics Implementation Coordinating Organization (ERTICO) [2] and the C2C Communication Consortium (C2C-CC) [3]. In this context, a large number of research projects focussing on C2C communications are currently being carried out throughout the world.

In C2C communication systems, the underlying radio channel differs from traditional fixed-to-mobile and mobile-to-fixed channels in the way that both the transmitter and the receiver are in motion. In this connection, robust and reliable traffic telematic systems have to be developed and tested, which calls for new channel models for C2C communication systems. Furthermore, MIMO communication systems can also be of great interest for C2C communications due to their higher throughput [4]. In this regard, several MIMO mobile-to-mobile (M2M) channel models have been developed and analyzed under different scattering conditions induced by, for example, the two-ring model [5], the elliptical model [6], the T-junction model [7], and the geometrical street model [8, 9]. A 2D reference model for narrowband single-input single-output (SISO) M2M Rayleigh fading channels has been proposed by Akki and Haber in [10, 11]. Simulation models for SISO M2M channels have been reported in [12, 13]. In [5, 14, 15], the 2D reference and simulation models have been presented for narrowband MIMO M2M channels. The proposed model in [15] combines the two-ring model and the elliptical model, where a combination of single- and double-bounce scattering in LOS propagation environments is assumed.

All aforementioned channel models are narrowband M2M channel models. In contrast with narrowband channels, a channel is called a wideband channel or frequency-selective channel if the signal bandwidth significantly exceeds the coherence bandwidth of the channel. Owing to increasing demands for high data rate wideband communication systems employing MIMO technologies, such as MIMO orthogonal frequency division multiplexing (OFDM) systems, it is of crucial importance to have accurate and realistic wideband MIMO M2M channel models. According to IEEE 802.11p [16], the dedicated frequency bands for short-range communications [17] will be between 5770 MHz and 5925 MHz depending on the region. The range 5795–5815 MHz will be devoted to Europe, while 5850–5925 MHz and 5770–5850 MHz will be assigned to North America and Japan, respectively. Consequently, a large number of C2C channel measurements have been carried out at different frequency bands, for example, at 2.4 GHz [18], 3.5 GHz [19], 5 GHz [20, 21], 5.2 GHz [22], and 5.9 GHz [23]. Real-world measurement campaigns for wideband C2C channels can be found in [24–27]. In the literature, there exist several papers [28–30] with the focus on the modeling of wideband MIMO M2M channels. A reference model derived from the geometrical T-junction scattering model has been proposed in [7] for wideband MIMO vehicle-to-vehicle (V2V) fading channels. In [29], a three-dimensional (3D) model for a wideband MIMO M2M channel has been studied. Its corresponding first- and second-order statistics have been investigated and validated on the basis of real-world measurement data. In the same paper, it has been shown that 3D scattering scenarios are more realistic than 2D scattering scenarios. However, 2D scattering models are more complexity efficient, and they provide a good approximation to 3D scattering models [31]. For those reasons, we propose in our paper a 2D street scattering model.

In the literature, numerous fundamental channel models with different scatterer distributions, such as the uniform, Gaussian, Laplacian, and von Mises distribution, have been proposed to characterize the angle-of-departure (AOD) and the angle-of-arrival (AOA) statistics. In [32], the author studied the effect of Gaussian distributed scatterers on the channel characteristics in a circular scattering region around a mobile station. The spatial and temporal properties of the first arrival path in multipath environments have also been analyzed in [32]. The authors of [9] assume rectangular scattering areas on both sides of the street, in which an infinite number of scatterers are uniformly distributed. It has been observed that the shape of the Doppler power spectral density (PSD) resembles a Gaussian function if the width of the scattering area is very large.

In contrast to our previous work in [9], where the focus was on the derivation of a reference channel model for narrowband SISO C2C channels, we design in this paper a wideband MIMO C2C channel model by starting from the same geometrical street scattering model. We focus on the statistical characterization of a wideband reference channel model assuming that an infinite number of scatterers are uniformly distributed within two rectangular areas. The radio propagation phenomena in street environments are modelled by a wide-sense stationary uncorrelated scattering process, where in addition a LOS component is taken into account. The reference model has been derived from the geometrical street scattering model assuming that the AOD

and the AOA are dependent due to single-bounce scattering. To account for the nature of C2C channels, we take the mobility of both the transmitter and the receiver for granted.

In our model, we consider a 2D street scattering environment to reduce the computational cost by still guaranteeing a good match between the reference model and measured channels. A typical propagation scenario for the proposed model is illustrated in Figure 1, where the buildings and the trees are considered as scattering objects. Such a typical dense urban environment scenario allows us to assume that the local scatterers are uniformly distributed in a specific area. An analytical expression will be derived for the STF-CCF from which the 2D space CCF, the TF-CCF, the temporal ACF, and the FCF can be obtained directly. To validate the proposed reference model, the mean Doppler shift and the Doppler spread of the reference model have been matched to the corresponding quantities of the measured channel described in [25] for different propagation environments, such as urban, rural, and highway areas. Furthermore, we have derived an SOC channel simulator from the reference model. It is shown that the designed channel simulator matches the underlying reference model with respect to the temporal ACF and the FCF.

The rest of this paper is organized as follows. Section 2 describes the geometrical street scattering model. In Section 3, the reference channel model is derived from the geometrical street model. Section 4 analyzes the correlation properties of the reference model, such as the STF-CCF, the 2D space CCF, the TF-CCF, the temporal ACF, and the FCF. The computation of the measurement-based model parameters and the characteristic quantities describing the Doppler effect are discussed in Section 5. Section 6 describes briefly the simulation model derived from the reference model. The illustration of some numerical results found for the correlation functions of the reference model and the corresponding simulation model is the topic of Section 7. Finally, Section 8 draws the conclusion of the paper.

2. The Geometrical Street Scattering Model

This section briefly describes the geometrical street scattering model for wideband MIMO C2C channels. The proposed geometrical model describes the scattering environment in an urban area, where the scatterers are located in two rectangular areas on both sides of the street as illustrated in Figure 2. We consider rectangular grids formed by rows and columns, where the length and the width of the rectangular grids are denoted by $L_A = A_1 + A_2$ and B_i ($i = 1, 2$), respectively. The scatterer located in the mth column of the nth row is denoted by $S^{(mn)}$ ($m = 1, 2, \ldots, M$, $n = 1, 2, \ldots, N$). It is assumed that the local scatterers $S^{(mn)}$ are uniformly distributed in the rectangles. The symbols MS_T and MS_R in Figure 2 stand for the mobile transmitter and the mobile receiver, respectively. The symbol D represents the scalar projection of the distance between the transmitter and the receiver onto the x-axis. The transmitter (receiver) is located at a distance y_{T_1} (y_{R_1}) from the left-hand side of the street and at a distance y_{T_2} (y_{R_2}) from the right-hand side of the street. Both the transmitter and the receiver are in motion

FIGURE 1: A typical propagation scenario along a straight street in urban areas.

and equipped with M_T transmitter antenna elements and M_R receiver antenna elements, respectively. The antenna element spacings at the transmitter and the receiver are denoted by δ_T and δ_R, respectively. The symbols $\alpha_T^{(mn)}$ and $\alpha_R^{(mn)}$ denote the AOD and the AOA, respectively. The angle $\gamma_T(\gamma_R)$ describes the tilt angle of the transmitter (receiver) antenna array. Moreover, it is assumed that the transmitter (receiver) moves with speed v_T (v_R) in the direction determined by the angle of motion φ_v^T (φ_v^R).

3. The Reference Model

3.1. Derivation of the Reference Model. In this section, we derive the reference model for the MIMO C2C channel under the assumption of LOS and NLOS propagation conditions. From Figure 2, we realize that the (mn)th homogeneous plane wave emitted from the lth antenna element $A_T^{(l)}$ ($l = 1, 2, \ldots, M_T$) of the transmitter travels over the local scatterer $S^{(mn)}$ before impinging on the kth antenna element $A_R^{(k)}$ ($k = 1, 2, \ldots, M_R$) of the receiver. The reference model is based on the assumption that the number of local scatterers within both rectangular areas is infinite, that is, $M, N \to \infty$. The temporal, spatial, and frequency characteristics of the reference model are determined by the $M_R \times M_T$ channel matrix $\mathbf{H}(f', t) = [H_{kl}(f', t)]_{M_R \times M_T}$, where $H_{kl}(f', t)$ denotes the time-variant transfer function (TVTF) of the channel for the link between the lth transmitter antenna element $A_T^{(l)}$ and the kth receiver antenna element $A_R^{(k)}$. The TVTF $H_{kl}(f', t)$ can be expressed as a superposition of the diffuse component and the LOS component as follows:

$$H_{kl}(f', t) = H_{kl}^{\mathrm{DIF}}(f', t) + H_{kl}^{\mathrm{LOS}}(f', t), \tag{1}$$

where $H_{kl}^{\mathrm{DIF}}(f', t)$ and $H_{kl}^{\mathrm{LOS}}(f', t)$ represent the diffuse and the LOS components of the channel, respectively.

Note that the single-bounce scattering components bear more energy than the double-bounce scattering components. Hence, in our analysis, we model the diffuse component $H_{kl}^{\mathrm{DIF}}(f', t)$ by only taking into account the single-bounce

scattering effects, which is in accordance with the assumptions made in [28, 33]. From the geometrical street scattering model shown in Figure 2, we can derive the TVTF of the diffuse component, which results in the following expression:

$$H_{kl}^{\mathrm{DIF}}(f', t) = \lim_{M,N \to \infty} \frac{1}{\sqrt{(c_R + 1)MN}} \sum_{m,n=1}^{M,N} a_{l,mn} b_{k,mn} c_{mn}$$
$$\cdot e^{j[\theta_{mn} + 2\pi(f_T^{(mn)} + f_R^{(mn)})t - 2\pi f' \tau_{kl}'^{(mn)}]}, \tag{2}$$

where

$$a_{l,mn} = e^{j\pi(\delta_T/\lambda)(M_T - 2l + 1)\cos(\alpha_T^{(mn)} - \gamma_T)}, \tag{3}$$

$$b_{k,mn} = e^{j\pi(\delta_R/\lambda)(M_R - 2k + 1)\cos(\alpha_R^{(mn)} - \gamma_R)}, \tag{4}$$

$$c_{mn} = e^{-j(2\pi/\lambda)(y_{T_1}/\sin(\alpha_T^{(mn)}) + y_{R_1}/\sin(\alpha_R^{(mn)}))}, \tag{5}$$

$$f_T^{(mn)} = f_{T_{\max}} \cos\left(\alpha_T^{(mn)} - \varphi_v^T\right), \tag{6}$$

$$f_R^{(mn)} = f_{R_{\max}} \cos\left(\alpha_R^{(mn)} - \varphi_v^R\right), \tag{7}$$

$$\tau_{kl}'^{(mn)} = \frac{1}{c_0}\left[D_T^{(l,mn)} + D_R^{(mn,k)}\right]. \tag{8}$$

In (6) and (7), the symbols $f_{T_{\max}} = v_T/\lambda$ and $f_{R_{\max}} = v_R/\lambda$ denote the maximum Doppler frequencies associated with the movement of the transmitter and the receiver, respectively, and λ is the wavelength. The symbol c_R in (2) represents the Rice factor, which is defined as the ratio of the power of the LOS component to the power of the diffuse component, that is, $c_R = E\{|H_{kl}^{\mathrm{LOS}}(f', t)|^2\}/E\{|H_{kl}^{\mathrm{DIF}}(f', t)|^2\}$. The phases θ_{mn} in (2) denote the phase shift introduced by the scatterer $S^{(mn)}$. It is assumed that the phases θ_{mn} are independent, identically distributed (i.i.d.) random variables, which are uniformly distributed over the interval $[0, 2\pi)$. The symbols $\tau_{kl}'^{(mn)}$ and c_0 represent the propagation delays of the diffuse component and the speed of light, respectively. In (8), the quantity $D_T^{(l,mn)}$

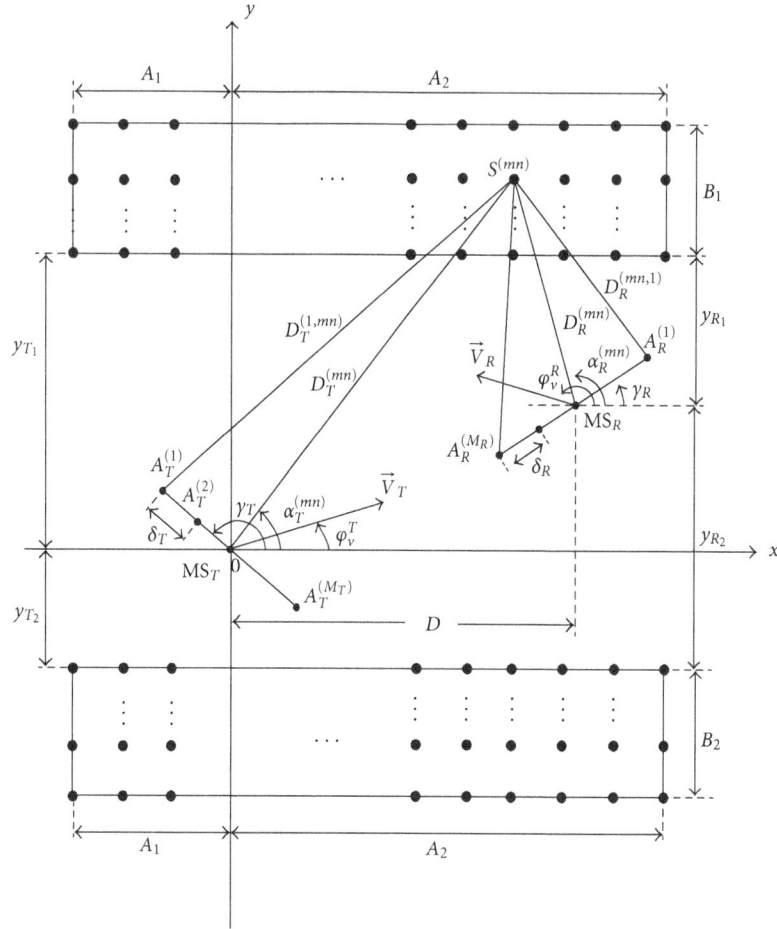

FIGURE 2: The geometrical street scattering model with local scatterers uniformly distributed in two rectangular areas on both sides of the street.

stands for the distance from the lth transmitter antenna element $A_T^{(l)}$ to the scatterer $S^{(mn)}$, whereas $D_R^{(mn,k)}$ is the distance between the scatterer $S^{(mn)}$ and the kth receiver antenna element $A_R^{(k)}$. It is assumed that $(M_T - 1)\delta_T \ll \min\{y_{T1}, y_{T2}\}$ and $(M_R - 1)\delta_R \ll \min\{y_{R1}, y_{R2}\}$. These assumptions, together with the approximation $\sqrt{1 + x} \approx 1 + x/2$ ($x \ll 1$), allow us to approximate the two distances $D_T^{(l,mn)}$ and $D_R^{(mn,k)}$ as follows:

$$D_T^{(l,mn)} \approx D_T^{(mn)}$$
$$- (M_T - 2l + 1)\left(\frac{\delta_T}{2}\right)\cos\left(\alpha_T^{(mn)} - \gamma_T\right), \quad (9)$$

$$D_R^{(mn,k)} \approx D_R^{(mn)} - (M_R - 2k + 1)\left(\frac{\delta_R}{2}\right)\cos\left(\alpha_R^{(mn)} - \gamma_R\right), \quad (10)$$

where $D_T^{(mn)}$ and $D_R^{(mn)}$ are given by $D_T^{(mn)} = y_{T1}/\sin(\alpha_T^{(mn)})$ and $D_R^{(mn)} = y_{R1}/\sin(\alpha_R^{(mn)})$, respectively.

It is noteworthy that one can also find articles [11, 34], in which only double-bounce scattering is assumed for M2M communications. However, by following a similar approach as in [15], one can easily extend our analysis on the basis of single-bounce scattering to the case of double-bounce

scattering, and thus also to a combination of single- and double-bounce scattering.

The TVTF of the LOS component is given by

$$H_{kl}^{\text{LOS}}(f', t) = \sqrt{\frac{c_R}{(c_R + 1)}} e^{j[2\pi(f_T^{(0)} + f_R^{(0)})t - (2\pi/\lambda)D_{kl} - 2\pi f' \tau_{kl}'^{(0)}]},$$
$$(11)$$

where

$$f_T^{(0)} = f_{T_{\max}}\cos\left(\alpha_T^{(0)} - \varphi_v^T\right), \quad (12)$$

$$f_R^{(0)} = f_{R_{\max}}\cos\left(\alpha_R^{(0)} - \varphi_v^R\right), \quad (13)$$

$$D_{kl} = D_0 - (M_T - 2l + 1)\frac{\delta_R}{2}\cos(\gamma_T)$$
$$+ (M_R - 2k + 1)\frac{\delta_R}{2}\cos(\gamma_R), \quad (14)$$

$$D_0 = \sqrt{D^2 + (y_{T_1} - y_{R_1})^2}. \quad (15)$$

In (11), $f_T^{(0)}$ and $f_R^{(0)}$ denote the Doppler shifts of the LOS component caused by the movement of the transmitter and the receiver, respectively. The symbols $\alpha_T^{(0)}$ and $\alpha_R^{(0)}$

in (12) and (13) represent the AOD and the AOA of the LOS component, respectively. Finally, $\tau_{kl}^{\prime (0)}$ denotes the propagation delay of the LOS component. The delay of the LOS component is defined by $\tau_{kl}^{\prime (0)} = D_{kl}/c_0$ with D_{kl} being the length of the direct path from the lth transmitter antenna element $A_T^{(l)}$ to the kth receiver antenna element $A_R^{(k)}$. The symbol D_0 in (14) denotes the Euclidean distance between the transmitter and the receiver. According to [35], the LOS component $H_{kl}^{\text{LOS}}(f', t)$ is assumed to be a deterministic process, while the diffuse component $H_{kl}^{\text{DIF}}(f', t)$ is a stochastic process.

3.2. Derivation of the AOD and the AOA.

The position of all local scatterers $S^{(mn)}$ is described by the Cartesian coordinates (x_m, y_n). In the reference model, the coordinates x_m and y_n are independent random variables, which are determined by the distribution of the local scatterers. With reference to Figure 2, we take into account that due to single-bounce scattering, the AOD $\alpha_T^{(mn)}$ and the AOA $\alpha_R^{(mn)}$ are dependent. By using the trigonometric identities, we can express the AOD $\alpha_T^{(mn)}$ and the AOA $\alpha_R^{(mn)}$ in terms of the coordinates (x_m, y_n) of the local scatterers $S^{(mn)}$ as follows:

$$\alpha_T^{(mn)}(x_m, y_n)$$
$$= \begin{cases} g(x_m, y_n), & \text{if } y_n \in J_i,\ x_m \in [0, A_2] \\ (-1)^{i+1}\pi + g(x_m, y_n), & \text{if } y_n \in J_i,\ x_m \in [-A_1, 0] \end{cases} \tag{16}$$

$$\alpha_R^{(mn)}(x_m, y_n)$$
$$= \begin{cases} f(x_m, y_n), & \text{if } y_n \in J_i,\ x_m \in [D, A_2] \\ (-1)^{i+1}\pi + f(x_m, y_n), & \text{if } y_n \in J_i,\ x_m \in [-A_1, D] \end{cases} \tag{17}$$

for $i = 1, 2$, where $J_1 = [y_{T_1}, y_{T_1} + B_1]$, $J_2 = [-y_{T_2} - B_2, -y_{T_2}]$, and

$$g(x_m, y_n) = \arctan \frac{y_n}{x_m}$$
$$f(x_m, y_n) = \arctan \frac{y_n - y_{T_1} + y_{R_1}}{x_m - D}. \tag{18}$$

4. Correlation Properties of the Reference Model

In this section, we derive a general analytical solution for the STF-CCF, from which other correlation functions, such as the 2D space CCF, the TF-CCF, the temporal ACF, and the FCF can easily be derived.

4.1. Derivation of the STF-CCF.

According to [10], the STF-CCF of the links $A_T^{(l)} - A_R^{(k)}$ and $A_T^{(l')} - A_R^{(k')}$ is defined as the correlation between the channel transfer functions $H_{kl}(f', t)$ and $H_{k'l'}(f', t)$, that is,

$$\rho_{kl,k'l'}(\delta_T, \delta_R, \nu', \tau) = E\{H_{kl}^*(f', t) H_{k'l'}(f' + \nu', t + \tau)\}$$
$$= \rho_{kl,k'l'}^{\text{DIF}}(\delta_T, \delta_R, \nu', \tau)$$
$$+ \rho_{kl,k'l'}^{\text{LOS}}(\delta_T, \delta_R, \nu', \tau), \tag{19}$$

where $(*)$ denotes the complex conjugate operator and $E\{\cdot\}$ stands for the expectation operator that applies to all random variables: the phases $\{\theta_{mn}\}$ and the coordinates (x_m, y_n) of the scatterers $S^{(mn)}$. The first term $\rho_{kl,k'l'}^{\text{DIF}}(\delta_T, \delta_R, \tau, \nu')$ represents the STF-CCF of the diffuse component. This correlation function can be expressed, after substituting (2) in (19), by

$$\rho_{kl,k'l'}^{\text{DIF}}(\delta_T, \delta_R, \nu', \tau)$$
$$= \lim_{M,N \to \infty} \frac{1}{(c_R + 1)MN} \tag{20}$$
$$\times \sum_{m,n=1}^{M,N} E\left\{c_{ll'}^{(mn)} d_{kk'}^{(mn)} e^{j2\pi[(f_T^{(mn)} + f_R^{(mn)})\tau - \nu' \tau_{kl}^{\prime (mn)}]}\right\},$$

where

$$c_{ll'}^{(mn)} = e^{j2\pi(\delta_T/\lambda)(l-l')\cos(\alpha_T^{(mn)} - \gamma_T)},$$
$$d_{kk'}^{(mn)} = e^{j2\pi(\delta_R/\lambda)(k-k')\cos(\alpha_R^{(mn)} - \gamma_R)}. \tag{21}$$

The quantities $f_T^{(mn)}$, $f_R^{(mn)}$, and $\tau_{kl}^{\prime (mn)}$ are given by (6), (7), and (8), respectively. We recall that the AOD $\alpha_T^{(mn)}$ and the AOA $\alpha_R^{(mn)}$ can be expressed in terms of the random variables x_m and y_n according to (16) and (17), respectively.

In Section 2, it has been mentioned that all scatterers are uniformly distributed in the two rectangular areas on both sides of the street, as illustrated in Figure 2. Hence, the random variables x_m and y_n are also uniformly distributed over the rectangular areas. If the number of scatterers tends to infinity, that is, $M, N \to \infty$, then the discrete random variables x_m and y_n become continuous random variables denoted by x and y, respectively. Thus, the probability density functions (PDFs) $p_x(x)$ and $p_y(y)$ of x and y, respectively, are given by

$$p_x(x) = \frac{1}{L_A}, \quad \text{if } x \in [-A_1, A_2],$$

$$p_y(y) = \begin{cases} \dfrac{1}{2B_1}, & \text{if } y \in [y_{T_1}, B_1 + y_{T_1}] \\ \dfrac{1}{2B_2}, & \text{if } y \in [-B_2 - y_{T_2}, -y_{T_2}], \end{cases} \tag{22}$$

where $L_A = A_1 + A_2$. Assuming that the random variables x and y are independent, the joint PDF $p_{xy}(x, y)$ of the random variables x and y can be expressed as a product of the marginal PDFs $p_x(x)$ and $p_y(y)$, that is,

$$
\begin{aligned}
&p_{xy}(x, y) \\
&= p_x(x) \cdot p_y(y) \\
&= \begin{cases} \dfrac{1}{2L_A B_1}, & \text{if } x \in [-A_1, A_2], \ y \in [y_{T_1}, B_1 + y_{T_1}] \\[2ex] \dfrac{1}{2L_A B_2}, & \text{if } x \in [-A_1, A_2], \ y \in [-B_2 - y_{T_2}, -y_{T_2}]. \end{cases}
\end{aligned}
\tag{23}
$$

The infinitesimal power of the diffuse component corresponding to the differential axes dx and dy is proportional to $p_{xy}(x, y)\,dx\,dy$. As $M, N \to \infty$, this infinitesimal contribution must be equal to $1/MN = p_{xy}(x, y)\,dx\,dy$. Consequently, it follows from (20) that the STF-CCF of the diffuse component can be expressed as

$$
\begin{aligned}
&\rho_{kl,k'l'}^{\text{DIF}}(\delta_T, \delta_R, \nu', \tau) \\
&= \frac{1}{2L_A B_1 (c_R + 1)} \int_{y_{T_1}}^{y_{T_1} + B_1} \int_{-A_1}^{A_2} c_{ll'}^{\text{DIF}}(\delta_T, x, y)\, d_{kk'}^{\text{DIF}}(\delta_R, x, y) \\
&\quad \times e^{j2\pi[(f_T(x,y)+f_R(x,y))\tau - \nu' \tau'_{kl}(x,y)]}\,dx\,dy \\
&\quad + \frac{1}{2L_A B_2 (c_R + 1)} \\
&\quad \times \int_{-B_2 - y_{T_2}}^{-y_{T_2}} \int_{-A_1}^{A_2} c_{ll'}^{\text{DIF}}(\delta_T, x, y)\, d_{kk'}^{\text{DIF}}(\delta_R, x, y) \\
&\quad \times e^{j2\pi[(f_T(x,y)+f_R(x,y))\tau - \nu' \tau'_{kl}(x,y)]}\,dx\,dy,
\end{aligned}
\tag{24}
$$

where

$$
c_{ll'}^{\text{DIF}}(\delta_T, x, y) = e^{j2\pi(\delta_T/\lambda)(l-l')\cos(\alpha_T(x,y)-\gamma_T)},
$$

$$
d_{kk'}^{\text{DIF}}(\delta_R, x, y) = e^{j2\pi(\delta_R/\lambda)(k-k')\cos(\alpha_R(x,y)-\gamma_R)},
$$

$$
f_T(x, y) = f_{T_{\max}}\cos\left(\alpha_T(x,y) - \varphi_\nu^T\right),
\tag{25}
$$

$$
f_R(x, y) = f_{R_{\max}}\cos\left(\alpha_R(x,y) - \varphi_\nu^R\right),
$$

$$
\tau'_{kl}(x, y) = \frac{1}{c_0}\left[D_T^{(l)}(x,y) + D_R^{(k)}(x,y)\right].
$$

Using the functions in (9) and (10), the distances $D_T^{(l)}(x, y)$ and $D_R^{(k)}(x, y)$ can be expressed as

$$
\begin{aligned}
D_T^{(l)}(x, y) &\approx \frac{y_{T_1}}{\sin(\alpha_T(x,y))} \\
&\quad - (M_T - 2l + 1)\left(\frac{\delta_T}{2}\right)\cos(\alpha_T(x,y) - \gamma_T),
\end{aligned}
$$

$$
\begin{aligned}
D_R^{(k)}(x, y) &\approx \frac{y_{R_1}}{\sin(\alpha_R(x,y))} \\
&\quad - (M_R - 2k + 1)\left(\frac{\delta_R}{2}\right)\cos(\alpha_R(x,y) - \gamma_R).
\end{aligned}
\tag{26}
$$

In (19), the quantity $\rho_{kl,k'l'}^{\text{LOS}}(\delta_T, \delta_R, \tau, \nu')$, which represents the STF-CCF of the LOS component, can be written as

$$
\begin{aligned}
\rho_{kl,k'l'}^{\text{LOS}}(\delta_T, \delta_R, \nu', \tau) &= \frac{c_R}{(c_R + 1)} c_{ll'}^{(0)}(\delta_T) \\
&\quad \times d_{kk'}^{(0)}(\delta_R) e^{j2\pi[(f_T^{(0)} + f_R^{(0)})\tau - \nu' \tau'^{(0)}_{kl}]},
\end{aligned}
\tag{27}
$$

where

$$
c_{ll'}^{(0)}(\delta_T) = e^{j2\pi(\delta_T/\lambda)(l-l')\cos(\gamma_T)},
\tag{28}
$$

$$
d_{kk'}^{(0)}(\delta_R) = e^{-j2\pi(\delta_R/\lambda)(k-k')\cos(\gamma_R)}.
\tag{29}
$$

The Doppler shifts $f_T^{(0)}$ and $f_R^{(0)}$ are given by (12) and (13), respectively.

4.2. Derivation of the 2D Space CCF. The 2D space CCF $\rho_{kl,k'l'}(\delta_T, \delta_R)$ is defined as $\rho_{kl,k'l'}(\delta_T, \delta_R) = E\{H_{kl}^*(f', t)H_{k'l'}(f', t)\}$, which is equal to the STF-CCF $\rho_{kl,k'l'}(\delta_T, \delta_R, \nu', \tau)$ in (19) by setting ν' and τ to zero, that is,

$$
\begin{aligned}
\rho_{kl,k'l'}(\delta_T, \delta_R) &= \rho_{kl,k'l'}(\delta_T, \delta_R, 0, 0) \\
&= \frac{1}{2L_A B_1 (c_R + 1)} \int_{y_{T_1}}^{y_{T_1} + B_1} \\
&\quad \times \int_{-A_1}^{A_2} c_{ll'}^{\text{DIF}}(\delta_T, x, y)\, d_{kk'}^{\text{DIF}}(\delta_R, x, y)\,dx\,dy \\
&\quad + \frac{1}{2L_A B_2 (c_R + 1)} \int_{-B_2 - y_{T_2}}^{-y_{T_2}} \\
&\quad \times \int_{-A_1}^{A_2} c_{ll'}^{\text{DIF}}(\delta_T, x, y)\, d_{kk'}^{\text{DIF}}(\delta_R, x, y)\,dx\,dy \\
&\quad + \frac{c_R}{(c_R + 1)} c_{ll'}^{(0)}(\delta_T)\, d_{kk'}^{(0)}(\delta_R).
\end{aligned}
\tag{30}
$$

4.3. Derivation of the TF-CCF. The TF-CCF of the transmission link from $A_T^{(l)}$ ($l = 1, 2, \ldots, M_T$) to $A_R^{(k)}$ ($k = 1, 2, \ldots, M_R$) is defined by $r_{kl}(\nu', \tau) := E\{H_{kl}^*(f', t)H_{kl}(f' + \nu', t+\tau)\}$ [36]. The TF-CCF can be obtained directly from the STF-CCF [see (19)] by setting the antenna element spacings δ_T and δ_R to zero, that is,

$$
\begin{aligned}
r_{kl}(\nu', \tau) &= \rho_{kl,k'l'}^{\text{DIF}}(0, 0, \nu', \tau) + \rho_{kl,k'l'}^{\text{LOS}}(0, 0, \nu', \tau) \\
&= \frac{1}{2L_A B_1 (c_R + 1)} \int_{y_{T_1}}^{y_{T_1} + B_1} \\
&\quad \times \int_{-A_1}^{A_2} e^{j2\pi[(f_T(x,y)+f_R(x,y))\tau - \nu' \tau'_{kl}(x,y)]}\,dx\,dy \\
&\quad + \frac{1}{2L_A B_2 (c_R + 1)} \int_{-B_2 - y_{T_2}}^{-y_{T_2}} \\
&\quad \times \int_{-A_1}^{A_2} e^{j2\pi[(f_T(x,y)+f_R(x,y))\tau - \nu' \tau'_{kl}(x,y)]}\,dx\,dy \\
&\quad + \frac{c_R}{(c_R + 1)} e^{j2\pi(f_T^{(0)} + f_R^{(0)})\tau} e^{-j2\pi\nu' \tau'^{(0)}_{kl}}.
\end{aligned}
\tag{31}
$$

4.4. Derivation of the Temporal ACF and the Doppler PSD. The temporal ACF of the transmission link from $A_T^{(l)}$ ($l = 1, 2, \ldots, M_T$) to $A_R^{(k)}$ ($k = 1, 2, \ldots, M_R$) is defined by $r_{kl}(\tau) := E\{H_{kl}^*(f', t)H_{kl}(f', t+\tau)\}$ [36, Page 376]. The temporal ACF can be obtained directly from the TF-CCF (see (31)) by setting at ν' to zero, that is, $r_{kl}(\tau) = r_{kl}(\tau, 0)$, which gives

$$r_{kl}(\tau) = \frac{1}{2L_A B_1(c_R + 1)} \int_{y_{T_1}}^{y_{T_1}+B_1}$$
$$\times \int_{-A_1}^{A_2} e^{j2\pi[f_T(x,y)+f_R(x,y)]\tau} dx dy$$
$$+ \frac{1}{2L_A B_2(c_R + 1)} \int_{-B_2-y_{T_2}}^{-y_{T_2}} \qquad (32)$$
$$\times \int_{-A_1}^{A_2} e^{j2\pi[f_T(x,y)+f_R(x,y)]\tau} dx dy$$
$$+ \frac{c_R}{(c_R + 1)} e^{j2\pi(f_T^{(0)}+f_R^{(0)})\tau}.$$

Notice that the expression in (32) reveals that the ACF $r_{kl}(\tau)$ is independent of k and l.

Computing the Fourier transform of the temporal ACF $r_{kl}(\tau)$ results in the Doppler PSD $S_{kl}(f)$, that is,

$$S_{kl}(f) = \int_{-\infty}^{\infty} r_{kl}(\tau) e^{-j2\pi f \tau} d\tau. \qquad (33)$$

The two most important statistical quantities characterizing the Doppler PSD $S_{kl}(f)$ are the average Doppler shift $B_{kl}^{(1)}$ and the Doppler spread $B_{kl}^{(2)}$ [35]. The average Doppler shift $B_{kl}^{(1)}$ is defined as the first moment of $S_{kl}(f)$, which can be expressed as follows:

$$B_{kl}^{(1)} = \frac{\int_{-\infty}^{\infty} f S_{kl}(f) df}{\int_{-\infty}^{\infty} S_{kl}(f) df}. \qquad (34)$$

The Doppler spread $B_{kl}^{(2)}$ is defined as the square root of the second central moment of $S_{kl}(f)$, which can be written as

$$B_{kl}^{(2)} = \sqrt{\frac{\int_{-\infty}^{\infty} \left(f - B_{kl}^{(1)}\right)^2 S_{kl}(f) df}{\int_{-\infty}^{\infty} S_{kl}(f) df}}. \qquad (35)$$

4.5. Derivation of the FCF. The frequency characteristics of the reference model are described by the FCF $r_{kl}(\nu')$. The FCF $r_{kl}(\nu')$ of the transmission link from $A_T^{(l)}$ to $A_R^{(k)}$ is defined by $r_{kl}(\nu') := E\{H_{kl}^*(f', t)H_{kl}(f'+\nu', t)\}$ for all $l = 1, 2, \ldots, M_T$ and $k = 1, 2, \ldots, M_R$. This function can be obtained directly

from the TF-CCF [see (31)] by setting τ to zero, that is, $r_{kl}(\nu') = r_{kl}(0, \nu')$, which results in

$$r_{kl}(\nu') = \frac{1}{2L_A B_1(c_R + 1)} \int_{y_{T_1}}^{y_{T_1}+B_1} \int_{-A_1}^{A_2} e^{-j2\pi\nu'\tau'_{kl}(x,y)} dx dy$$
$$+ \frac{1}{2L_A B_2(c_R + 1)} \int_{-B_2-y_{T_2}}^{-y_{T_2}} \int_{-A_1}^{A_2} e^{-j2\pi\nu'\tau'_{kl}(x,y)} dx dy$$
$$+ \frac{c_R}{(c_R + 1)} e^{-j2\pi\nu'\tau'^{(0)}_{kl}}. \qquad (36)$$

In contrast to the temporal ACF $r_{kl}(\tau)$, the FCF $r_{kl}(\nu')$ depends on k and l.

5. Measurement-Based Computation of the Model Parameters

The objective of this section is to determine the set of model parameters $\mathcal{P} = \{A_1, A_2, B_1, B_2, y_{T_1}, y_{T_2}, y_{R_1}, y_{R_2}, D, f_{T_{max}}, f_{R_{max}}, c_R\}$ describing the reference model in such a way that the average Doppler shift $B_{kl}^{(1)}$ and the Doppler spread $B_{kl}^{(2)}$ of the reference model match the corresponding quantities ($B_{kl}^{\star(1)}$ and $B_{kl}^{\star(2)}$) of the measured channel reported in [25]. To determine the set of model parameters $\mathcal{P}$, we minimize the following error:

$$E_{min} = W_1 E_{B_{kl}^{(1)}} + W_2 E_{B_{kl}^{(2)}}, \qquad (37)$$

where W_1 and W_2 denote the weighting factors. The symbols $E_{B_{kl}^{(1)}}$ and $E_{B_{kl}^{(2)}}$ in (37) stand for the absolute errors of the average Doppler shift and Doppler spread, respectively, which are defined as

$$E_{B_{kl}^{(1)}} = \arg\min_{\mathcal{P}} \left| B_{kl}^{\star(1)} - B_{kl}^{(1)} \right|, \qquad (38)$$

$$E_{B_{kl}^{(2)}} = \arg\min_{\mathcal{P}} \left| B_{kl}^{\star(2)} - B_{kl}^{(2)} \right|. \qquad (39)$$

In (38) and (39), the notation $\arg\min_x f(x)$ stands for the argument of the minimum, which is the set of points of the given argument for which $f(x)$ reaches its minimum value. At the beginning of the optimization procedure, the weighting factors W_1 and W_2 are selected arbitrarily, but such that they satisfy the equality $W_1 + W_2 = 1$. If the error $E_{B_{kl}^{(i)}}$ ($i = 1, 2$) in (37) is large, then we reduce the corresponding weighting factor W_i and vice versa. We continue the optimization procedure until the result in (37) reaches an error floor, meaning that the average Doppler shift and the Doppler spread of the reference model best match the measured average Doppler shift and the measured Doppler spread, respectively.

For the measured channels in [25], the resulting optimized model parameters and the corresponding average Doppler shift and Doppler spread are listed in Table 1. The results found for the reference model demonstrate an excellent fitting to real-world measured channels for rural, urban, and highway propagation areas, which validates the usefulness of the proposed reference model. It is worth

TABLE 1: Measurement-based parameters of the geometrical street scattering model and the resulting average Doppler shift and the Doppler spread.

Model parameters	Propagation environment				
	Urban LOS	Urban NLOS	Rural LOS	Highway LOS	Highway NLOS
$A_1(A_2)$ (m)	546.28 (1249)	537.03 (908.3)	546.52 (1236)	547.69 (1207)	546.88 (1193)
$B_1(B_2)$ (m)	198.96 (198.77)	76.46 (1.1113)	20.89 (18.25)	199.8 (200)	0.01 (0.01)
$f_{T_{\max}}(f_{R_{\max}})$ (Hz)	223.55 (219.77)	262.1 (209.97)	463.72 (491.65)	511.68 (442.62)	491.67 (481.97)
$y_{T1}(y_{T2})$ (m)	10.42 (7)	2.12 (1.18)	15.28 (4.63)	17.62 (19.78)	1.3 (1.3)
$y_{R1}(y_{R2})$ (m)	19.82 (6.6)	20 (7.06)	14.57 (9.4)	19.63 (25)	20 (9.4)
D (m)	238.6	236.7	186.77	896.7	749.6
c_R	0.485	0	0.27	0.4	0
Measured average Doppler shift $B_{kl}^{\star(1)}$ (Hz) [25]	−20	103	201	209	−176
Theoretical average Doppler shift $B_{kl}^{(1)}$ (Hz)	−20	102.67	200.55	208.8	−110
Measured Doppler spread $B_{kl}^{\star(2)}$ (Hz) [25]	341	298	782	761	978
Theoretical Doppler spread $B_{kl}^{(2)}$ (Hz)	341	298	782.03	760.88	941

mentioning that the computed average Doppler shift $B_{kl}^{(1)} = -110$ Hz and the Doppler spread $B_{kl}^{(2)} = 941$ Hz do not closely agree with the measured channel ($B_{kl}^{\star(1)} = -176$ Hz and $B_{kl}^{\star(2)} = 978$ Hz) in case of the highway NLOS scenario. For this scenario, a close agreement can be found for sufficiently small values of $c_R \neq 0$.

6. The Simulation Model

The reference model described above is a theoretical model, which is based on the assumption that the number of scatterers (M, N) is infinite. Owing to an infinite realization complexity, the reference model is non-realizable. However, the reference model can serve as a ground for the derivation of stochastic and deterministic simulation models. According to the generalized principle of deterministic channel modeling [35, Sec. 8.1], a stochastic simulation model can be derived from the reference model introduced in (1) by using only a finite number of scatterers. In the literature, several different models exist that allow for a proper simulation of mobile channels. The SOC model is an appropriate simulation model for mobile radio channels under non-isotropic scattering conditions. A detailed description and the design of SOC models can be found in [37, 38], respectively. In [38], several parametrization techniques for SOC models have been discussed and analyzed. Here, we use the L_p-norm method (LPNM), which is a high-performance parameter computation method for the design of SOC channel simulators.

7. Numerical Results

This section illustrates the analytical results given by (30), (31), (32), and (36). The correctness of the analytical results will be verified by simulations. The performance of the channel simulator has been assessed by comparing its temporal ACF and the FCF to the corresponding system functions of the reference model (see (32) and (36)).

As an example for our geometrical street scattering model, we consider rectangular scattering areas on both sides of the street with a length of $L_A = A_1 + A_2$, where $A_1 = 50$ m and $A_2 = 450$ m, and a width of $B_1 = B_2 = 100$ m. With reference to Figure 2, the position of the transmitter and the receiver are defined by the distances $D = 400$ m, $y_{T_1} = y_{R_2} = 20$ m, and $y_{T_2} = y_{R_1} = 10$ m. For the reference model, all theoretical results have been obtained by choosing the following parameters: $\gamma_T = 90°$, $\gamma_R = 90°$, $\varphi_v^T = 0°$, $\varphi_v^R = 180°$, and $f_{T_{\max}} = f_{R_{\max}} = 91$ Hz. The Rice factor c_R was chosen from the set $\{0, 0.5, 1\}$. The scatterers are uniformly distributed over the considered rectangular areas. The L_p-norm method has been applied to optimize the simulation model parameters by using a finite number of scatterers (cisoids). For the simulation model, we use $M \times N = 50 \times 25$ scatterers (cisoids) within the rectangle on the left-hand side as well as on the right-hand side.

In Figure 3, the absolute value of the 2D space CCF $|\rho_{11,22}(\delta_T, \delta_R)|$ of the reference model is presented for the NLOS propagation scenario ($c_R = 0$). The results have been obtained by using (30). From Figure 3, we can observe that the 2D space CCF decreases as the antenna element spacings

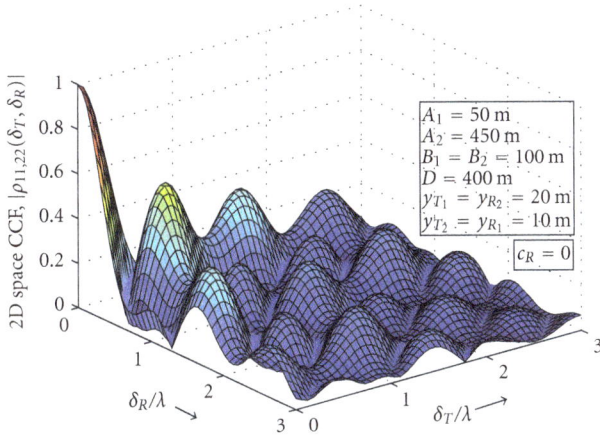

FIGURE 3: Absolute value of the 2D space CCF $|\rho_{11,22}(\delta_T, \delta_R)|$ of the reference model for a NLOS propagation scenario ($c_R = 0$).

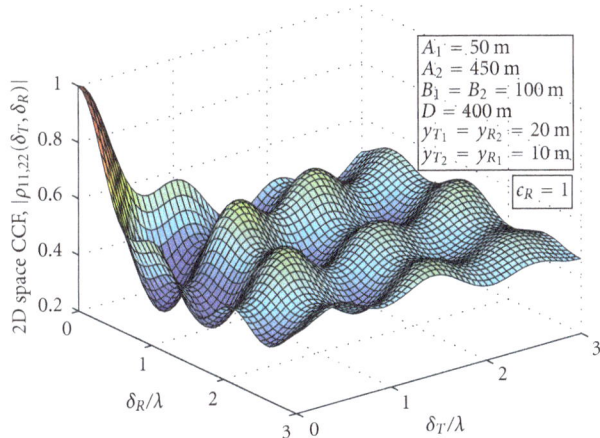

FIGURE 5: Absolute value of the TF-CCF $|r_{11}(\nu', \tau)|$ of the reference model for a NLOS propagation scenario ($c_R = 0$).

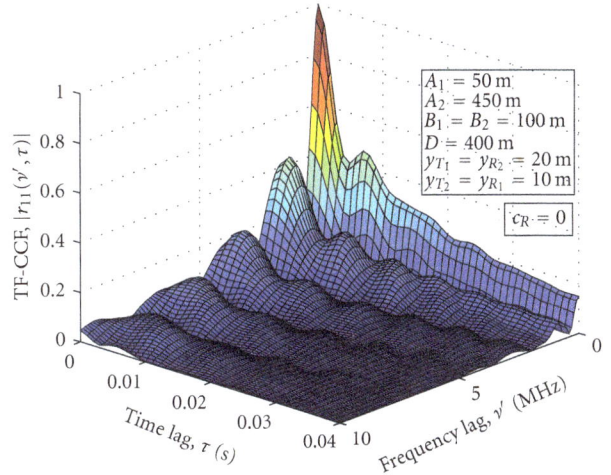

FIGURE 4: Absolute value of the 2D space CCF $|\rho_{11,22}(\delta_T, \delta_R)|$ of the reference model for a LOS propagation scenario ($c_R = 1$).

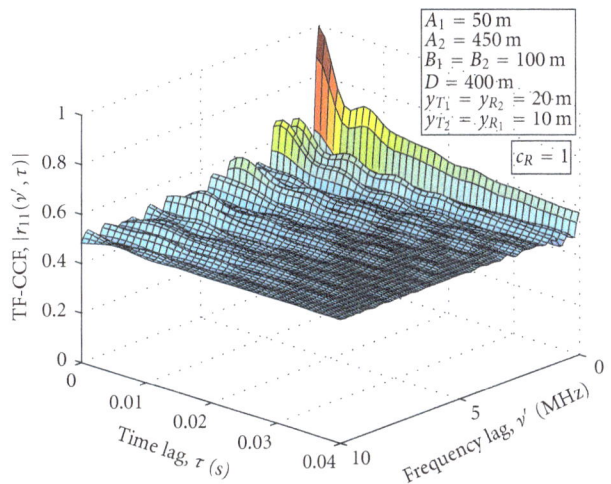

FIGURE 6: Absolute value of the TF-CCF $|r_{11}(\nu', \tau)|$ of the reference model for a LOS propagation scenario ($c_R = 1$).

increase. For comparison reasons, the absolute value of the 2D space CCF $|\rho_{11,22}(\delta_T, \delta_R)|$ is depicted in Figure 4 for a LOS propagation scenario ($c_R = 1$). From Figure 4, one can see that the channel transfer functions $H_{kl}(f', t)$ and $H_{k'l'}(f', t)$ are highly correlated over a large range of antenna element spacings δ_T and δ_R. This can be concluded from the fact that even for large antenna element spacings, for example, $\delta_T = \delta_R = 3\lambda$, the absolute value of the 2D space CCF $|\rho_{11,22}(\delta_T, \delta_R)|$ equals approximately one half of its maximum value. Comparing Figures 3 and 4 shows that by increasing the Rice factor c_R, the 2D space CCF also increases.

Figures 5 and 6 illustrate the TF-CCFs of the reference model under NLOS and LOS propagation conditions, respectively. From Figure 5, we can observe that the TF-CCF decreases as the time and frequency lags increase in NLOS propagation environments. A comparison of Figures 5 and 6 shows that the absolute value of the TF-CCF under LOS conditions is in general higher than under NLOS.

Figure 7 depicts the absolute value of the temporal ACF $|r_{kl}(\tau)|$ according to (32) if both the transmitter and the receiver are moving towards each other. A good match

between the temporal ACF of the reference model and that of the simulation model can be observed in Figure 7. This figure demonstrates also that the experimental simulation results of the temporal ACF match very well with the theoretical results.

Finally, Figure 8 illustrates the absolute value of the FCF $|r_{kl}(\nu')|$ for different Rice factors $c_R = \{0, 0.5, 1\}$ if both the transmitter and the receiver are moving towards each other. A close agreement between the reference model and the simulation model can be seen in Figure 8 for all chosen Rice factors. One can realize that the experimental simulation results of the FCF match very well with the theoretical results.

8. Conclusion

In this paper, a reference model for a wideband MIMO C2C channel has been derived by starting from the geometrical street scattering model. Taking both LOS and NLOS

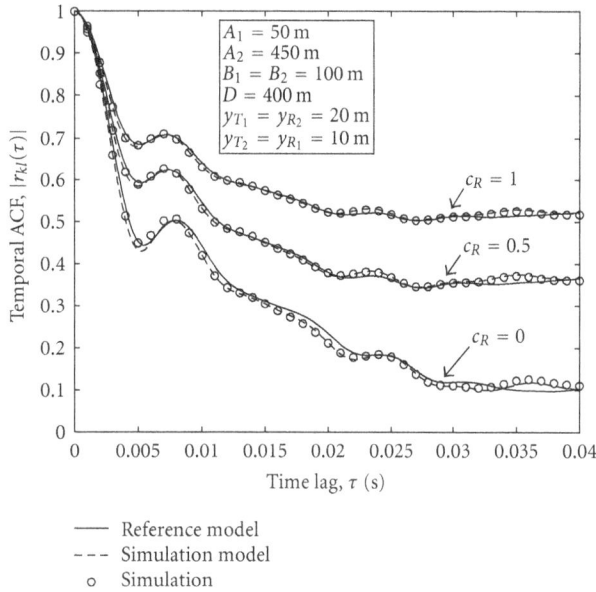

FIGURE 7: Absolute values of the ACFs $|r_{kl}(\tau)|$ (reference model) and $|\hat{r}_{kl}(\tau)|$ (simulation model) for different values of the Rice factor $c_R \in \{0, 0.5, 1\}$.

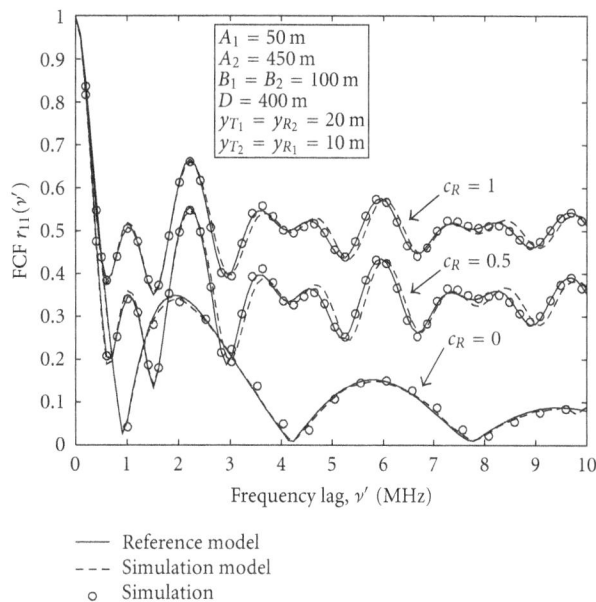

FIGURE 8: Absolute values of the FCFs $|r_{11}(\nu')|$ (reference model) and $|\hat{r}_{11}(\nu')|$ (simulation model) for different values of the Rice factor $c_R \in \{0, 0.5, 1\}$.

propagation conditions into account, we have analyzed the 2D space CCF and the TF-CCF of the reference model. To find a proper simulation model, the SOC principle has been applied. It has been shown that the SOC channel simulator approximates the reference model with high accuracy with respect to the temporal ACF and the FCF. An excellent fitting of the average Doppler shift and the Doppler spread of the reference model to the corresponding quantities of measured channels has validated the usefulness of the

proposed reference model. Further extensions of the proposed wideband MIMO C2C channel model incorporating the nonstationarity properties of real-world C2C channels are planned for future work.

References

[1] F. Qu, F. Y. Wang, and L. Yang, "Intelligent transportation spaces: vehicles, traffic, communications, and beyond," *IEEE Communications Magazine*, vol. 48, no. 11, pp. 136–142, 2010.

[2] http://www.ertico.com/.

[3] http://www.car-to-car.org/.

[4] E. Telatar, "Capacity of multi-antenna Gaussian channels," *European Transactions on Telecommunications*, vol. 10, no. 6, pp. 585–595, 1999.

[5] M. Pätzold, B. O. Hogstad, and N. Youssef, "Modeling, analysis, and simulation of MIMO mobile-to-mobile fading channels," *IEEE Transactions on Wireless Communications*, vol. 7, no. 2, pp. 510–520, 2008.

[6] M. Pätzold and B. O. Hogstad, "A wideband MIMO channel model derived from the geometric elliptical scattering model," *Wireless Communications and Mobile Computing*, vol. 8, no. 5, pp. 597–605, 2008.

[7] H. Zhiyi, C. Wei, Z. Wei, M. Pätzold, and A. Chelli, "Modelling of MIMO vehicle-to-vehicle fading channels in T-junction scattering environments," in *Proceedings of the 3rd European Conference on Antennas and Propagation (EuCAP '09)*, pp. 652–656, Berlin, Germany, March 2009.

[8] A. Chelli and M. Pätzold, "A MIMO mobile-to-mobile channel model derived from a geometric street scattering model," in *Proceedings of the 4th IEEE International Symposium on Wireless Communication Systems (ISWCS '07)*, pp. 792–797, Trondheim, Norway, October 2007.

[9] N. Avazov and M. Pätzold, "A geometric street scattering channel model for car-to-car communication systems," in *Proceedings of the International Conference on Advanced Technologies for Communications (ATC '11)*, pp. 224–230, Da Nang City, Vietnam, August 2011.

[10] A. S. Akki and F. Haber, "A statistical model of mobile-to-mobile land communication channel," *IEEE Transactions on Vehicular Technology*, vol. VT-35, no. 1, pp. 2–7, 1986.

[11] A. S. Akki, "Statistical properties of mobile-to-mobile land communication channels," *IEEE Transactions on Vehicular Technology*, vol. 43, no. 4, pp. 826–831, 1994.

[12] R. Wang and D. Cox, "Channel modeling for ad hoc mobile wireless networks," in *Proceedings of the 55th IEEE Vehicular Technology Conference (VTC '02)*, vol. 1, pp. 21–25, Birmingham, AL, USA, May 2002.

[13] C. S. Patel, G. L. Stüber, and T. G. Pratt, "Simulation of Rayleigh-faded mobile-to-mobile communication channels," *IEEE Transactions on Communications*, vol. 53, no. 11, pp. 1876–1884, 2005.

[14] A. G. Zajić and G. L. Stüber, "Space-time correlated mobile-to-mobile channels: modelling and simulation," *IEEE Transactions on Vehicular Technology*, vol. 57, no. 2, pp. 715–726, 2008.

[15] X. Cheng, C. X. Wang, D. I. Laurenson, H. H. Ghent, and A. V. Vasilakos, "A generic geometrical-based MIMO mobile-to-mobile channel model," in *Proceedings of the International Wireless Communications and Mobile Computing Conference (IWCMC '08)*, pp. 1000–1005, August 2008.

[16] "IEEE 802.11p, Part 11: wireless LAN medium access control (MAC) and physical layer (PHY) specifications amendment 6: wireless access in vehicular environments, IEEE standards association," June 2010.

[17] "Standard specification for telecommunications and information exchange between roadside and vehicle systems—5 GHz band dedicated short range communications (DSRC) medium access control (MAC) and physical layer (PHY) specifications," ASTM E2213-03, September 2003.

[18] G. Acosta, K. Tokuda, and M. A. Ingram, "Measured joint Doppler-delay power profiles for vehicle-to-vehicle communications at 2.4 GHz," in *Proceedings of IEEE Global Telecommunications Conference (GLOBECOM '04)*, pp. 3813–3817, Dallas, Tex, USA, December 2004.

[19] P. C. F. Eggers, T. W. C. Brown, K. Olesen, and G. F. Pedersen, "Assessment of capacity support and scattering in experimental high speed vehicle-to-vehicle MIMO links," in *Proceedings of the IEEE 65th Vehicular Technology Conference (VTC '07)*, pp. 466–470, April 2007.

[20] I. Sen and D. W. Matolak, "Vehicle-to-vehicle channel models for the 5-GHz band," *IEEE Transactions on Intelligent Transportation Systems*, vol. 9, no. 2, pp. 235–245, 2008.

[21] Q. Wu, D. W. Matolak, and I. Sen, "5-GHz-band vehicle-to-vehicle channels: models for multiple values of channel bandwidth," *IEEE Transactions on Vehicular Technology*, vol. 59, no. 5, pp. 2620–2625, 2010.

[22] J. Maurer, T. Fügen, and W. Wiesbeck, "Narrow-band measurement and analysis of the inter-vehicle transmission channel at 5.2 GHz," in *Proceedings of the 55th Vehicular Technology Conference (VTC '02)*, vol. 3, pp. 1274–1278, May 2002.

[23] L. Cheng, B. E. Henty, D. D. Stancil, F. Bai, and P. Mudalige, "Mobile vehicle-to-vehicle narrow-band channel measurement and characterization of the 5.9 GHz dedicated short range communication (DSRC) frequency band," *IEEE Journal on Selected Areas in Communications*, vol. 25, no. 8, pp. 1501–1516, 2007.

[24] A. Paier, J. Karedal, N. Czink, H. Hofstetter, and C. Dumard, "Car-to-car radio channel measurements at 5 GHz: pathloss, power-delay profile, and delay-Doppler spectrum," in *Proceedings of the 4th IEEE International Symposium on Wireless Communication Systems (ISWCS '07)*, pp. 224–228, Trondheim, Norway, October 2007.

[25] I. Tan, W. Tang, K. Laberteaux, and A. Bahai, "Measurement and analysis of wireless channel impairments in DSRC vehicular communications," in *Proceedings of IEEE International Conference on Communications (ICC '08)*, pp. 4882–4888, May 2008.

[26] J. Kunisch and J. Pamp, "Wideband car-to-car radio channel measurements and model at 5.9 GHz," in *Proceedings of the 68th IEEE Vehicular Technology (VTC '08)*, pp. 1–5, September 2008.

[27] O. Renaudin, V. M. Kolmonen, P. Vainikainen, and C. Oestges, "Wideband measurement-based modeling of inter-vehicle channels in the 5 GHz band," in *Proceedings of the 5th European Conference on Antennas and Propagation (EUCAP '11)*, pp. 2881–2885, April 2011.

[28] A. Chelli and M. Pätzold, "A wideband multiple-cluster MIMO mobile-to-mobile channel model based on the geometrical street model," in *Proceedings of IEEE 19th International Symposium on Personal, Indoor and Mobile Radio Communications (PIMRC '08)*, pp. 1–6, Cannes, France, September 2008.

[29] A. G. Zajić, G. L. Stüber, T. G. Pratt, and S. T. Nguyen, "Wideband MIMO mobile-to-mobile channels: geometry-based statistical modeling with experimental verification," *IEEE Transactions on Vehicular Technology*, vol. 58, no. 2, pp. 517–534, 2009.

[30] X. Cheng, C. X. Wang, D. I. Laurenson, S. Salous, and A. V. Vasilakos, "An adaptive geometry-based stochastic model for non-isotropic MIMO mobile-to-mobile channels," *IEEE Transactions on Wireless Communications*, vol. 8, no. 9, pp. 4824–4835, 2009.

[31] G. D. Durgin, *Space-Time Wireless Channels*, Prentice Hall, 2002.

[32] S. H. Kong, "TOA and AOD statistics for down link Gaussian scatterer distribution model," *IEEE Transactions on Wireless Communications*, vol. 8, no. 5, pp. 2609–2617, 2009.

[33] Y. Ma and M. Pätzold, "A wideband one-ring MIMO channel model under non-isotropic scattering conditions," in *Proceedings of IEEE 67th Vehicular Technology Conference (VTC '08)*, pp. 424–429, Singapore, May 2008.

[34] F. Vatalaro, "Doppler spectrum in mobile-to-mobile communications in the precense of three-dimensional multipath scattering," *IEEE Transactions on Vehicular Technology*, vol. 46, no. 1, pp. 213–219, 1997.

[35] M. Pätzold, *Mobile Radio Channels*, John Wiley & Sons, Chichester, UK, 2nd edition, 2011.

[36] A. Papoulis and S. U. Pillai, *Probability, Random Variables and Stochastic Processes*, McGraw-Hill, New York, NY, USA, 4th edition, 2002.

[37] M. Pätzold and B. Talha, "On the statistical properties of sum-of-cisoids-based mobile radio channel simulators," in *Proceedings of the 10th International Symposium on Wireless Personal Multimedia Communications (WPMC '07)*, pp. 394–400, Jaipur, India, December 2007.

[38] C. A. Gutiérrez and M. Pätzold, "The design of sum-of-cisoids Rayleigh fading channel simulators assuming non-isotropic scattering conditions," *IEEE Transactions on Wireless Communications*, vol. 9, no. 4, pp. 1308–1314, 2010.

10

Mathematical Model and Matlab Simulation of Strapdown Inertial Navigation System

Wen Zhang,[1] Mounir Ghogho,[2,3] and Baolun Yuan[1]

[1] *College of Opto-Electronic Science and Technology, National University of Defense Technology, Changsha 410073, China*
[2] *School of Electronic and Electrical Engineering, University of Leeds, Leeds LS2 9JT, UK*
[3] *International University of Rabat, Rabat 11 100, Morocco*

Correspondence should be addressed to Wen Zhang, zhangwendaisy@hotmail.com

Academic Editor: Ahmed Rachid

Basic principles of the strapdown inertial navigation system (SINS) using the outputs of strapdown gyros and accelerometers are explained, and the main equations are described. A mathematical model of SINS is established, and its Matlab implementation is developed. The theory is illustrated by six examples which are static status, straight line movement, circle movement, s-shape movement, and two sets of real static data.

1. Introduction

Many navigation books and papers on inertial navigation system (INS) provide readers with the basic principle of INS. Some also superficially describe simulation methods and rarely provide the free code which can be used by new INS users to help them understand the theory and develop INS applications. Commercial simulation software is available but is not free. The objective of this paper is to develop an easy-to-understand step-by-step development method for simulating INS. Here we consider the most popular INS which is the strapdown inertial navigation system (SINS). The mathematical operations required in our work are mostly matrix manipulations and more generally basic linear algebra [1]. In this paper, Matlab [2] is chosen as the simulation environment. It is a popular computing environment to perform complex matrix calculations and to produce sophisticated graphics in a relatively easy manner. A large collection of Matlab scripts are now available for a wide variety of applications and are often used for university courses. Matlab is also becoming more and more popular in industrial research centers in the design and simulation stages.

The main purposes of this paper are to establish a mathematical model and to develop a comprehensive Matlab implementation for SINS. The structure of the proposed mathematical model and Matlab simulation of SINS is

shown in Figure 1. In Section 2, the INS-related orthogonal coordinates (the body frame, the inertial frame, the Earth frame, the navigation frame, the *ENU*-frame, and the wander azimuth navigation frame) are described and figures to illustrate the relationship between the frames are provided. The basic principle of SINS is described in the wander azimuth navigation frame (*p*-frame). In Section 3, two important direction cosine matrices (DCMs), the vehicle attitude DCM and the position DCM, and the related important attitude and position angles are defined. In Section 4, the simulation for data generation of gyros and accelerometers is described in *ENU*-frame. Instead of *p*-frame, *ENU*-frame is chosen because the outputs of gyros and accelerometers are easier to obtain in this frame. The Matlab implementation is given and described step by step. Four kinds of scenarios (static, straight, circle, and s-shape) are set as examples of different kinds of vehicle trajectories. In Section 5, the mathematical model of SINS is set up and the calculation steps in *p*-frame are provided. In Section 6, the required initial parameters and other initial data calculation for the SINS model are given for the different simulation scenarios. In Section 7, Matlab implementation code functions are listed and described. Further, simulation results for the four above-mentioned scenarios are presented; two examples from real SINS experiment data are also provided to verify the validity of the developed codes. Finally, conclusions

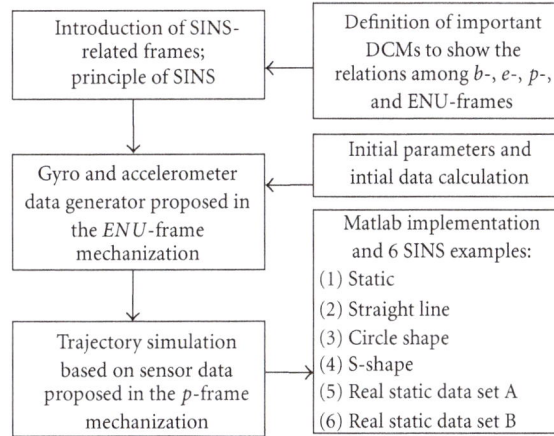

Figure 1: The schema of the proposed mathematical model and Matlab simulation of SINS.

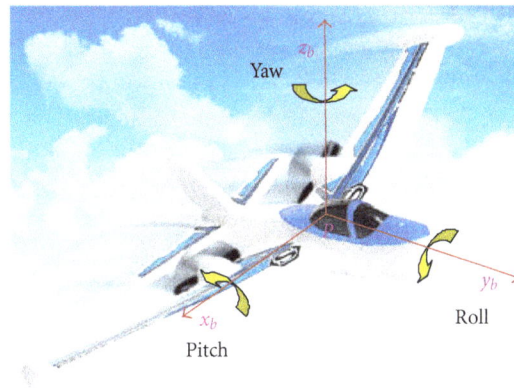

Figure 2: The b-frame illustration and the definition of axis rotations.

are drawn. Mathematical details are given in Appendices A–D.

2. Principles

A fundamental aspect of inertial navigation is the precise definition of a number of Cartesian coordinate reference frames. Each frame is an orthogonal, right-handed, coordinate frame or axis set. For all the coordinate frames used in this paper, a positive rotation about each axis is taken to be in a clockwise direction looking along the axis from the origin, as indicated in Figure 2. A negative rotation corresponds to an anticlockwise direction. This convention is used throughout this paper. It is also worth pointing out that a change in attitude of a body, which is subjected to a series of rotations about different axes, is not only a function of the rotation angles, but also on the order in which the rotations occur. In this paper, the following coordinate frames are used [3].

(1) The body frame (b-frame): the b-frame, depicted in Figure 2, is an orthogonal axis set which has its origin at the center of the vehicle, point P, and is aligned

with the pitch Px_b axis, roll Py_b axis, and yaw Pz_b axis of the vehicle in which the navigation system is installed.

(2) The inertial frame (i-frame): the i-frame, depicted in Figure 3, has its origin at the center of the Earth and its axes nonrotating with respect to fixed stars; these axes are denoted by Ox_i, Oy_i, and Oz_i, with Oz_i being coincident with the Earth polar axis.

(3) The Earth frame (e-frame): the e-frame, depicted in Figure 3, has its origin at the center of the Earth and axes nonrotating with respect to the Earth; these axes are denoted by Ox_e, Oy_e, and Oz_e. The axis Oz_e is the Earth polar axis. The axis Ox_e is along the intersection of the plane of the Greenwich meridian and the Earth equatorial plane. The Earth frame rotates with respect to the inertial frame at a rate ω_{ie} about the axis Oz_i.

(4) The navigation frame (n-frame): the n-frame, depicted in Figure 3, is a local geographic navigation

frame which has its origin at the location of the navigation system, point P (the navigation system is fixed inside the vehicle and we assume that the navigation system is located exactly at the center of the vehicle), and axes aligned with the directions of east PE, north PN and the local vertical up PU. When the n-frame is defined in this way, it is called the "ENU-frame." The turn rate of the navigation frame with respect to the Earth-fixed frame, $\boldsymbol{\omega}_{en}$, is governed by the motion of the point P with the Earth. This is often referred to as the transport rate.

(5) The wander azimuth navigation frame (p-frame): the p-frame, depicted in Figure 3, may be used to avoid the singularities in the computation which occur at the poles of the n-frame. Like the n-frame, it is of a local level but rotates through the wander angle about the local vertical. Here we do not call this frame w-frame (w for wander) for notation clarity since w and ω may look similar when printed. Letter p in p-frame stands for platform; indeed the wander azimuth navigation frame is of a local level and thus forms a horizontal platform.

In this paper, we choose the p-frame as the navigation frame for vehicle trajectory calculation, for the following reason. In the local geographic navigation frame mechanization, the n-frame is required to rotate continuously as the system moves over the surface of the Earth in order to keep its Py_N axis pointing to the true north. In order to achieve this condition worldwide, the n-frame must rotate at much greater rates about its Pz_U axis as the navigation system moves over the surface of the Earth in the polar regions, compared to the rates required at lower latitudes. It is clear that near the polar areas the local geographic navigation frame must rotate about its Pz_U axis rapidly in order to maintain the Py_N axis pointing to the pole. The heading direction will abruptly change by 180° when moving past the pole. In the most extreme case, the turn rate becomes infinite when passing over the pole. One way of avoiding the singularity, and also providing a navigation system with worldwide capability, is to adopt a wander azimuth mechanization in which the z-component of $\boldsymbol{\omega}_{ep}^p$ is always set to zero, that is, $\omega_{epz}^p = 0$. A wander axis system is a local level frame which moves over the Earth surface with the moving vehicle, as depicted in Figure 3. However, as the name implies, the azimuth angle α between Py_N axis and Py_p axis varies with the vehicle position on Earth. This variation is chosen in order to avoid discontinuities in the orientation of the wander frame with respect to Earth as the vehicle passes over either the north or south pole.

In the remainder of this section, the main principle of SINS in the p-frame is described.

Along the same lines as in [3], a navigation equation for a wander azimuth system can be constructed as follows:

$$\dot{\mathbf{v}}_e^p = \mathbf{C}_b^p \mathbf{f}^b - \left(2\mathbf{C}_e^p \boldsymbol{\omega}_{ie}^e + \boldsymbol{\omega}_{ep}^p\right)\mathbf{v}_e^p + \mathbf{g}^p, \qquad (1)$$

where $\mathbf{C}_b^p$ is the direction cosine matrix used to transform the measured specific force vector in b-frame into p-frame.

This matrix propagates in accordance with the following equation:

$$\dot{\mathbf{C}}_b^p = \mathbf{C}_b^p \boldsymbol{\Omega}_{pb}^b, \qquad (2)$$

where $\boldsymbol{\Omega}_{pb}^b$ is the skew symmetric form of $\boldsymbol{\omega}_{pb}^b$, the b-frame angular rate with respect to the p-frame.

Equation (1) is integrated to generate estimates of the vehicle speed in the wander azimuth frame, $\mathbf{v}_e^p$. This is then used to generate the turn rate of the wander frame with respect to the Earth frame, $\boldsymbol{\omega}_{ep}^p$. The direction cosine matrix which relates the wander frame to the Earth frame, $\mathbf{C}_e^p$, may be updated using the following equation:

$$\dot{\mathbf{C}}_p^e = \mathbf{C}_p^e \boldsymbol{\Omega}_{ep}^p, \qquad (3)$$

$$\left(\dot{\mathbf{C}}_p^e\right)^T = \left(\boldsymbol{\Omega}_{ep}^p\right)^T \left(\mathbf{C}_p^e\right)^T = -\boldsymbol{\Omega}_{ep}^p \left(\mathbf{C}_p^e\right)^T, \qquad (4)$$

where the superscript T means matrix transposition.

Since $(\dot{\mathbf{C}}_p^e)^T = \dot{\mathbf{C}}_e^p$, $(\mathbf{C}_p^e)^T = \mathbf{C}_e^p$ and skew symmetric matrix is $(\boldsymbol{\Omega}_{ep}^p)^T = -\boldsymbol{\Omega}_{ep}^p$ (see Appendix A), (4) can be rewritten as

$$\dot{\mathbf{C}}_e^p = -\boldsymbol{\Omega}_{ep}^p \mathbf{C}_e^p, \qquad (5)$$

where $\boldsymbol{\Omega}_{ep}^p$ is a skew symmetric matrix formed from the elements of the angular rate vector $\boldsymbol{\omega}_{ep}^p$; we could have $\boldsymbol{\omega}_{ep}^p = -\boldsymbol{\omega}_{pe}^e$ when the rotation angles are reciprocal. Because the z-component of $\boldsymbol{\omega}_{ep}^p$ is set to zero, $\omega_{epz}^p = 0$, the matrix expression of $\boldsymbol{\omega}_{ep}^p$ is $\boldsymbol{\omega}_{ep}^p = [\omega_{epx}^p \ \omega_{epy}^p \ 0]^T$. This process is implemented iteratively and enables any singularities to be avoided.

In the next section, the two important DCMs, the vehicle attitude DCM and vehicle position DCM, are defined, as well as the vehicle-attitude-related attitude angles and vehicle-position-related position angles.

3. Direction Cosine Matrices (DCMs)

In this section, the vehicle attitude DCM with the corresponding attitude angles and the vehicle position DCM with the corresponding position angles are described separately.

3.1. Vehicle Attitude DCM $\mathbf{C}_b^p$. The definition of the rotation sequence from p-frame to b-frame is (see Figure 4)

$$x_p y_p z_p \xrightarrow{z_p, \psi_G} x_e' y_e' z_e' \xrightarrow{y_p'', \theta} x_e'' y_e'' z_e'' \xrightarrow{y_p'', \gamma} x_b y_b z_b, \qquad (6)$$

where ψ_G is the gird azimuth angle (0–360°), θ is the pitch angle ($-90°$–$90°$), and γ is the roll angle ($-180°$–$180°$). The above rotation can be written in the following matrix form:

$$\mathbf{C}_p^b = \mathbf{C}_3\mathbf{C}_2\mathbf{C}_1$$

$$= \begin{bmatrix} \cos\gamma & 0 & -\sin\gamma \\ 0 & 1 & 0 \\ \sin\gamma & 0 & \cos\gamma \end{bmatrix} \begin{bmatrix} 1 & 0 & 0 \\ 0 & \cos\theta & \sin\theta \\ 0 & -\sin\theta & \cos\theta \end{bmatrix}$$

$$\times \begin{bmatrix} \cos\psi_G & \sin\psi_G & 0 \\ -\sin\psi_G & \cos\psi_G & 0 \\ 0 & 0 & 1 \end{bmatrix}. \tag{7}$$

The vehicle attitude DCM T_b^p is then obtained as

$$\mathbf{C}_b^p = \left(\mathbf{C}_p^b\right)^T = \begin{bmatrix} \cos\gamma\cos\psi_G - \sin\gamma\sin\theta\sin\psi_G & -\cos\theta\sin\psi_G & \sin\gamma\cos\psi_G + \cos\gamma\sin\theta\sin\psi_G \\ \cos\gamma\sin\psi_G + \sin\gamma\sin\theta\cos\psi_G & \cos\theta\cos\psi_G & \sin\gamma\sin\psi_G - \cos\gamma\sin\theta\cos\psi_G \\ -\sin\gamma\cos\theta & \sin\theta & \cos\gamma\cos\theta \end{bmatrix}. \tag{8}$$

For the p-frame system, the angle between the grid north y_p and the true north y_N is the wander azimuth angle α. So the angle between the horizontal projection along y_p' axis of the vehicle's vertical axis z_b and the real north y_N is the heading angle ψ. We have that

$$\psi = \psi_G + \alpha. \tag{9}$$

So the direction cosine matrix $\mathbf{C}_b^n$ from b-frame to n-frame is

$$\mathbf{C}_b^n = \begin{bmatrix} \cos\gamma\cos\psi - \sin\gamma\sin\theta\sin\psi & -\cos\theta\sin\psi & \sin\gamma\cos\psi + \cos\gamma\sin\theta\sin\psi \\ \cos\gamma\sin\psi + \sin\gamma\sin\theta\cos\psi & \cos\theta\cos\psi & \sin\gamma\sin\psi - \cos\gamma\sin\theta\cos\psi \\ -\sin\gamma\cos\theta & \sin\theta & \cos\gamma\cos\theta \end{bmatrix}. \tag{10}$$

The gimbal angles ψ, θ, and γ and the gimbal rates $\dot{\psi}$, $\dot{\theta}$, and $\dot{\gamma}$ are related to the body rate ω_{nb}^b, which is the turn rate of the b-frame with respect to n-frame and measured in b-frame as follows:

$$\begin{bmatrix} \omega_{nbx}^b \\ \omega_{nby}^b \\ \omega_{nbz}^b \end{bmatrix} = \begin{bmatrix} 0 \\ \dot{\gamma} \\ 0 \end{bmatrix} + \mathbf{C}_3 \begin{bmatrix} \dot{\theta} \\ 0 \\ 0 \end{bmatrix} + \mathbf{C}_3\mathbf{C}_2 \begin{bmatrix} 0 \\ 0 \\ \dot{\psi} \end{bmatrix}$$

$$= \begin{bmatrix} \cos\gamma\dot{\theta} - \sin\gamma\cos\theta\dot{\psi} \\ \dot{\gamma} + \sin\theta\dot{\psi} \\ \sin\gamma\dot{\theta} + \cos\gamma\cos\theta\dot{\psi} \end{bmatrix}. \tag{11}$$

3.2. *Vehicle Position DCM* $\mathbf{C}_e^p$. Position matrix $\mathbf{C}_e^p$ is the DCM from e-frame to p-frame. It has the following rotating sequence (see Figure 5):

$$x_e y_e z_e \xrightarrow{z_e, \lambda} x_e' y_e' z_e' \xrightarrow{y_p'', 90° - \varphi} x_e'' y_e'' z_e''$$

$$\xrightarrow{z_p'', 90°} x_E y_N z_U \xrightarrow{z_U, \alpha} x_p y_p z_p, \tag{12}$$

where λ is the longitude angle $(-180°-180°)$, φ is the latitude angle $(-90°-90°)$, and α is the wander azimuth angle $(0-360°)$. The above rotation can be written in the following matrix form:

$$\mathbf{C}_e^p = \begin{bmatrix} -\sin\alpha\sin\varphi\cos\lambda - \cos\alpha\sin\lambda & -\sin\alpha\sin\varphi\sin\lambda + \cos\alpha\cos\lambda & \sin\alpha\cos\varphi \\ -\cos\alpha\sin\varphi\cos\lambda + \sin\alpha\sin\lambda & -\cos\alpha\sin\varphi\sin\lambda - \sin\alpha\cos\lambda & \cos\alpha\cos\psi \\ \cos\varphi\cos\lambda & \cos\varphi\sin\lambda & \sin\varphi \end{bmatrix}. \tag{13}$$

In Section 4, a trajectory simulation method in the *ENU*-frame is described step by step to generate sensor data. In Section 5, a trajectory and attitude simulator method in the p-frame is described step by step to derive the desired trajectory and attitude from the simulated sensor data or real sensor data; Section 6 provides the initial parameters and initial data calculation.

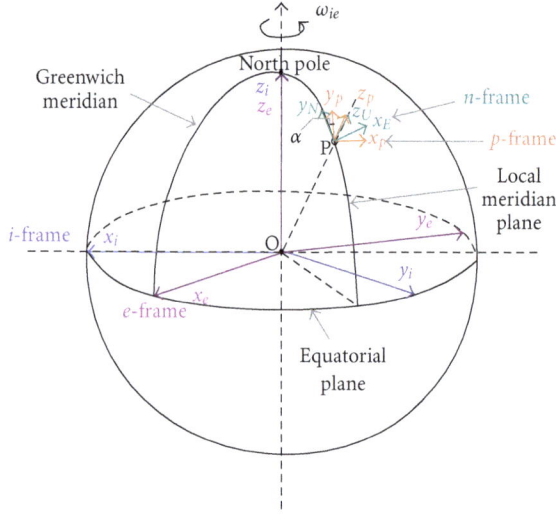

FIGURE 3: The reference frames.

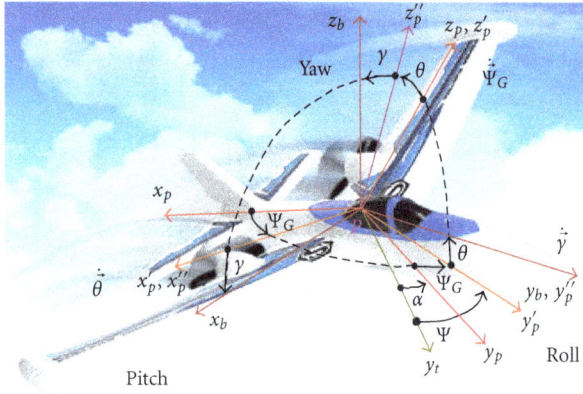

FIGURE 4: The relation between b-frame and p-frame.

4. Sensor Data Generator

The purpose of trajectory simulation is to generate data of the 3 orthogonal gyros and the 3 orthogonal accelerometers according to the designed trajectory. It is mentioned in Section 2 that p-frame is set up to avoid the singularities when the vehicle passes over either the north or south pole. But in most applications, the SINS systems are seldom operated under this extreme environment. The *ENU*-frame can be implemented easier than p-frame, so it is chosen as the navigation frame. Figure 6 shows the whole process of the SINS principal in the *ENU*-frame mechanization. First, the vehicle trajectory in the *ENU*-frame is set. Then, the sensor ideal output is derived using the inverse principle of INS. The sensor simulation data can be obtained by adding noise to the ideal data. Then, we use the simulated sensor data to derive the noise-corrupted simulated trajectory. Besides, the difference between the ideal and simulated state vectors can be set as the input for the observed measurements in the Kalman filter.

4.1. The Initial Parameters. For the designed trajectory, the initial parameters are

(1) initial position, latitude φ_0, longitude λ_0, height h_0;

(2) initial velocity $\mathbf{v} = [v_{E0}, v_{N0}, v_{U0}]$;

(3) the designed variation of acceleration $\mathbf{a}$, which varies with time according to the designed trajectory;

(4) the designed variations of the attitude angles, pitch θ, roll γ, and heading ψ, and attitude angle rates, $\dot{\theta}$, $\dot{\gamma}$, and $\dot{\psi}$, which vary with time according to the designed trajectory.

4.2. The Update of Velocity. The velocity is updated as

$$\mathbf{v} \longleftarrow \mathbf{v} + \mathbf{a}\Delta t, \qquad (14)$$

where Δt is the time step.

4.3. The Update of Position. The position is updated as

$$\text{latitude: } L \longleftarrow L + \frac{v_N \Delta t}{R_N},$$

$$\text{longitude: } \lambda \longleftarrow \lambda + \frac{v_E \Delta t \sec L}{R_E}, \qquad (15)$$

$$\text{altitude: } h \longleftarrow h + v_U \Delta t.$$

4.4. The Update of Attitude. The attitude angles are updated as

$$\text{pitch: } \theta \longleftarrow \theta + \Delta\theta,$$

$$\text{roll: } \gamma \longleftarrow \gamma + \Delta\gamma, \qquad (16)$$

$$\text{heading: } \psi \longleftarrow \psi + \Delta\psi.$$

The attitude rates are updated as

$$\text{pitch: } \dot{\theta} \longleftarrow \dot{\theta} + \Delta\dot{\theta},$$

$$\text{roll: } \dot{\gamma} \longleftarrow \dot{\gamma} + \Delta\dot{\gamma}, \qquad (17)$$

$$\text{heading: } \dot{\psi} \longleftarrow \dot{\psi} + \Delta\dot{\psi}.$$

The expressions for $\Delta\theta$, $\Delta\gamma$, $\Delta\psi$, $\Delta\dot{\theta}$, $\Delta\dot{\gamma}$, and $\Delta\dot{\psi}$ depend on the designed trajectory.

The direction cosine matrix $\mathbf{C}_b^n$ can be calculated using matrix expression (10). We have that $\mathbf{C}_n^b = (\mathbf{C}_b^n)^T$.

4.5. Gyro Data Generator. The output of the gyros is

$$\boldsymbol{\omega}_{ib}^b = \left(\mathbf{I} - \mathbf{S}_g\right)\left(\mathbf{C}_n^b\left(\boldsymbol{\omega}_{ie}^n + \boldsymbol{\omega}_{en}^n\right) + \boldsymbol{\omega}_{nb}^b\right) + \varepsilon^b, \qquad (18)$$

where $\boldsymbol{\omega}_{ib}^b$ is the simulated actual output, $\mathbf{I}$ is the 3×3 unit matrix, $\mathbf{S}_g$ is the 3×3 diagonal matrix whose diagonal elements correspond to the 3 gyros' scale factor errors, and

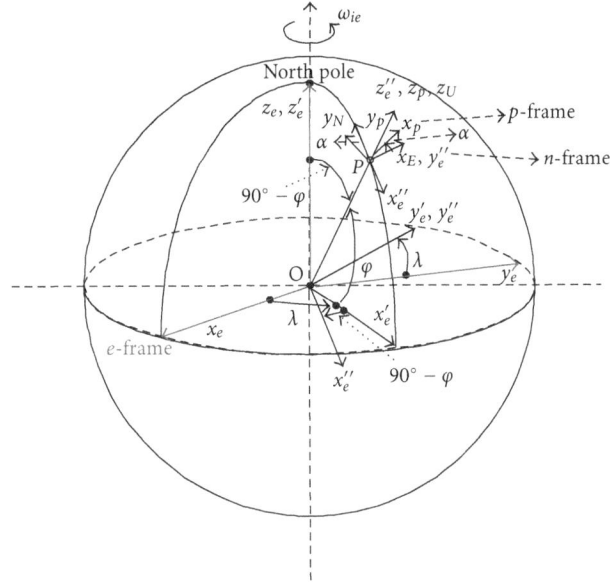

FIGURE 5: The relation between e-frame and p-frame.

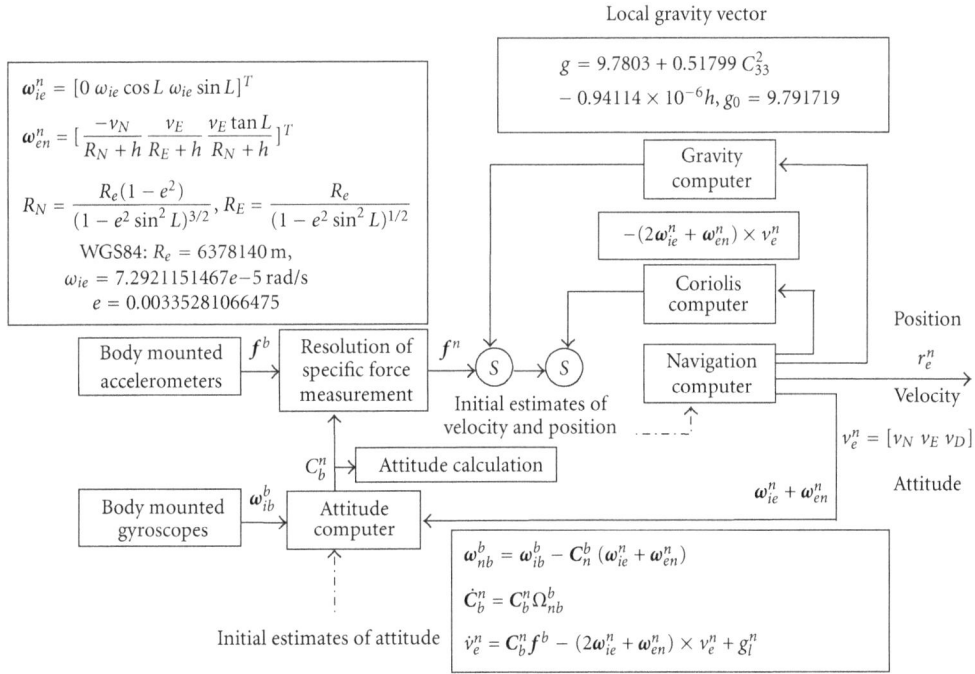

FIGURE 6: SINS ENU-frame mechanization.

ε^b is the gyro's drift and can be simulated as the sum of a constant noise and a random white noise: $\varepsilon^b = \varepsilon_{const}^b + \varepsilon_{random}^b$:

$$\boldsymbol{\omega}_{ie}^n = \begin{bmatrix} 0 \\ \omega_{ie} \cos L \\ \omega_{ie} \sin L \end{bmatrix}. \qquad (19)$$

In a static base, $\boldsymbol{\omega}_{nb}^b$ is equal to zero, whereas, in a moving base it is obtained as

$$\boldsymbol{\omega}_{nb}^b = \begin{bmatrix} \cos\gamma\,\dot{\theta} - \sin\gamma\cos\theta\,\dot{\psi} \\ \dot{\gamma} + \sin\theta\,\dot{\psi} \\ \sin\gamma\,\dot{\theta} + \cos\gamma\cos\theta\,\dot{\psi} \end{bmatrix}. \qquad (20)$$

$$\dot{C}_e^p = -\Omega_{ep}^p C_e^p$$

$$\boldsymbol{\omega}_{ie}^p = \boldsymbol{C}_e^p \boldsymbol{\omega}_{ie}^e = \omega_{ie}\,[C_{e13}^p\ C_{e23}^p\ C_{e33}^p]^T$$

$$\boldsymbol{\omega}_{ep}^p = [\omega_{epx}^p\ \omega_{epy}^p\ 0]^T$$

$$
\begin{matrix}
\omega_{epx}^p \\
\omega_{epy}^p
\end{matrix}
=
\begin{matrix}
-1/\tau_a & -1/R_{yp} & v_{ex}^p \\
1/R_{xp} & 1/\tau_a & v_{ey}^p
\end{matrix}
$$

$$1/R_{yp} = (1/R_e)(1 - eC_{33}^2 + 2eC_{23}^2)$$

$$1/R_{xp} = (1/R_e)(1 - eC_{33}^2 + 2eC_{13}^2)$$

$$1/\tau_a = (2e/R_e)C_{13}C_{23}$$

WGS84: $R_e = 6378140\,\text{m}$, $\omega_{ie} = 7.2921151467e{-}5\,\text{rad/s}$ $e = 0.00335281066475$

$g = 9.7803 + 0.51799\,C_{33}^2 - 0.94114 \times 10^{-6} h$, $g_0 = 9.791719$

C_e^p

Position $[\lambda\ \varphi\ \alpha]$

T_b^p

Attitude $[\Psi_G\ \theta\ \gamma]$

$\Psi = \Psi_G + \alpha$

Gravity computer

$-(2\boldsymbol{\omega}_{ie}^p + \boldsymbol{\omega}_{ep}^p)\times v_e^p$

Coriolis computer

Body-mounted accelerometers → $\boldsymbol{f}^b$ → Resolution of specific force measurement → $\boldsymbol{f}^p$ → S → S → Navigation computer → Results

Initial velocity and position

T_b^p → Attitude calculation

Body-mounted gyroscopes → $\boldsymbol{\omega}_{ib}^b$ → Attitude computer

$\boldsymbol{\omega}_{ie}^p + \boldsymbol{\omega}_{ep}^p$

Initial attitude

$$\dot{C}_b^p = C_b^p \Omega_{pb}^b$$
$$\boldsymbol{\omega}_{pb}^b = \boldsymbol{\omega}_{ib}^b - (C_b^p)^{-1}(\boldsymbol{\omega}_{ie}^p + \boldsymbol{\omega}_{ep}^p)$$
$$\dot{v}_e^p = C_b^p f^b - (2\boldsymbol{\omega}_{ie}^p + \boldsymbol{\omega}_{ep}^p) \times v_e^p + g^p$$

Velocity
$$v_e^p = [v_{ex}^p\ v_{ey}^p\ v_{ez}^p]$$
$$v_e^n = [v_N\ v_E\ v_D]^T$$
$$=
\begin{matrix}
v_{ey}^p\cos(\alpha) - v_{ex}^p\sin(\alpha) \\
v_{ey}^p\sin(\alpha) + v_{ex}^p\cos(\alpha) \\
v_{ez}^p
\end{matrix}$$

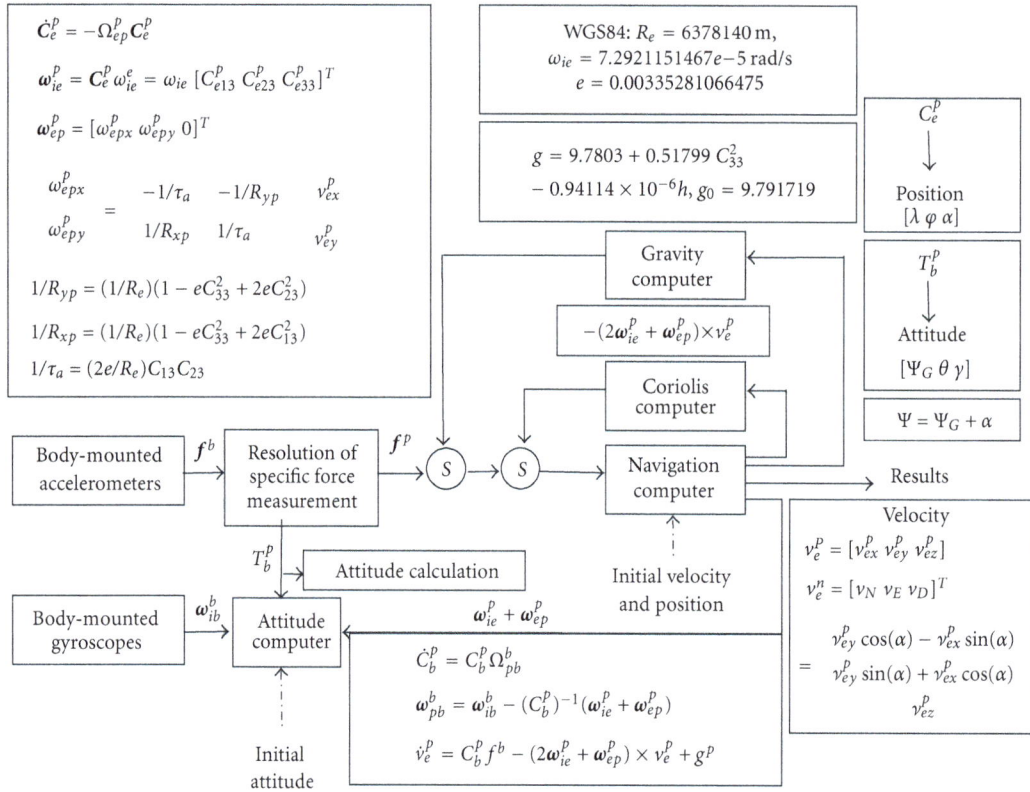

FIGURE 7: SINS p-frame mechanization.

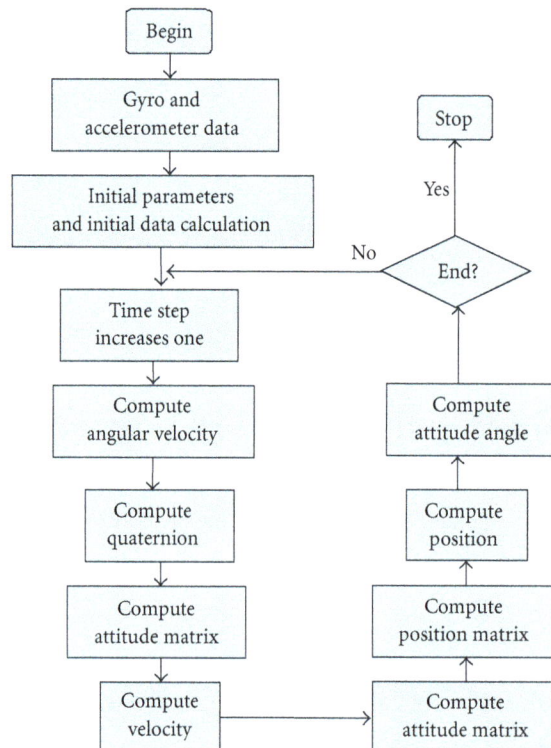

Begin → Gyro and accelerometer data → Initial parameters and initial data calculation → Time step increases one → Compute angular velocity → Compute quaternion → Compute attitude matrix → Compute velocity → Compute attitude matrix → Compute position matrix → Compute position → Compute attitude angle → End? — No → Time step; Yes → Stop

FIGURE 8: SINS program structure.

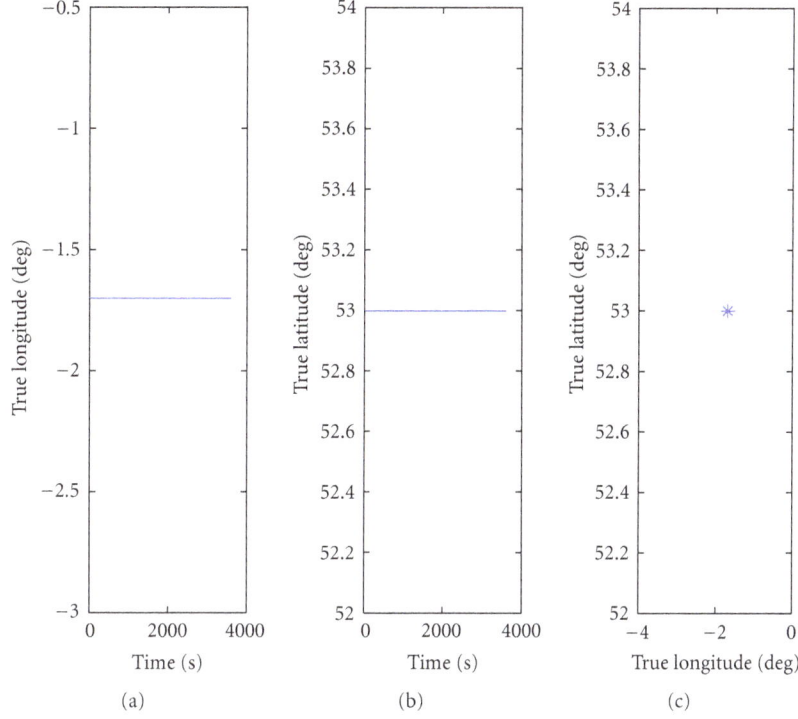

FIGURE 9: The designed trajectory of static simulation.

$\boldsymbol{\omega}_{en}^n$ is related to velocity $\mathbf{v} = [v_E, v_N, v_U]^T$ and can be expressed as

$$\boldsymbol{\omega}_{en}^n = \begin{bmatrix} \dfrac{-v_N}{R_N} \\ \dfrac{v_E}{R_E} \\ \dfrac{v_E \tan L}{R_N} \end{bmatrix}. \tag{21}$$

4.6. Accelerometer Data Generator. The measurement of the accelerometer is the specific force:

$$\mathbf{f}^b = (\mathbf{I} - \mathbf{S}_a)\mathbf{C}_n^b \mathbf{f}^n + \boldsymbol{\eta}^b,$$
$$\mathbf{f}^n = \mathbf{a} + (2\boldsymbol{\omega}_{ie}^n + \boldsymbol{\omega}_{en}^n) \times \mathbf{v} - \mathbf{g}, \tag{22}$$

where $\mathbf{f}^b$ is the simulated actual output, $\mathbf{I}$ is the 3×3 unit matrix. $\mathbf{S}_a$ is the 3×3 diagonal matrix whose diagonal elements correspond to the 3 accelerometers' scale factor errors, $\boldsymbol{\eta}^b$ is the bias considered as the sum of a constant noise and a random white noise $\boldsymbol{\eta}^b = \boldsymbol{\eta}_{\text{const}}^b + \boldsymbol{\eta}_{\text{random}}^b$. $\mathbf{g} = [0\ 0\ g]^T$, and $g = 9.7803 + 0.051799 C_{33}^2 - 0.94114 \times 10^{-6} h\,(\text{m/s}^2)$, where C_{33} is the 9th element of $\mathbf{C}_e^p$ and h is the vehicle altitude.

4.7. Examples. For four examples of static, straight line, circle, and s-shape situations, details will be given next under the conditions that the vehicle is moving on the surface of the Earth with no attitude change except for the heading angle,

which means that the pitch angle, roll angle, and altitude are constants during the simulation process:

$$\begin{aligned} \Delta\theta &= 0, \\ \Delta\gamma &= 0, \\ \Delta\dot{\theta} &= 0, \\ \Delta\dot{\gamma} &= 0, \end{aligned} \tag{23}$$

The calculation method for the other parameters for the four situations is described as follows.

(1) Static:

$$\begin{aligned} L &= \text{constant}, \\ \lambda &= \text{constant}, \\ v_E &= \text{constant}, \\ v_N &= \text{constant}, \\ \Delta\psi &= 0, \\ \Delta\dot{\psi} &= 0. \end{aligned} \tag{24}$$

(2) Straight line:

$$\begin{aligned} a_E &= \text{constant}, \\ a_N &= \text{constant}, \\ v_E &= v_E + a_E\Delta t, \\ v_N &= v_N + a_N\Delta t, \end{aligned}$$

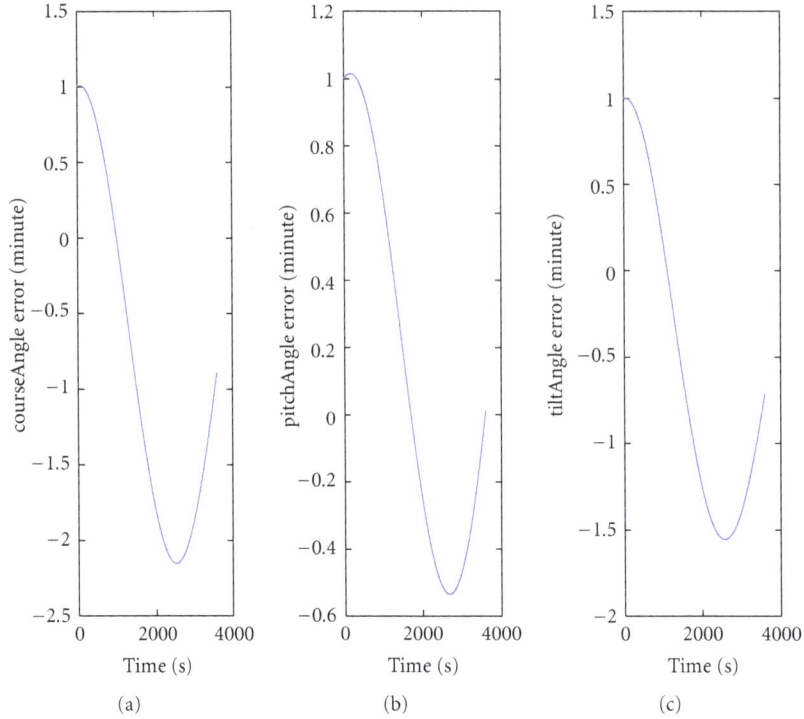

FIGURE 10: Angle error of static simulation.

$$\psi = \psi_0 + \arctan\left(\frac{v_E}{v_N}\right),$$

$$\dot{\psi} = \frac{a_N v_E - a_E v_N}{v_E^2 + v_N^2}.$$

(25)

(3) Circle:

$$v_g = \text{constant},$$

$$\Delta\psi = \text{mod}\left(\frac{2\pi\Delta t}{T_{\text{circle}}}, 2\pi\right),$$

$$\Delta\dot{\psi} = \frac{2\pi}{T_{\text{circle}}},$$

$$a_E = -\frac{2\pi v_g \cos\psi}{T_{\text{circle}}},$$

$$a_N = -\frac{2\pi v_g \sin\psi}{T_{\text{circle}}}.$$

(26)

(4) S-shape:

$$v_g = \text{constant},$$

$$a_E = -\frac{v_g \cos\left(\psi_0 + A_{\text{sshape}}\sin\left(2\pi t/T_{\text{sshape}}\right)\right)}{T_{\text{sshape}}},$$

$$\cdot \frac{2\pi A_{\text{sshape}}\cos\left(2\pi t/T_{\text{sshape}}\right)}{T_{\text{sshape}}},$$

$$a_N = -\frac{v_g \sin\left(\psi_0 + A_{\text{sshape}}\sin\left(2\pi t/T_{\text{sshape}}\right)\right)}{T_{\text{sshape}}}$$

$$\cdot \frac{2\pi A_{\text{sshape}}\cos\left(2\pi t/T_{\text{sshape}}\right)}{T_{\text{sshape}}},$$

$$\psi = \psi_0 + A_{\text{sshape}}\sin\left(\frac{2\pi t}{T_{\text{sshape}}}\right).$$

$$\dot{\psi} = \frac{2\pi A_{\text{sshape}}\cos\left(2\pi t/T_{\text{sshape}}\right)}{T_{\text{sshape}}}.$$

(27)

5. Mathematical Model and Trajectory Calculation Steps

After the gyro and accelerometer data are simulated using the method described in the previous section under the designed scenario, the next step we have to do is to figure out the mathematical model of SINS and the calculation steps to process the sensor data to get the calculated trajectories. Based on the basic principles of strapdown inertial navigation system [4], we draw the mathematical model in the p-frame mechanization in Figure 7. The calculation steps are described below. Although the situation that the vehicle passes over either the north or south pole seldom happens, the universal p-frame is still chosen instead of the simpler ENU-frame to give a navigation illustration in a different frame.

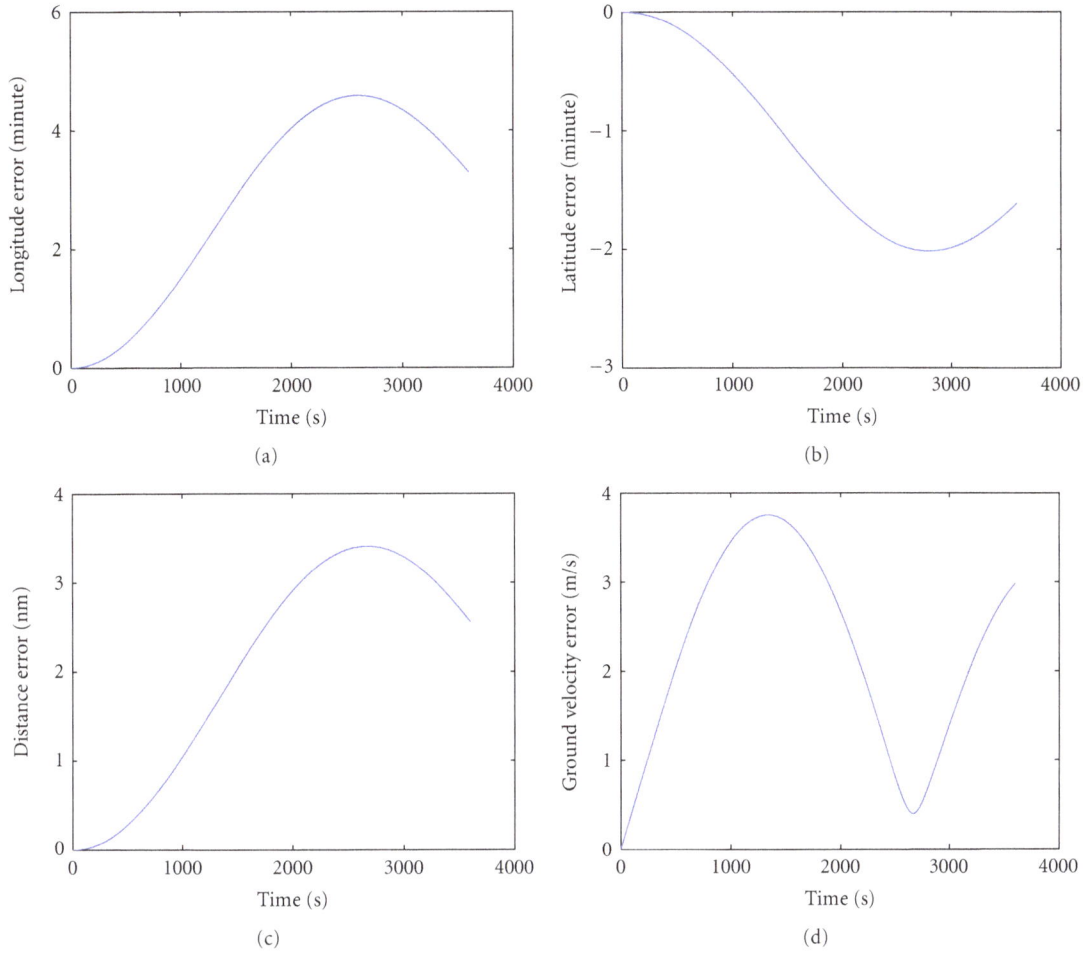

FIGURE 11: Position and velocity error of static simulation.

5.1. Quaternion **Q** Update and Optimal Normalization.

There are three kinds of strapdown attitude representations: DCM, Euler angle, and quaternion. In this paper, we choose quaternion. The reason why quaternion is chosen is explained in [3].

The quaternion formed by a rotating body frame around the platform frame is

$$\mathbf{Q} = q_0 + q_1 i_b + q_2 j_b + q_3 k_b. \tag{28}$$

The update for the quaternion can be obtained by solving the following quaternion differential equation:

$$\begin{bmatrix} \dot{q}_0 \\ \dot{q}_1 \\ \dot{q}_2 \\ \dot{q}_3 \end{bmatrix} = \frac{1}{2} \begin{bmatrix} 0 & -\omega_{pbx}^b & -\omega_{pby}^b & -\omega_{pbz}^b \\ \omega_{pbx}^b & 0 & \omega_{pbz}^b & -\omega_{pby}^b \\ \omega_{pby}^b & -\omega_{pbz}^b & 0 & \omega_{pbx}^b \\ \omega_{pbz}^b & \omega_{pby}^b & -\omega_{pbx}^b & 0 \end{bmatrix} \begin{bmatrix} q_0 \\ q_1 \\ q_2 \\ q_3 \end{bmatrix}. \tag{29}$$

Based on the Euclide norm minimized indicator [4], the optimal normalization for the quaternion is

$$\mathbf{Q} \leftarrow \frac{\mathbf{Q}}{\sqrt{q_0^2 + q_1^2 + q_2^2 + q_3^2}}. \tag{30}$$

5.2. $\mathbf{C}_b^p$ Calculation.

$\mathbf{C}_b^p$ is vehicle attitude DCM which transforms the measured angle in the b-frame to the p-frame, with its 9 components T_{ij}, $i, j = 1, 2, 3$. (Here we use T_{ij} to distinguish it from the components C_{ij}, $i, j = 1, 2, 3$ of $\mathbf{C}_e^p$ which is used below.)

After obtaining q_0, q_1, q_2, and q_3 using (29), $\mathbf{C}_b^p$ can be calculated as

$$\mathbf{C}_b^p$$

$$= \begin{bmatrix} T_{11} & T_{12} & T_{13} \\ T_{21} & T_{22} & T_{23} \\ T_{31} & T_{32} & T_{33} \end{bmatrix}$$

$$= \begin{bmatrix} q_0^2 + q_1^2 - q_2^2 - q_3^2 & 2(q_1 q_2 - q_0 q_3) & 2(q_1 q_3 + q_0 q_2) \\ 2(q_1 q_2 + q_0 q_3) & q_0^2 - q_1^2 + q_2^2 - q_3^2 & 2(q_2 q_3 - q_0 q_1) \\ 2(q_1 q_3 - q_0 q_2) & 2(q_2 q_3 + q_0 q_1) & q_0^2 - q_1^2 - q_2^2 + q_3^2 \end{bmatrix}. \tag{31}$$

5.3. Specific Force Transformation from $\mathbf{f}^b$ in b-Frame to $\mathbf{f}^p$ in p-Frame.

The specific force $\mathbf{f}^b$ in the b-frame can be transformed to $\mathbf{f}^p$ in the p-frame by multiplication with

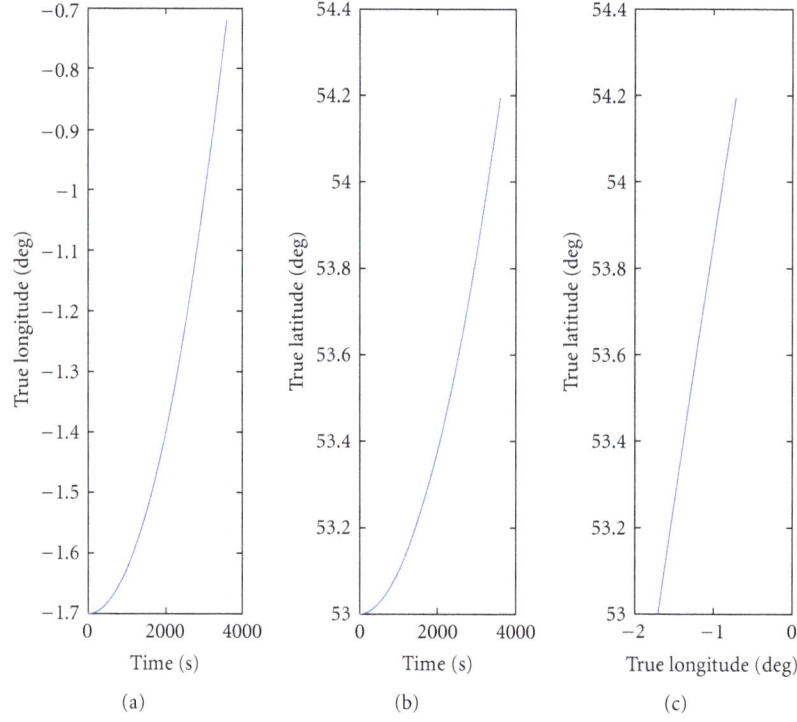

FIGURE 12: The designed trajectory of straight line simulation.

DCM $\mathbf{C}_b^p$:

$$\mathbf{f}^p = \mathbf{C}_b^p \mathbf{f}^b,$$

$$\begin{bmatrix} f_x^p \\ f_y^p \\ f_z^p \end{bmatrix} = \mathbf{C}_b^p \begin{bmatrix} f_x^b \\ f_y^b \\ f_z^b \end{bmatrix}, \tag{32}$$

5.4. Velocity $\mathbf{v}_e^p$ Calculation. The velocity $\mathbf{v}_e^p$ update can be obtained by solving the following differential equation:

$$\dot{\mathbf{v}}_e^p = \mathbf{f}^p - \left(2\boldsymbol{\omega}_{ie}^p + \boldsymbol{\omega}_{ep}^p\right)\mathbf{v}_e^p + \mathbf{g}^p,$$

$$\begin{bmatrix} \dot{v}_x \\ \dot{v}_y \\ \dot{v}_z \end{bmatrix} = \begin{bmatrix} f_x^p \\ f_y^p \\ f_z^p \end{bmatrix} - \begin{bmatrix} 0 \\ 0 \\ g \end{bmatrix}$$

$$+ \begin{bmatrix} 0 & 2\omega_{iez}^p & -\left(2\omega_{iey}^p + \omega_{epy}^p\right) \\ -2\omega_{iez}^p & 0 & 2\omega_{iex}^p + \omega_{epx}^p \\ 2\omega_{iey}^p + \omega_{epy}^p & -\left(2\omega_{iex}^p + \omega_{epx}^p\right) & 0 \end{bmatrix}$$

$$\times \begin{bmatrix} v_x \\ v_y \\ v_z \end{bmatrix}. \tag{33}$$

The ground speed is the vehicle velocity projection on the horizontal plane:

$$v_g = \sqrt{v_x^2 + v_y^2}. \tag{34}$$

5.5. Position Matrix $\mathbf{C}_e^p$ Update. The update for the position matrix $\mathbf{C}_e^p$ can be obtained by solving the following differential equation, noticing that $\omega_{epz}^p = 0$:

$$\dot{\mathbf{C}}_e^p = -\boldsymbol{\Omega}_{ep}^p \mathbf{C}_e^p,$$

$$\mathbf{C}_e^p = \begin{bmatrix} C_{11} & C_{12} & C_{13} \\ C_{21} & C_{22} & C_{23} \\ C_{31} & C_{32} & C_{33} \end{bmatrix},$$

$$\begin{bmatrix} \dot{C}_{11} & \dot{C}_{12} & \dot{C}_{13} \\ \dot{C}_{21} & \dot{C}_{22} & \dot{C}_{23} \\ \dot{C}_{31} & \dot{C}_{32} & \dot{C}_{33} \end{bmatrix} = \begin{bmatrix} 0 & 0 & -\omega_{epy}^p \\ 0 & 0 & \omega_{epx}^p \\ \omega_{epy}^p & -\omega_{epx}^p & 0 \end{bmatrix} \begin{bmatrix} C_{11} & C_{12} & C_{13} \\ C_{21} & C_{22} & C_{23} \\ C_{31} & C_{32} & C_{33} \end{bmatrix}. \tag{35}$$

5.6. Position Angular Velocity $\boldsymbol{\omega}_{ep}^p$ Update. In the chosen wander azimuth navigation frame, we have $\omega_{epz}^p = 0$, and

$$\begin{bmatrix} \omega_{epx}^p \\ \omega_{epy}^p \end{bmatrix} = \begin{bmatrix} -\dfrac{1}{\tau_a} & -\dfrac{1}{R_{yp}} \\ \dfrac{1}{R_{xp}} & \dfrac{1}{\tau_a} \end{bmatrix} \begin{bmatrix} v_{ex}^p \\ v_{ey}^p \end{bmatrix}, \tag{36}$$

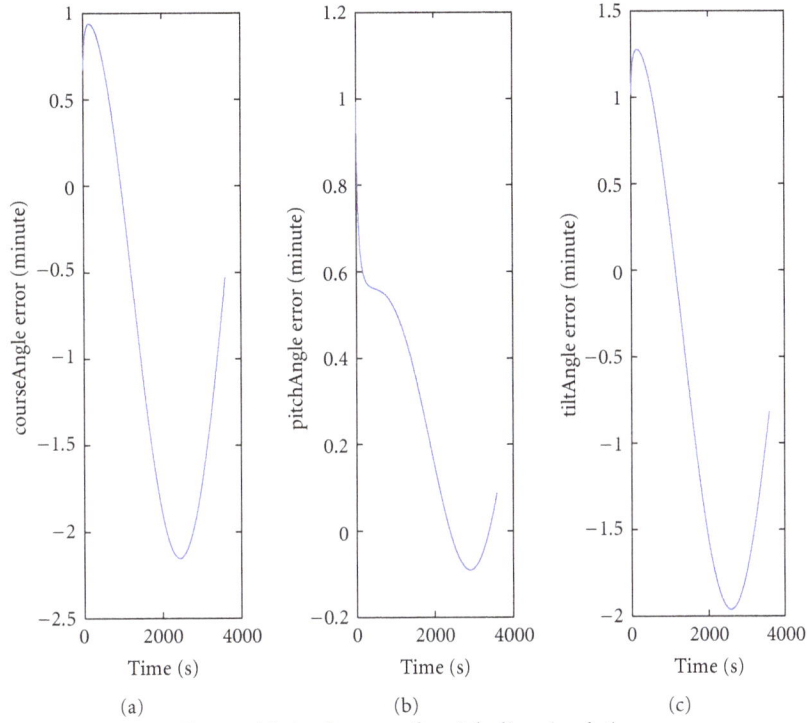

FIGURE 13: Angle error of straight line simulation.

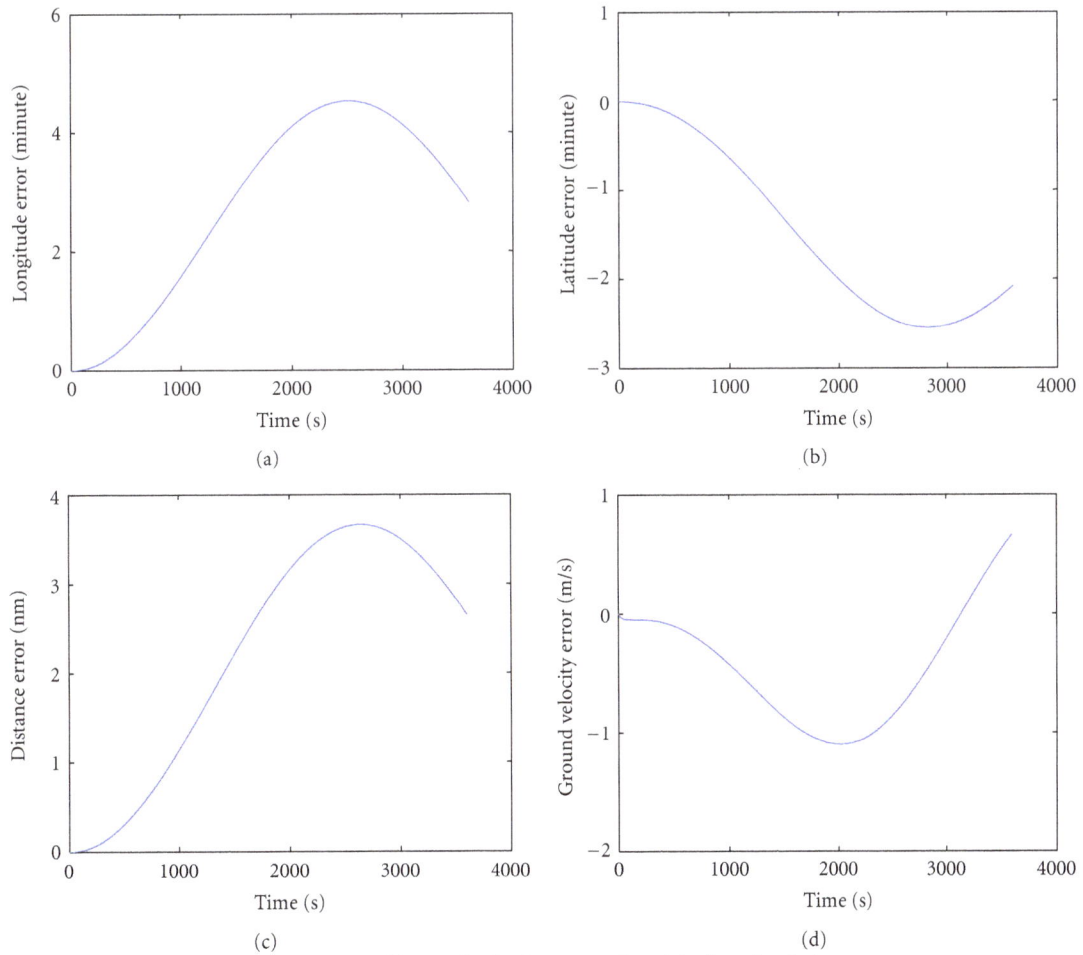

FIGURE 14: Position and velocity error of straight line simulation.

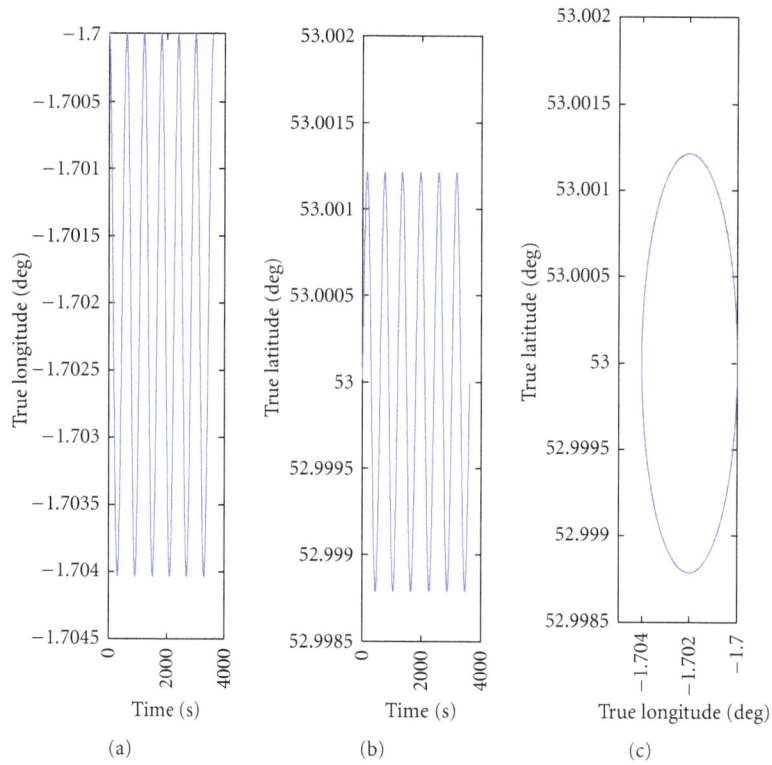

FIGURE 15: The designed trajectory of circle simulation.

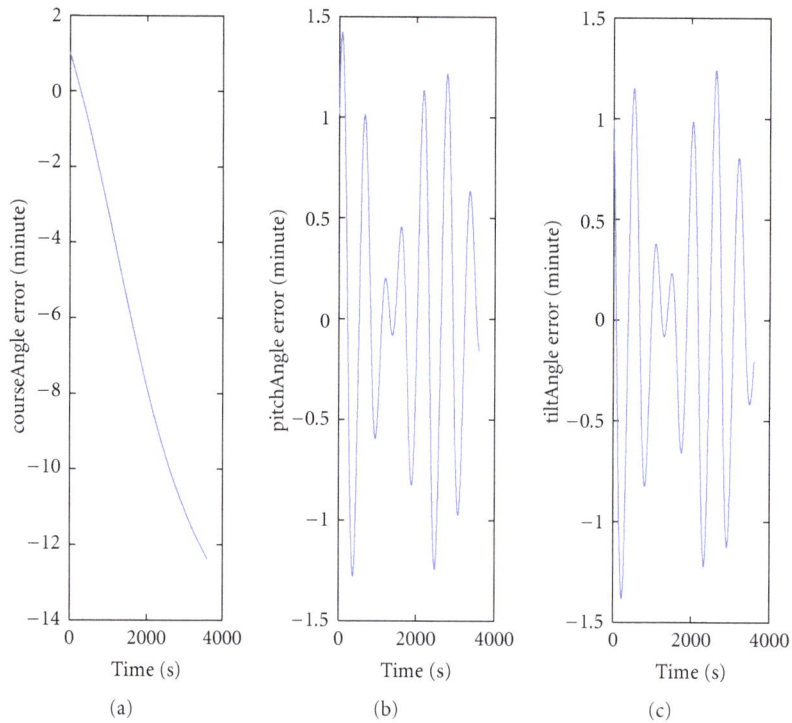

FIGURE 16: Angle error of circle simulation.

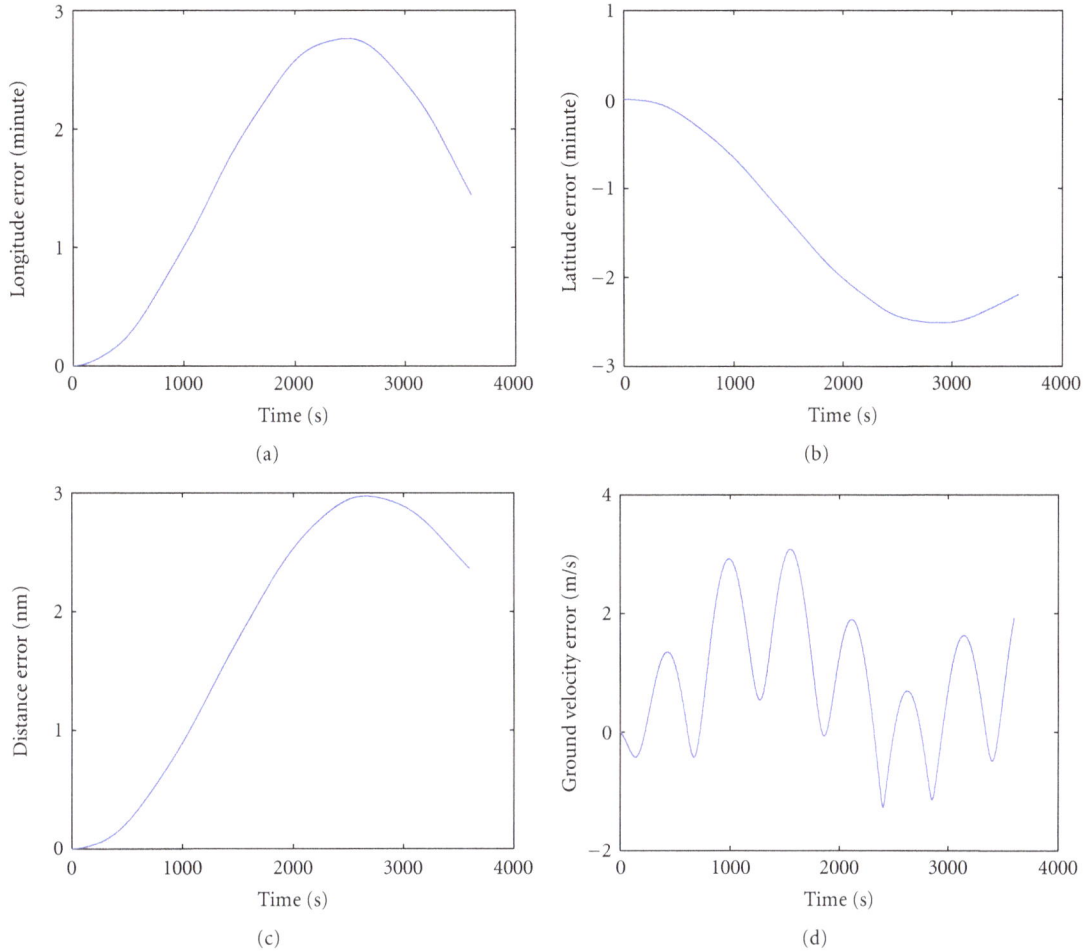

FIGURE 17: Position and velocity error of circle simulation.

where

$$\frac{1}{R_{yp}} = \frac{1}{R_e}\left(1 - eC_{33}^2 + 2eC_{23}^2\right),$$

$$\frac{1}{R_{xp}} = \frac{1}{R_e}\left(1 - eC_{33}^2 + 2eC_{13}^2\right), \quad (37)$$

$$\frac{1}{\tau_a} = \frac{2e}{R_e}C_{13}C_{23},$$

where the elements of position matrix $\mathbf{C}_e^p$ can be obtained using (35).

5.7. Earth Angular Velocity $\boldsymbol{\omega}_{ie}^p$ and Attitude Angular Velocity $\boldsymbol{\omega}_{pb}^b$ Calculation. We have that

$$\boldsymbol{\omega}_{ie}^p = \mathbf{C}_e^p \boldsymbol{\omega}_{ie}^e = \begin{bmatrix} C_{11} & C_{12} & C_{13} \\ C_{21} & C_{22} & C_{23} \\ C_{31} & C_{32} & C_{33} \end{bmatrix} \begin{bmatrix} 0 \\ 0 \\ \omega_{ie} \end{bmatrix} = \begin{bmatrix} \omega_{ie}C_{13} \\ \omega_{ie}C_{23} \\ \omega_{ie}C_{33} \end{bmatrix}, \quad (38)$$

$$\boldsymbol{\omega}_{pb}^b = \boldsymbol{\omega}_{ib}^b - \boldsymbol{\omega}_{ip}^b = \boldsymbol{\omega}_{ib}^b - \left(\mathbf{C}_b^p\right)^{-1}\left(\boldsymbol{\omega}_{ie}^p + \boldsymbol{\omega}_{ep}^p\right), \quad (39)$$

5.8. Attitude Angle Calculation. The relation between attitude matrix $\mathbf{C}_b^p$ and the three attitude angles, grid azimuth angle ψ_G, pitch angle θ, and roll angle γ, is

$$\mathbf{C}_b^p = \begin{bmatrix} \cos\gamma\cos\psi_G - \sin\gamma\sin\theta\sin\psi_G & -\cos\theta\sin\psi_G & \sin\gamma\cos\psi_G + \cos\gamma\sin\theta\sin\psi_G \\ \cos\gamma\sin\psi_G + \sin\gamma\sin\theta\cos\psi_G & \cos\theta\cos\psi_G & \sin\gamma\sin\psi_G - \cos\gamma\sin\theta\cos\psi_G \\ -\sin\gamma\cos\theta & \sin\theta & \cos\gamma\cos\theta \end{bmatrix}. \quad (40)$$

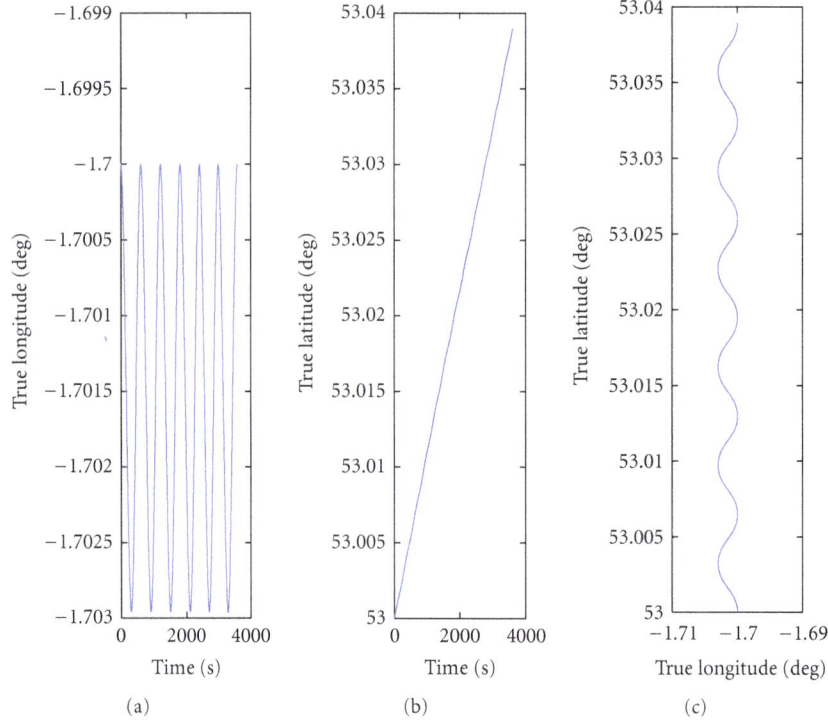

FIGURE 18: The designed trajectory of s-shape simulation.

Thus, the principal values of ψ_G, θ, and γ are

$$\theta_{principal} = \sin^{-1} T_{32},$$

$$\gamma_{principal} = \tan^{-1}\frac{-T_{31}}{T_{33}},$$

$$\varphi_{Gprincipal} = \tan^{-1}\frac{-T_{12}}{T_{22}}. \tag{41}$$

Considering the defined range of the angles, the expressions of the real values of ψ_G, θ, γ, and are

$$\theta \longleftarrow \theta_{principal},$$

$$\gamma \longleftarrow \begin{cases} \gamma_{principal}, & \text{if } T_{33} > 0, \\ \gamma_{principal} + 180°, & \text{if } T_{33} < 0, \ \gamma_{principal} < 0, \\ \gamma_{principal} - 180°, & \text{if } T_{33} < 0, \ \gamma_{principal} > 0, \end{cases}$$

$$\psi_G \longleftarrow \begin{cases} \psi_{Gprincipal}, & \text{if } T_{22} > 0, \ \psi_{Gprincipal} > 0, \\ \psi_{Gprincipal} + 360°, & \text{if } T_{22} > 0, \ \psi_{Gprincipal} < 0, \\ \psi_{Gprincipal} + 180°, & \text{if } T_{22} < 0. \end{cases} \tag{42}$$

5.9. *Position Angle Calculation.* The relation between position matrix $\mathbf{C}_e^p$ and the 3 position angles, longitude λ, latitude φ, and wander azimuth angle α, is

$$\mathbf{C}_e^p = \begin{bmatrix} -\sin\alpha\sin\varphi\cos\lambda - \cos\alpha\sin\lambda & -\sin\alpha\sin\varphi\sin\lambda + \cos\alpha\cos\lambda & \sin\alpha\cos\varphi \\ -\cos\alpha\sin\varphi\cos\lambda + \sin\alpha\sin\lambda & -\cos\alpha\sin\varphi\sin\lambda - \sin\alpha\cos\lambda & \cos\alpha\cos\psi \\ \cos\varphi\cos\lambda & \cos\varphi\sin\lambda & \sin\varphi \end{bmatrix}. \tag{43}$$

Thus, the principal values of φ, λ, and α are

$$\varphi_{principal} = \sin^{-1} C_{33},$$

$$\lambda_{principal} = \tan^{-1}\frac{C_{32}}{C_{31}},$$

$$\alpha_{principal} = \tan^{-1}\frac{C_{13}}{C_{23}}. \tag{44}$$

Considering the defined range of the angles, the expressions of the real values of φ, λ, and α are

$$\varphi \longleftarrow \varphi_{principal},$$

$$\lambda \longleftarrow \begin{cases} \lambda_{principal}, & \text{if } C_{31} > 0, \\ \lambda_{principal} + 180°, & \text{if } C_{31} < 0, \ \lambda_{principal} < 0, \\ \lambda_{principal} - 180°, & \text{if } C_{31} < 0, \ \lambda_{principal} > 0, \end{cases}$$

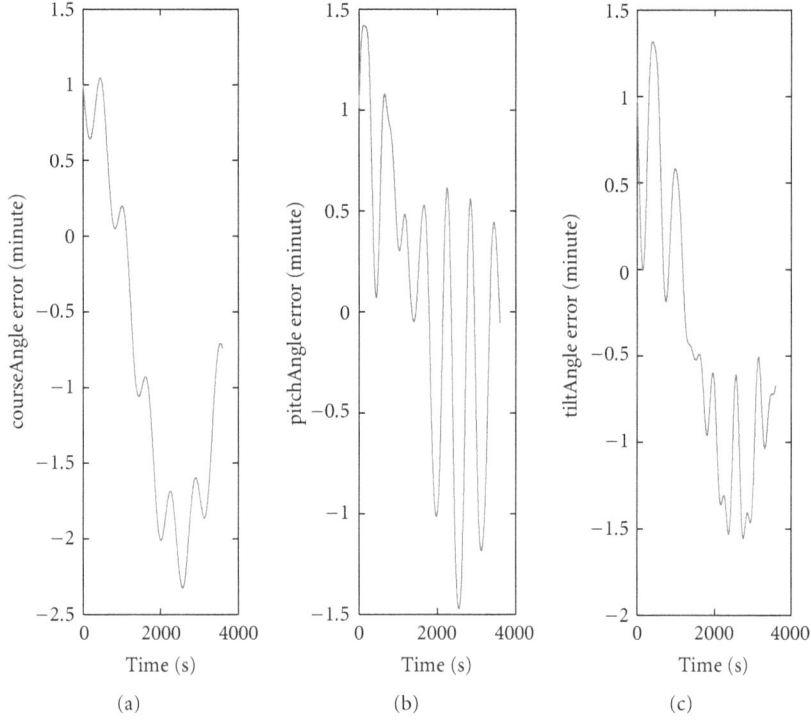

FIGURE 19: Angle error of s-shape simulation.

$$\alpha \longleftarrow \begin{cases} \alpha_{\text{principal}}, & \text{if } C_{23} > 0, \ \alpha_{\text{principal}} > 0, \\ \alpha_{\text{principal}} + 360°, & \text{if } C_{23} > 0, \ \alpha_{\text{principal}} < 0, \\ \alpha_{\text{principal}} + 180°, & \text{if } C_{23} < 0. \end{cases}$$

$$(45)$$

5.10. Heading Angle Calculation. The heading angle ψ is calculated as

$$\psi = \psi_G + \alpha. \tag{46}$$

To make sure that ψ will not be out of range, we should determine it according to

$$\psi \longleftarrow \begin{cases} \psi, & \text{if } \psi < 360°, \\ \psi - 360°, & \text{if } \psi \geq 360°. \end{cases} \tag{47}$$

5.11. Velocity $\mathbf{v}_e^n$ in n-Frame Calculation. We have that

$$\mathbf{v}_e^n = \begin{bmatrix} v_E \\ v_N \\ v_U \end{bmatrix} = \begin{bmatrix} v_{ey}^p \cos \alpha - v_{ex}^p \sin \alpha \\ v_{ey}^p \sin \alpha + v_{ex}^p \cos \alpha \\ v_{ez}^p \end{bmatrix}. \tag{48}$$

5.12. Altitude Calculation. For the calculation of the altitude, damped methods should be used because it diverges with time. To simplify problems, in our simulations, we set the altitude to zero, that is, surface of the Earth.

5.13. Local Gravity g Calculation. The local gravity g is calculated as [5]

$$g = 9.7803 + 0.051799C_{33}^2 - 0.94114 \times 10^{-6}h \ (\text{m/s}^2), \tag{49}$$

where $C_{33} = \sin \varphi$, φ is the latitude and h is the altitude above sea level.

Before we carry out the implementation of the above described mathematical model of SINS, we have to know the initial parameters of the system, which will be described in the following Section.

6. Initial Parameters and Initial Data Calculation

For the calculations in Section 5, we first need to know the given initial parameters and the corresponding initial data.

6.1. Initial Parameters

(1) Initial position, latitude φ_0, longitude λ_0, height h_0. The values of these parameters should be the same as the corresponding ones in Section 4.1.

(2) Initial wander azimuth angle α_0. We could choose $\alpha_0 = 0$ at the very beginning. The value should be the same as the corresponding ones in Section 4.1.

(3) Initial velocity v_{E0}, v_{N0}, v_{U0}.

(4) If barometric altimeter applied, initial external reference height h_{ref0} can be supplied.

6.2. Initial Alignment Data

(1) Initial attitude matrix is determined by initial alignment process $\mathbf{C}_{b0}^p$. $\mathbf{C}_{b0}^p = \mathbf{C}_{b0}^n$ when $\alpha_0 = 0$.

(2) Initial position matrix is determined by initial alignment process $\mathbf{C}_{e0}^p$. $\mathbf{C}_{e0}^p = \mathbf{C}_{e0}^n$ when $\alpha_0 = 0$.

6.3. Initial Data Calculation.

(1) Initial attitude angles φ_0, λ_0, and α_0 determination: The initial attitude angles ψ_{G0}, θ_0, and γ_0 can be calculated using (41) and (42). Because $\alpha_0 = 0$, heading angle $\psi_0 = \psi_{G0}$.

(2) Initial quaternion calculation: From the diagonal elements in (31) and the quaternion constraint equation, we have that

$$q_0^2 + q_1^2 - q_2^2 - q_3^2 = T_{11},$$
$$q_0^2 - q_1^2 + q_2^2 - q_3^2 = T_{22},$$
$$q_0^2 - q_1^2 - q_2^2 + q_3^2 = T_{33},$$
$$q_0^2 + q_1^2 + q_2^2 + q_3^2 = 1,$$

(50)

The solution to (50)

$$|q_1| = \frac{1}{2}\sqrt{1 + T_{11} - T_{22} - T_{33}},$$
$$|q_2| = \frac{1}{2}\sqrt{1 - T_{11} + T_{22} - T_{33}},$$
$$|q_3| = \frac{1}{2}\sqrt{1 - T_{11} - T_{22} + T_{33}},$$
$$|q_0| = \sqrt{1 - q_1^2 - q_2^2 - q_3^2}.$$

(51)

Assuming q_0 to be positive, according to (31), we have that

$$\text{sign}(q_0) = \text{sign}(1),$$
$$\text{sign}(q_1) = \text{sign}(T_{32} - T_{23}),$$
$$\text{sign}(q_2) = \text{sign}(T_{13} - T_{31}),$$
$$\text{sign}(q_3) = \text{sign}(T_{21} - T_{12}).$$

(52)

(3) Initial position matrix $\mathbf{C}_{e0}^p$: Substituting initial position, latitude φ_0, longitude λ_0 and initial wander azimuth $\alpha_0 = 0$ into (43), we can obtain the initial position matrix $\mathbf{C}_{e0}^p$.

(4) Initial Earth angular velocity $\boldsymbol{\omega}_{ie0}^p$ and initial attitude angular velocity $\boldsymbol{\omega}_{pb0}^b$ calculations: use (38) and (39).

(5) Initial position angular velocity $\boldsymbol{\omega}_{ep0}^p$ calculation: use (36) and (37).

(6) Initial gravity g_0 calculation: use (49) and element C_{33} in $\mathbf{C}_{e0}^p$.

(7) Initial ground velocity v_{g0} calculation: use (34).

At this point, the whole SINS model, including sensor data generator and initial parameters, is fully described. The following Section will provide a Matlab implementation of the SINS theory.

7. Matlab Implementation and Simulation Examples

First, the Matlab program structure and the main codes are given. The Matlab implementation is illustrated using six examples: static, straight, circle, s-shape, and the other two from real SINS experimental data.

7.1. Matlab Implementation and Codes. The program structure is given in Figure 8. The program starts from "Begin" and ends at "Stop." The gyro and accelerometer data are obtained either from a sensor data generator described in Section 4 or from the real SINS experiment logged files. Processing the sensor data with the initial parameters, using the method described in Section 5, we get the attitude, velocity and position values of the system at specific times. After all data are processed, the program will stop and the results will be provided.

The main Matlab codes are presented next.

(1) Initial settings:

   (a) initSettings.m contains initial parameters and constants used in the simulation project.

(2) Trajectory part:

   (a) initialCalculation_static.m gives the initial calculation for the static situation;
   (b) trajectorySimulater_static.m simulates gyro and accelerometer data for the static situation;
   (c) initialCalculation_straight.m gives the initial calculation for the straight line situation;
   (d) trajectorySimulater_straight.m simulates gyro and accelerometer data for the straight line situation;
   (e) initialCalculation_cirlce.m gives the initial calculation for the circle situation;
   (f) trajectorySimulater_circle.m simulates gyro and accelerometer data for the circle situation;
   (g) initialCalculation_Sshape.m gives the initial calculation for the s-shape situation;
   (h) trajectorySimulater_Sshape.m simulates gyro and accelerometer data for the s-shape situation;

(3) Simulation part:

   (a) INSmain.m is the main program; the simulation starts from here;
   (b) AltitudeParamete.m calculates the four damping parameters to damp the altitude error according to the input parameters k_4 and τ, to be used with the external reference altitude;

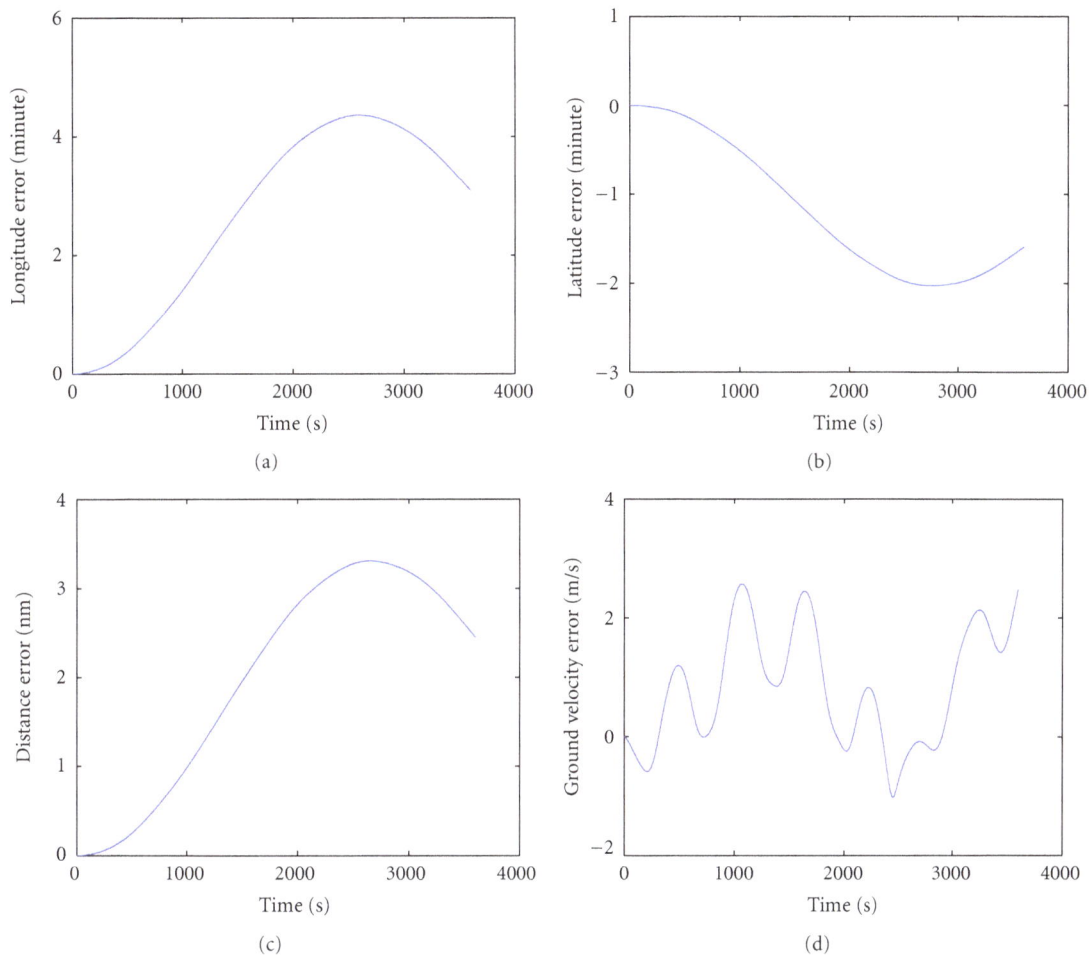

FIGURE 20: Position and velocity error of s-shape simulation.

(c) *InitializePosition.m* gives the initial position *initLong, initLat, initAlt*, the external reference altitude *extAlt*, and the wander azimuth angle *wanderAzimuth*; it calculates the initial position matrix and then orthogonalizes the matrix;

(d) *InitializeAttitude.m* gives the initial alignment error and calculates the attitude matrix (strapdown matrix);

(e) *InitializeQuaternion.m* calculates the quaternion according to the input attitude matrix;

(f) *ComputeAngularVelocity.m* calculates the position angular velocity, earth angular velocity, and position angle increment in the *p*-frame and resets the gyroscopes and accelerometers;

(g) *ComputeQuaternionRungeKutta.m* computes the quaternion using Runge-Kutta method [6]; see Appendices B and C;

(h) *ComputeAttitudeMatrix.m* computes the attitude matrix and transfers the raw data of the accelerometers to the *p*-frame;

(i) *ComputeVelocity.m* computes the velocity, in the wander azimuth frame (*p*-frame) and *ENU*-frame, the ground velocity and altitude;

(j) *ComputePositionMatrix.m* computes the position matrix.

(k) *ComputePosition.m* computes *latitude, longitude* and *wanderAzimuth*;

(l) *ComputeAttitudeAngle.m* computes the attitude angle of *pitchAngle, tiltAngle, gridAzimuth* and *courseAngle*;

(m) *OrthogonalizeMatrix.m* computes matrix orthogonalization; see Appendix D;

(n) *QuaCofMatrix.m* is called by *ComputeQuaternionRungeKutta.m*;

(o) *PlotResult.m* plots the results of the simulation project.

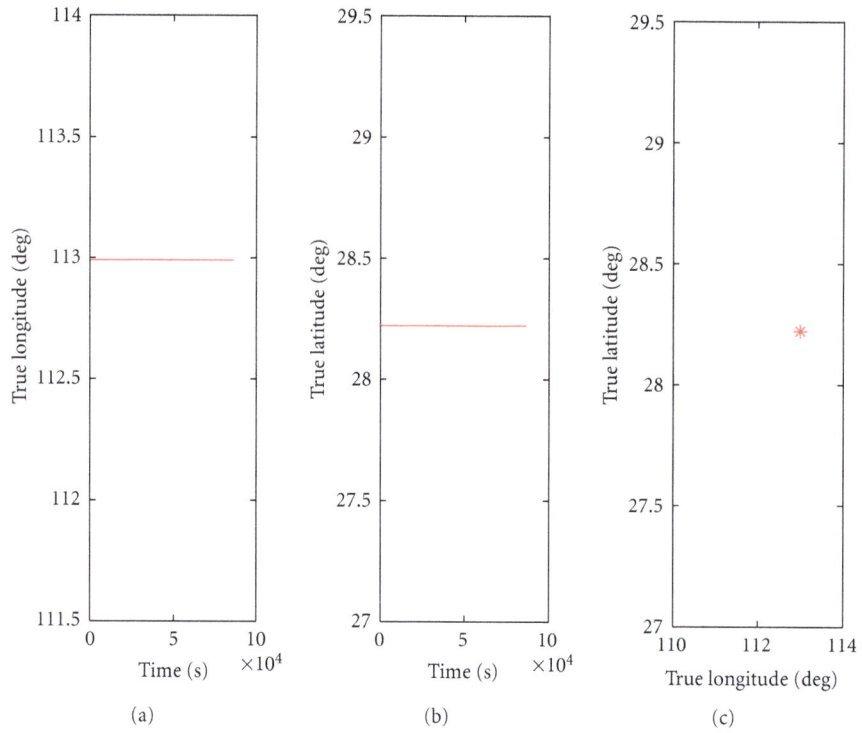

FIGURE 21: The trajectory of real data set A.

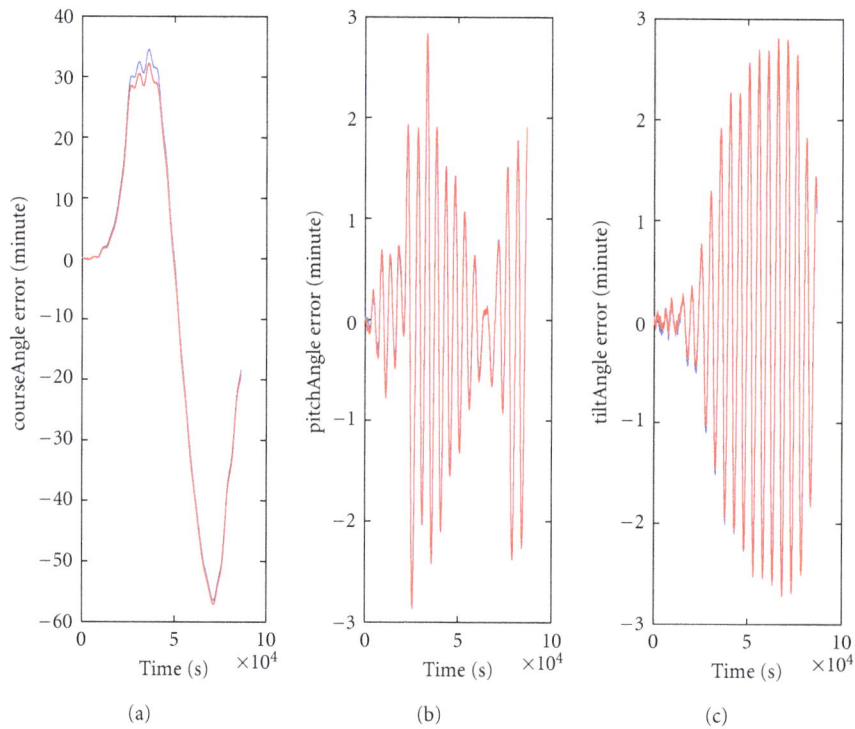

FIGURE 22: Angle error of real data set A.

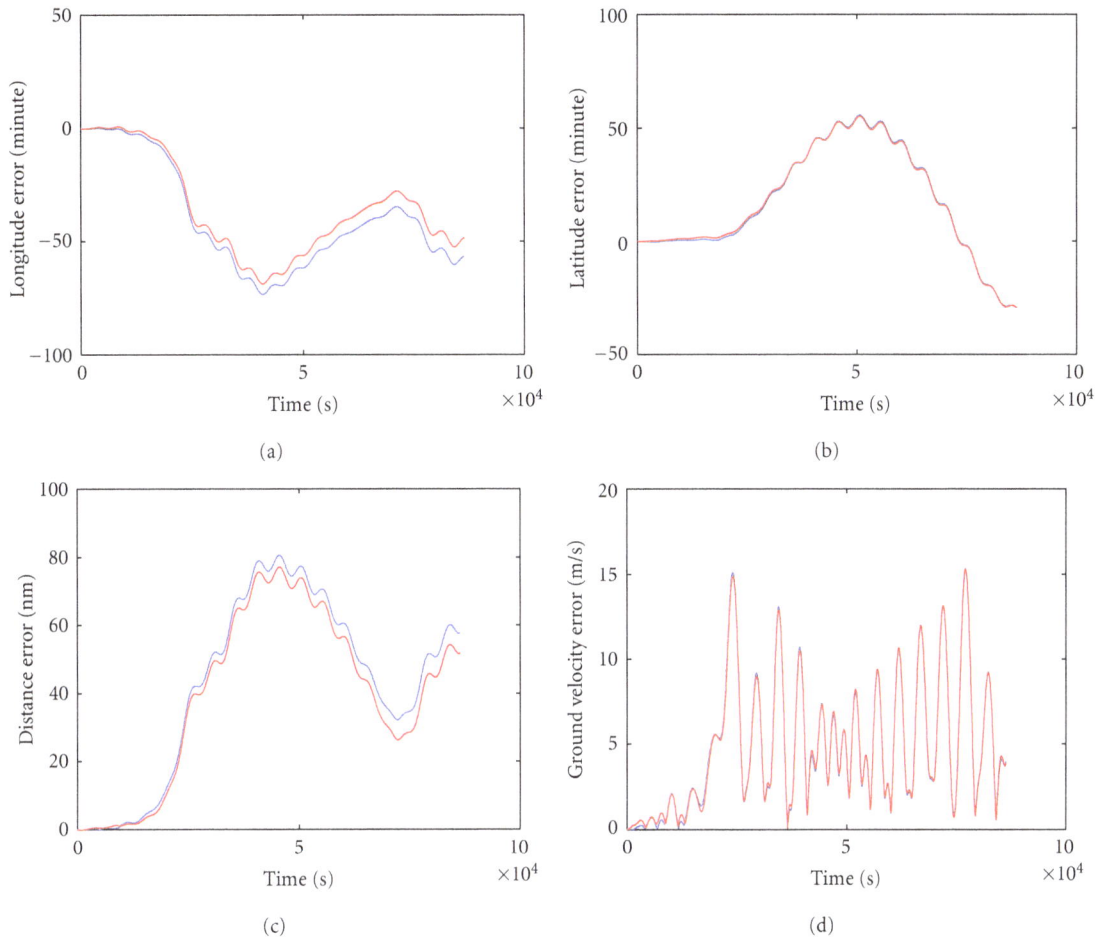

FIGURE 23: Position and velocity error of real data set A.

7.2. Simulation Examples. In this subsection, there are 6 SINS simulation examples. Example 1 is the static situation simulation, where the vehicle trajectory in the n-frame is a fixed point. Example 2 is the straight line situation simulation, where the vehicle trajectory in the n-frame is a straight line. Example 3 is the circle situation simulation, where the vehicle trajectory in the n-frame is a circle. Example 4 is the s-shape situation simulation, where the vehicle trajectory in the n-frame is an s-shape line. Here, high-accuracy SINS simulation is applied to the four situations. The initial latitude and longitude errors are set to be 1 minute. The simulation time is set to 3600 seconds. The initial positions are dependent on the designed trajectories.

In order to verity the validity of the Matlab codes further, two sets of real static data are used, and we refer to these as Examples 5 and 6. The two sets of real data, set A and set B, are collected from the same SINS in the same place but at different times. The 2 data sets are 24 hours long.

All the errors (the angle error, the velocity error, and the position error) will contribute to the distance error in the INS trajectory calculation. Thus, the distance error is a key index of an INS system. The distance error will increase with time, so it is always associated with a time stamp.

Example 1 (Static situation simulation). The static situation is the most basic and simple situation where the output of the gyro is the Earth rotating angular velocity and the output of the accelerometer is the gravity. Figure 9 shows the designed true trajectory. Figure 10 shows the difference between the calculated angle and the true angle. Figure 11 shows the differences between the calculated PV (position and velocity) and the true PV. The maximum value of the distance error in 1 hour is 3.5 nm (nautical mile).

Example 2 (Straight line situation simulation). The straight line situation corresponds to a vehicle moving along the northwest direction. Figure 12 shows the designed true trajectory. Figure 13 shows the difference between the calculated angle and the true angle. Figure 14 shows the differences between the calculated PV and the true PV. The maximum value of the distance error in 1 hour is 3.7 nm.

Example 3 (Circle situation simulation). The circle situation corresponds to a vehicle moving along a circle. Figure 15 shows the designed true trajectory. Figure 16 shows the difference between the calculated angle and the true angle. Figure 17 shows the difference between the calculated PV and the true PV. The maximum value of the distance error in 1 hour is 3.0 nm.

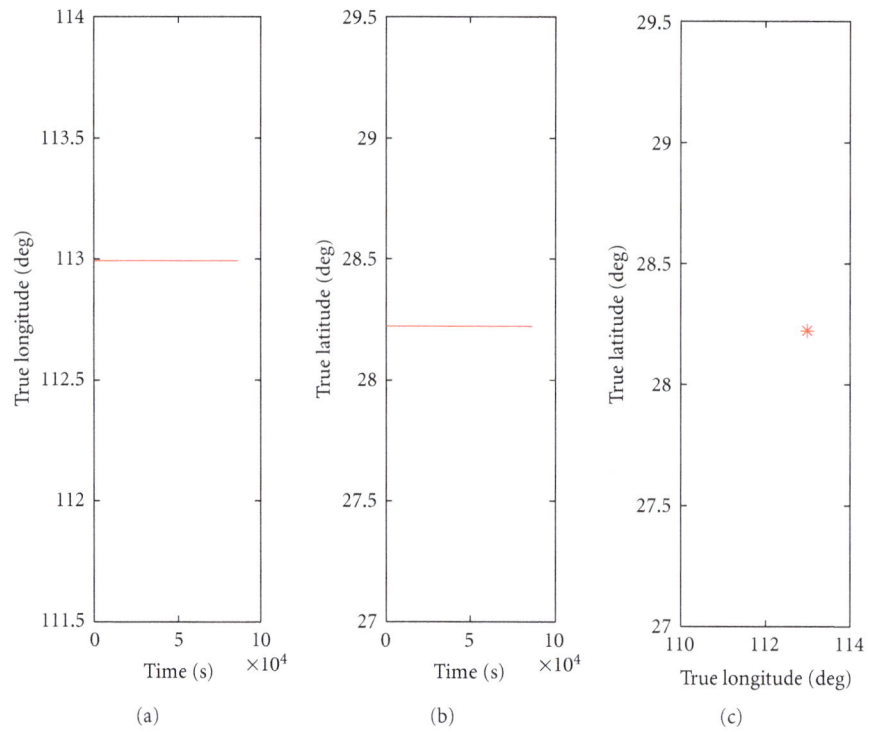

FIGURE 24: The trajectory of real data set B.

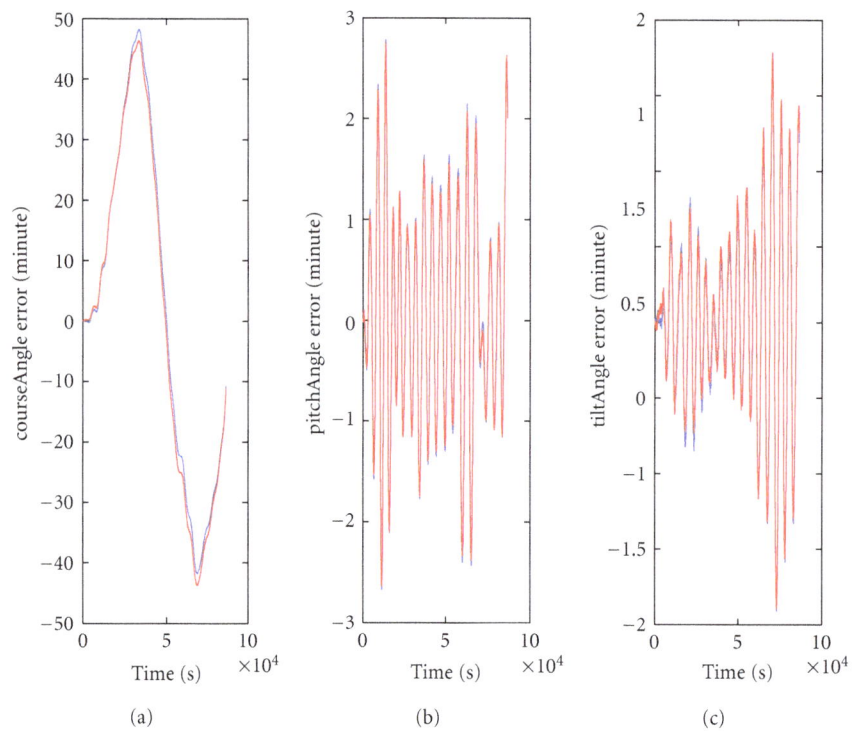

FIGURE 25: Angle error of real data set B.

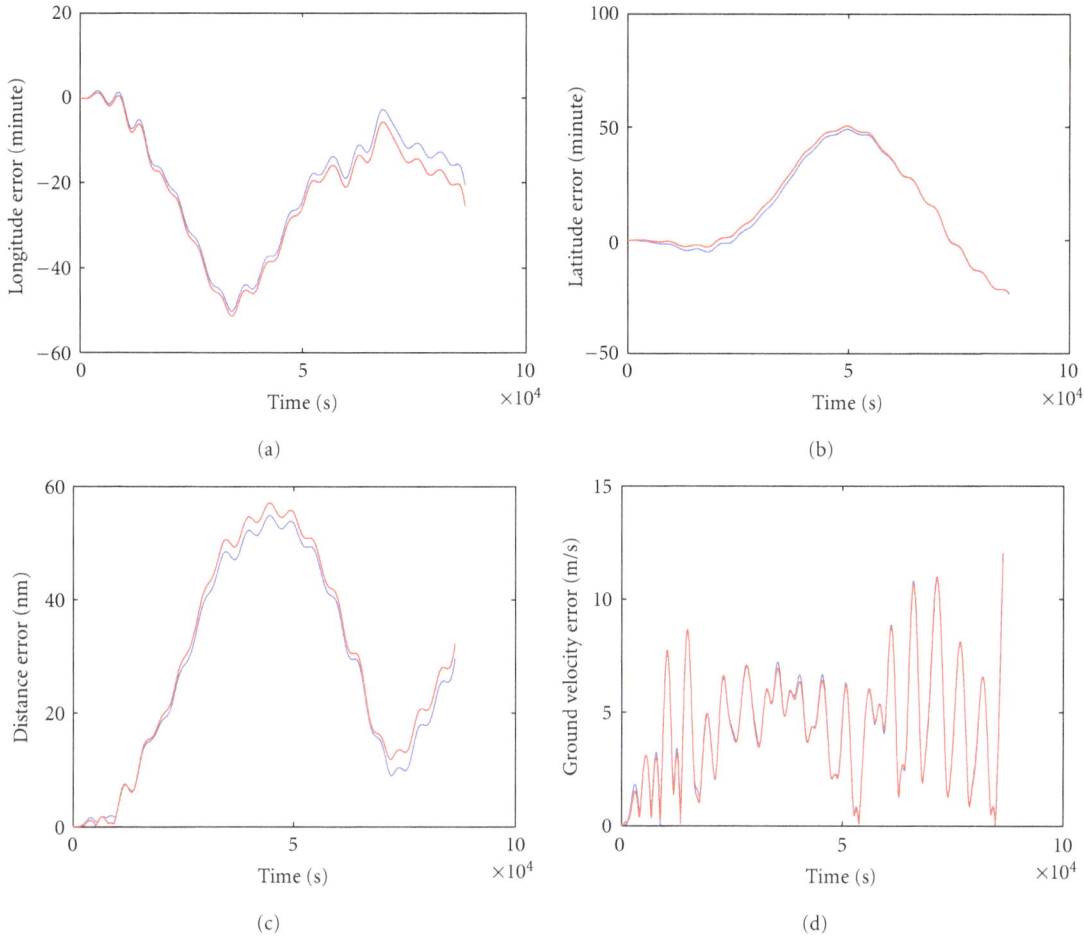

FIGURE 26: Position and velocity error of real data set B.

Example 4 (S-shape situation simulation). The s-shape situation corresponds to a vehicle moving along an s-shaped line. Figure 18 shows the designed true trajectory of s-shape situation simulation. Figure 19 shows the difference between the calculated angle and the true angle. Figure 20 shows the differences between the calculated PV and the true PV. The maximum value of the distance error in 1 hour is 3.3 nm.

Example 5 (Real static data set A simulation). First, we process data set A [7]. Figure 21 shows the trajectory for the real data set A; from the figure we can conclude that the system is static. In Figure 22, the red line corresponds to the three attitude angle errors of the real system, while the blue line corresponds to the three attitude angle errors processed by the Matlab code. We can also show that the difference between the red and blue lines is negligible. In Figure 23, the red line corresponds to the position and velocity errors of the real system, while the blue line corresponds to the position and velocity errors processed by the Matlab code. We can also see that the difference between the red and blue lines is negligible and this validates the correctness of the Matlab code. The error described by the red lines (output from the real system) is slightly smaller than that described by the blue lines (simulation). This is due to the fact that the real system

is processed in a much higher rate and thus its input is more accurate than the simulated system.

Example 6 (Real static data set B simulation). Figure 24 shows the trajectory of the real data set B; from the figure we can conclude that the system is static too. In Figure 25, the red line corresponds to the three attitude angle errors of the real system, while the blue line corresponds to the three attitude angle errors obtained by the Matlab code when applied to the real raw sensor data set B. We can also see that the difference between the red and blue lines is negligible. In Figure 26, the red line corresponds to the position and velocity errors of the real system, while the blue line corresponds to the position and velocity errors obtained by the Matlab code when applied to the real raw sensor data set B. We can also see that the difference between the red and blue lines is negligible, and this further validates the correctness of the Matlab code.

8. Conclusions

In this paper, a mathematical model for the strapdown inertial navigation system (SINS) is built and its Matlab implementation is developed. First, a number of Cartesian coordinate reference frames that relate to SINS are introduced, the basic principle of SINS in the wander azimuth navigation frame (*p*-frame) is explained, and the main equations are described. Second, the important attitude direction cosine matrix and position direction cosine matrix in the *p*-frame are defined in detail. Third, the mathematical model for SINS simulation is described in detail. Fourth, a trajectory simulator model is set up to generate data from three orthogonal gyros and three orthogonal accelerometers. The initial parameters and initial data calculations for the mathematical model are also carried out. Finally, a Matlab implementation of SINS is developed. The proposed simulation method is illustrated with four examples, static, straight line, circle, and s-shape trajectories; details are given under the condition that the pitch angle, roll angle, and altitude are constant during the simulation process. Further, two sets of real experimental data are processed to verify the validity of the Matlab code.

Appendices

A. Symmetric Matrix Basic Operation

For a vector $\boldsymbol{\omega} = [\omega_x\ \omega_y\ \omega_z]^T$, its skew symmetric matrix $\boldsymbol{\Omega}$ is

$$\boldsymbol{\Omega} = \begin{bmatrix} 0 & -\omega_z & \omega_y \\ \omega_z & 0 & -\omega_x \\ -\omega_y & \omega_x & 0 \end{bmatrix}. \tag{A.1}$$

We can easily show that

$$\boldsymbol{\Omega}^T = -\boldsymbol{\Omega}. \tag{A.2}$$

B. Fourth-Order Runge-Kutta Method

For numerical analysis, the fourth-order Runge-Kutta method is an important iterative method for the approximation of solutions of ordinary differential equations. Here, in this paper, the fourth-order Runge-Kutta method is adopted to update the quaternion.

The steps for the fourth-order Runge-Kutta method are the following.

(1) Calculate slope k_1, the slope at the beginning of the interval, to determine the value of $y_{i+1/2}$ at the point $t_{i+1/2}$ using the Euler method:

$$k_1 = f(\omega_i, y_i, t_i),$$
$$y_{i+1/2} = y_i + \frac{\tau}{2}k_1, \tag{B.1}$$

where τ is the time step between time t_i and time t_{i+1}, $\tau = t_{i+1} - t_i$.

(2) Calculate slope k_2, the slope at the midpoint of the interval, to determine the value of $y'_{i+1/2}$ at the point $t_{i+1/2}$ using Euler's method:

$$k_2 = f(\omega_{i+1/2}, y_{i+1/2}, t_{i+1/2}),$$
$$y'_{i+1/2} = y_i + \frac{\tau}{2}k_2. \tag{B.2}$$

(3) Calculate slope k_3, again the slope at the midpoint, to determine the y_{i+1} value:

$$k_3 = f(\omega_{i+1/2}, y'_{i+1/2}, t_{i+1/2}),$$
$$y_{i+1} = y_i + \tau k_3. \tag{B.3}$$

(4) Calculate slope k_4, the slope at the end of the interval, with its y_{i+1} value determined using k_3:

$$k_4 = f(\omega_{i+1}, y_{i+1}, t_{i+1}). \tag{B.4}$$

(5) Average the four slopes; greater weights are given to the slopes at the midpoint:

$$k = \frac{1}{6}(k_1 + 2k_2 + 2k_3 + k_4). \tag{B.5}$$

(6) Finally, using the average slope k, the value of y_{i+1} is

$$y_{i+1} = y_i + \tau k. \tag{B.6}$$

C. Angular Velocity Extraction

From Appendix B and (29), we need to provide the attitude angular velocity ω_{pb}^b in a period of $\tau/2$ to update the quaternion. By inspecting the expression of ω_{pb}^b in (39), we know that the variations of ω_{ep}^p and ω_{ie}^p are slow, while ω_{ib}^b changes quickly. So, only ω_{ib}^b needs to be given in a period of $\tau/2$. We know that ω_{ib}^b (we next use ω to simplify notation) is the output of gyro which gives data in the form of angle increment $\Delta\theta_i$ during the time interval τ. For first-order angular velocity extraction, we have that

$$\omega = \frac{\Delta\theta_i}{\tau}. \tag{C.1}$$

In order to provide $\omega(t_i)\omega(t_{i+1/2})$ and $\omega(t_{i+1})$, we need to do second-order angular velocity extraction:

$$\omega(t_i) = \frac{3\Delta\theta_{i1} - \Delta\theta_{i2}}{\tau},$$
$$\omega(t_{i+1/2}) = \frac{\Delta\theta_{i1} + \Delta\theta_{i2}}{\tau}, \tag{C.2}$$
$$\omega(t_{i+1}) = \frac{-\Delta\theta_{i1} + 3\Delta\theta_{i2}}{\tau},$$

where $\Delta\theta_{i1}$ is the angle increment from time t_i to $t_{i+1/2}$ and $\Delta\theta_{i2}$ is the angle increment from time $t_{i+1/2}$ to time t_{i+1}.

D. Matrix Orthogonalization Method

For the direction cosine matrix $\mathbf{C}$, the optimal orthogonalization method is to get $\hat{\mathbf{C}}$ which makes the following Euclidian function have the minimum value [8]:

$$D = \left[\sum_{i=1}^{3} \sum_{j=1}^{3} \left(C_{ij} - \hat{C}_{ij} \right) \right]^{1/2}. \tag{D.1}$$

The expression for $\hat{\mathbf{C}}$ is thus

$$\hat{\mathbf{C}} = \pm \mathbf{C} \left(\mathbf{C}^T \mathbf{C} \right)^{-3/2}, \tag{D.2}$$

where the superscript T means the transpose operator. It is difficult to solve the above equation directly. Instead, we use an iterative method. Assume $\mathbf{C}_0$ to be initial matrix, and $\mathbf{C}_n$ to be the matrix obtained after n iterations. The iteration process is as follows:

$$\mathbf{C}_0 = \mathbf{C}, \\ \vdots \\ \mathbf{C}_{n+1} = \mathbf{C}_n - \frac{1}{2} \left(\mathbf{C}_n \mathbf{C}^T \mathbf{C}_n - \mathbf{C} \right). \tag{D.3}$$

If at the $n + 1$ step, the following function:

$$f_n = \sum_{i=1}^{3} \sum_{j=1}^{3} \left(C_{ij} - C_{ijn} \right)^2 \tag{D.4}$$

satisfies $f_{n+1} - f_n \leq \epsilon$ (e.g., $\epsilon = 10^{-10}$), then the iteration procedure can be stopped and $\mathbf{C}_{n+1}$ is taken to be the final result.

Abbreviations and Symbols

SINS: Strapdown inertial navigation system
DCM: Direction cosine matrix
O: Center of the Earth
P: Center of the vehicle
x, y, z: 3 orthogonal axes or the 3 components of a Cartesian coordinate
b: Body frame
i: Inertial frame
e: Earth frame
n: Navigation frame
ENU: East-North-UP navigation frame, which is identical to the n-frame in this paper
p: Wander azimuth frame

$\mathbf{A}^T$: Transpose of matrix $\mathbf{A}$
$\mathbf{v}_e^p$: Velocity vector measured in p-frame with respect to e-frame
$\mathbf{C}_b^p$: Vehicle attitude DCM used to transform the measured angle in b-frame to p-frame, with its 9 components T_{ij}, $i, j = 1, 2, 3$
$\mathbf{C}_p^b$: Transpose of $\mathbf{C}_b^p$ is used to transform the measured vector in p-frame to b-frame
$\mathbf{C}_e^p$: Vehicle position DCM used to transform the measured vector in e-frame to p-frame, with its 9 components C_{ij}, $i, j = 1, 2, 3$
$\mathbf{C}_p^e$: Transpose of $\mathbf{C}_e^p$ is used to transform the measured vector in p-frame to e-frame
$\mathbf{C}_b^n$: Vehicle attitude DCM used to transform the measured angle in b-frame to n-frame
$\mathbf{f}^p$: Specific force vector measured in p-frame
$\mathbf{f}^n$: Specific force vector measured in n-frame
$\mathbf{f}^b$: Specific force vector measured in b-frame; the output of the 3 accelerometers
ω_{ie}: Constant value of the turn rate of the Earth, $\omega_{ie} = 7.2921151467 \times 10^{-5}$ rad/s
ω_{ie}^n: Turn rate of the Earth measured in n-frame
ω_{ib}^b: Turn rate of the b-frame with respect to i-frame, which is measured in b-frame; the output of the 3 gyros
ω_{en}^n: Transport rate of the n-frame with respect to e-frame, which is measured in n-frame
ω_{ie}^e: Turn rate of the e-frame with respect to i-frame, which is measured in e-frame
ω_{ep}^p: Turn rate of the p-frame with respect to e-frame, which is measured in p-frame
ω_{pe}^e: Turn rate of the e-frame with respect to p-frame, which is measured in e-frame
ω_{pb}^b: Turn rate of the b-frame with respect to p-frame, which is measured in b-frame
ω_{nb}^b: Turn rate of the b-frame with respect to n-frame, which is measured in b-frame
$\mathbf{\Omega}_{pb}^b$: Skew matrix form of ω_{pb}^b
$\mathbf{\Omega}_{ep}^p$: Skew matrix form of ω_{ep}^p
$\mathbf{g}^p$: Gravity vector measured in p-frame
g: Local gravity scalar
g_0: Local gravity scalar at sea level

ψ_G: Grid azimuth angle of the vehicle in b-frame with respect to p-frame

α: Wander azimuth angle of p-frame with respect to n-frame

ψ: Heading angle of the vehicle in b-frame with respect to n-frame

θ: Grid pitch angle of the vehicle in b-frame with respect to n-frame or p-frame

γ: Grid roll angle of the vehicle in b-frame with respect to n-frame or p-frame

$\Delta\psi$: Increase of the heading angle ψ

$\Delta\theta$: Increase of the grid pitch angle θ

$\Delta\gamma$: Increase of the grid roll angle γ

λ: Longitude of the vehicle

φ, L: Latitude of the vehicle

h: Altitude of the vehicle above the sea level of the Earth

$\varphi_0, \lambda_0, h_0$: Initial vehicle position (latitude, longitude, height)

Δt: Time step

$\mathbf{a}$: Vehicle acceleration

$\mathbf{v}_0 = [v_{E0}, v_{N0}, v_{U0}]$: Initial vehicle velocity (east, north, up)

$\mathbf{v} = [v_E, v_N, v_U]$: Vehicle velocity (east, north, up)

v_g: Vehicle ground velocity

$\mathbf{r}_e^n$: Vehicle position measured in n-frame with respect to e-frame

e: Major eccentricity of the ellipsoid of the Earth

R_e: Length of the semi-major axis of the Earth

R_N: Meridian radius of curvature of the Earth

R_E: Transverse radius of curvature of the Earth

R_{xp}, R_{yp}: Free curvature radiuses

$1/\tau_a$: Turn torsion

$\mathbf{Q}$: Quaternion

q_1, q_2, q_3, q_4: Four components of the quaternion $\mathbf{Q}$

T_{circle}: Period of the circle trajectory in simulation

T_{sshape}: Period of the s-shape trajectory in simulation

A_{sshape}: Amplitude of the s-shape trajectory in simulation

PV: Position and velocity.

References

[1] H. Schneider and N. E. George Philip Barker, *Matrices and Linear Algebra*, Dover Publications, New York, NY, USA, 1989.

[2] A. Gilat, *Matlab: An Introduction with Applications*, John Wiley & Sons, New York, NY, USA, 3rd edition, 2008.

[3] D. H. Titterton and J. L. Weston, *Strapdown Inertial Navigation Technology*, Institution of Engineering and Technology, Stevenage, UK, 2004.

[4] Z. Chen, *Strapdown Inertial Navigation System Principles*, China Astronautic Publishing House, Beijng, China, 1986.

[5] P. S. Maybeck, "Wander azimuth implimentation algorithm for a strapdown inertial system," Air Force Flight Dynamics Laboratory AFFDL-TR-73-80, Tech. Rep., Ohio, USA, 1973.

[6] J. C. Butcher, *Numerical Methods for Ordinary Differencial Equations*, John Wiley & Sons, New York, NY, USA, 2003.

[7] B. Yuan, *Research on Rotating Inertial Navigation System with Four-Frequency Differential Laser Gyroscope*, Graduate School of National University of Defense Technology, Changsha, China, 2007.

[8] I. Y. Bar-Itzhack, "Iterative optimal orthogonalization of the strapdowm matrix," *IEEE Transactions on Aerospace and Electronic Systems*, vol. 11, no. 1, pp. 30–37, 1975.

11

Online Projective Integral with Proper Orthogonal Decomposition for Incompressible Flows Past NACA0012 Airfoil

Sirod Sirisup[1] and Montri Maleewong[2, 3]

[1] *Large-Scale Simulation Research Laboratory, National Electronics and Computer Technology Center, Prathum Thani 12120, Thailand*
[2] *Department of Mathematics, Faculty of Science, Kasetsart University, Bangkok 10900, Thailand*
[3] *Centre of Excellence in Mathematics, CHE, Si Ayutthaya Road, Bangkok 10400, Thailand*

Correspondence should be addressed to Montri Maleewong, montri.m@ku.ac.th

Academic Editor: Javier Otamendi

The projective integration method based on the Galerkin-free framework with the assistance of proper orthogonal decomposition (POD) is presented in this paper. The present method is applied to simulate two-dimensional incompressible fluid flows past the NACA0012 airfoil problem. The approach consists of using high-accuracy direct numerical simulations over short time intervals, from which POD modes are extracted for approximating the dynamics of the primary variables. The solution is then projected with larger time steps using any standard time integrator, without the need to recompute it from the governing equations. This is called the online projective integration method. The results by the projective integration method are in good agreement with the full scale simulation with less computational needs. We also study the individual function of each POD mode used in the projective integration method. It is found that the first POD mode can capture basic flow behaviors but the overall dynamic is rather inaccurate. The second and the third POD modes assist the first mode by correcting magnitudes and phases of vorticity fields. However, adding the fifth POD mode in the model leads to some incorrect results in phase-shift forms for both drag and lift coefficients. This suggests the optimal number of POD modes to use in the projective integration method.

1. Introduction

The projective integration methodology based on the "equation-free" framework has been successfully applied to analyze various kinds of engineering problems [1–3]. The key idea involves using short-duration initial simulations to estimate the time dynamics of the primary variables, without explicitly solving the governing equations either in full or reduced form. The data set obtained from the short-duration simulations can be effectively used to extract proper orthogonal decomposition (POD) basis set, which can then be used to estimate the time dynamics of the needed variables. The most efficient way to extract such a basis set is to employ the snapshots method proposed by Sirovich [4]. This "equation-free" framework has many advantages over the typical Galerkin approach. First, the reduced systems resulting from the Galerkin approach may result in spurious asymptotic states [5, 6]. Also it can be difficult to obtain an explicit form of the reduced system dynamics for a given set of governing equations. Due to these limitations, the equation-free framework of POD methods presents an attractive alternative for solving the problem, since the governing equations are not needed in explicit form during time marching.

The approach presented in this paper is called the "online" method which means that POD bases used in projective integration with "equation-free" framework are constructed from the snapshot data set collected from the full direct numerical simulation (DNS) during each projective integration step. Thus, the approach presented in this paper is different from the "offline" approach presented previously [2, 7]. In the offline approach, the global underlying POD modes are known prior to the projective integration. The online method is more practical than the offline method

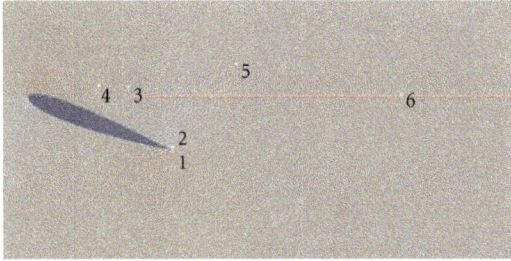

FIGURE 1: Computational domain and location of probes.

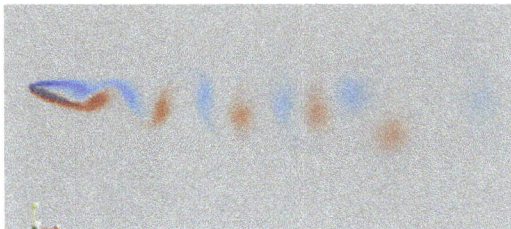

FIGURE 2: Karman vortex street for Re = 400 and t = 13.

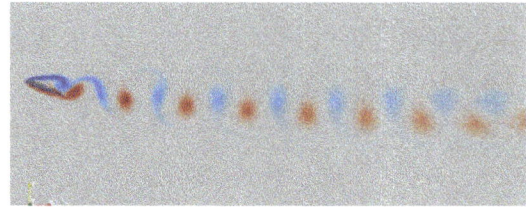

FIGURE 3: Karman vortex street for Re = 800 and t = 13.

2. Methodology

2.1. Navier-Stokes Equations. In this paper, we consider two-dimensional incompressible flow past the NACA0012 airfoil. The governing equations are the set of incompressible Navier-Stokes equations in the form of

$$\nabla \cdot \mathbf{v} = 0,$$
$$\frac{\partial \mathbf{v}}{\partial t} + (\mathbf{v} \cdot \nabla)\mathbf{v} = -\nabla \mathbf{p} + \frac{1}{\text{Re}}\nabla^2 \mathbf{v}, \tag{1}$$

where $\mathbf{v} \equiv \mathbf{v}(t, \mathbf{x})$ is the velocity field and $\mathbf{p} \equiv \mathbf{p}(t, \mathbf{x})$ the pressure term for $\mathbf{x} \in \Omega \subset R^2$ and $t \in R^+$. The equations are in dimensionless forms scaled by a characteristic velocity U and a characteristic length L. The Reynolds number is defined as Re $= UL/\nu$, where ν is the kinematic viscosity. The computational domain is shown in Figure 1. The object is the NACA0012 airfoil positioned at an angle of attack of $20°$ with respect to the horizontal axis. The points labeled 1–6 are the locations of probes to check the accuracy and efficiency of our method as discussed later. We impose the boundary conditions as follows. Uniform inflow is prescribed on the left boundary while periodic boundary conditions are specified on the top and bottom boundaries. We apply outflow condition on the right boundary, and no-slip condition on the surface of the airfoil. The full direct numerical simulations (DNS) are performed by solving the Navier-Stokes equations using the spectral/hp element method [12]. We have checked that the computational domain is long enough so that there is no effect from truncated domain. To obtain high accuracy of solutions, we apply the ninth-order Jacobi polynomial basis in each element. The full DNS results can then be regarded as the exact solutions in our problem.

because we cannot know in advance the underlying POD basis as well as the appropriate number of POD modes used in the projective integration approach. Besides, in order to obtain that information, the representations of each POD mode should be known so that the number of POD modes can be choosen appropriately. To this end, we also propose a method and provide such information for the current projective integration.

In this paper, we apply the POD-assisted projective integration based on the online approach as an accelerated scheme for simulating incompressible fluid flows past the NACA0012 airfoil. This is one of the most popular two-dimensional flow benchmark problems. There are many numerical and experimental results [8, 9] that provide us data to investigate the accuracy and efficiency of the present scheme. Recently, Siegel et al. [10] use an alternative approach of artificial neural network to study the problem. A similar problem of NACA0015 airfoil has also been studied by the POD based method [11]. The primary objective of our studies is to investigate numerically the efficiency of the POD-assisted projective integration based on the online method. Another objective is to explore the functions of each POD mode employed in the POD-assisted projective integration.

This paper is organized as follows. Data acquisition and the methodology of the projective integration method are presented in Section 2. Numerical results from the application of the method to incompressible flows past the NACA0012 airfoil for both transient and periodic cases are shown in Section 3. The functions for each dominant POD mode in the online method are demonstrated in the POD mode interplay in Section 4. Finally, conclusions are made in Section 5.

2.2. Proper Orthogonal Decomposition. In the equation-free or Galerkin-free framework, proper orthogonal decomposition (POD) is the main procedure used to extract the global behavior of flow fields (snapshots) during calculations in time. Some basic concepts of POD have been summarized in this subsection.

The POD procedure extracts empirical orthogonal features from any ensemble of data from the full set of DNS. The POD method is linear procedure that produces a reduced basis set which is optimal in L^2 sense. For continuous

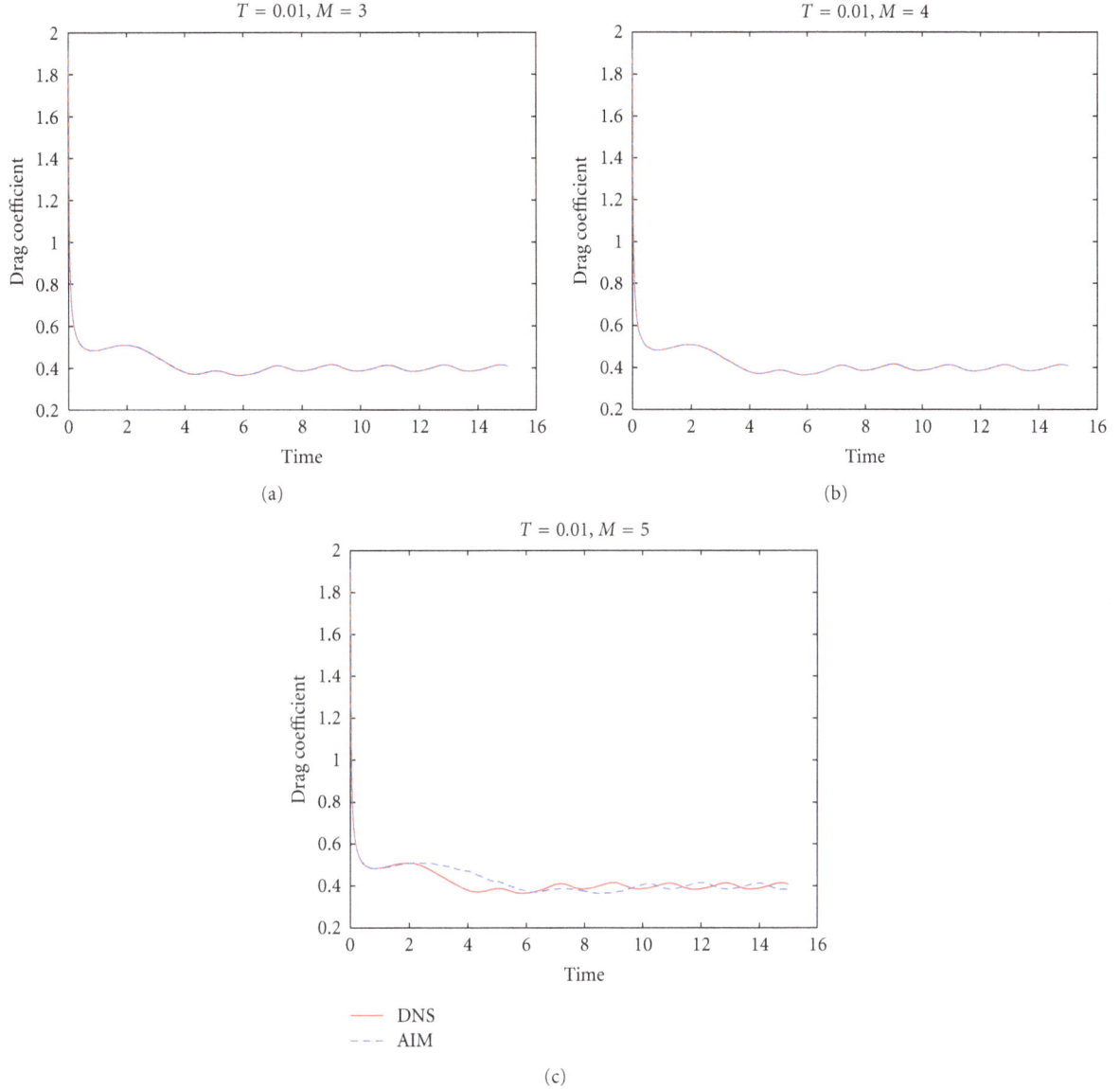

FIGURE 4: Transient dynamics: Re = 400, drag coefficient for $T = 0.01$ with various number of M. From top to bottom: $M = 3, 4,$ and 5.

problems [13], one can assume that the flow field $\mathbf{u}(t, \mathbf{x})$ can be represented as

$$\mathbf{u}(t, \mathbf{x}) = \sum_{k=0}^{\infty} a_k(t)\phi_k(\mathbf{x}), \qquad (2)$$

where $\{\phi_k(\mathbf{x})\}$ is the set of POD basis determined by the eigenvalue problem

$$\int_A C(t, t')a_k(t')dt' = \hat{\lambda}_k a_k(t), \quad t \in A, \qquad (3)$$

where $\{a_k(t)\}$ is the set of temporal modes, A is a specified time interval, and $C(t, t')$ is the correlation function defined by

$$C(t, t') = \int_{\Omega} \mathbf{u}(t, \mathbf{x}) \cdot \mathbf{u}(t', \mathbf{x})d\mathbf{x}. \qquad (4)$$

The POD basis is thus defined by

$$\phi_k(\mathbf{x}) = \int_A a_k(t)\mathbf{u}(t, \mathbf{x})dt, \quad \forall k. \qquad (5)$$

The nonnegative definiteness of the correlation function (4) allows us to order the eigenvalues and the corresponding POD modes by $\hat{\lambda}_k \geq \hat{\lambda}_{k+1}$. The POD expansion coefficients for (2) can be found by $a_k = \langle \mathbf{u}(t, \mathbf{x}), \phi_k(\mathbf{x}) \rangle$. Here $\langle , \rangle$ denotes the inner product operator in L^2 sense.

2.3. The Equation-Free POD-Assisted Model. The equation-free form of the POD-assisted model in conjunction with the projective integration technique allows us to integrate numerical solutions forward in time by using two main processes: restriction and lifting. We introduce the definitions of

$$T = 0.01, M = 3$$

$$T = 0.01, M = 4$$

(a)

(b)

$$T = 0.01, M = 5$$

—— DNS
- - - AIM

(c)

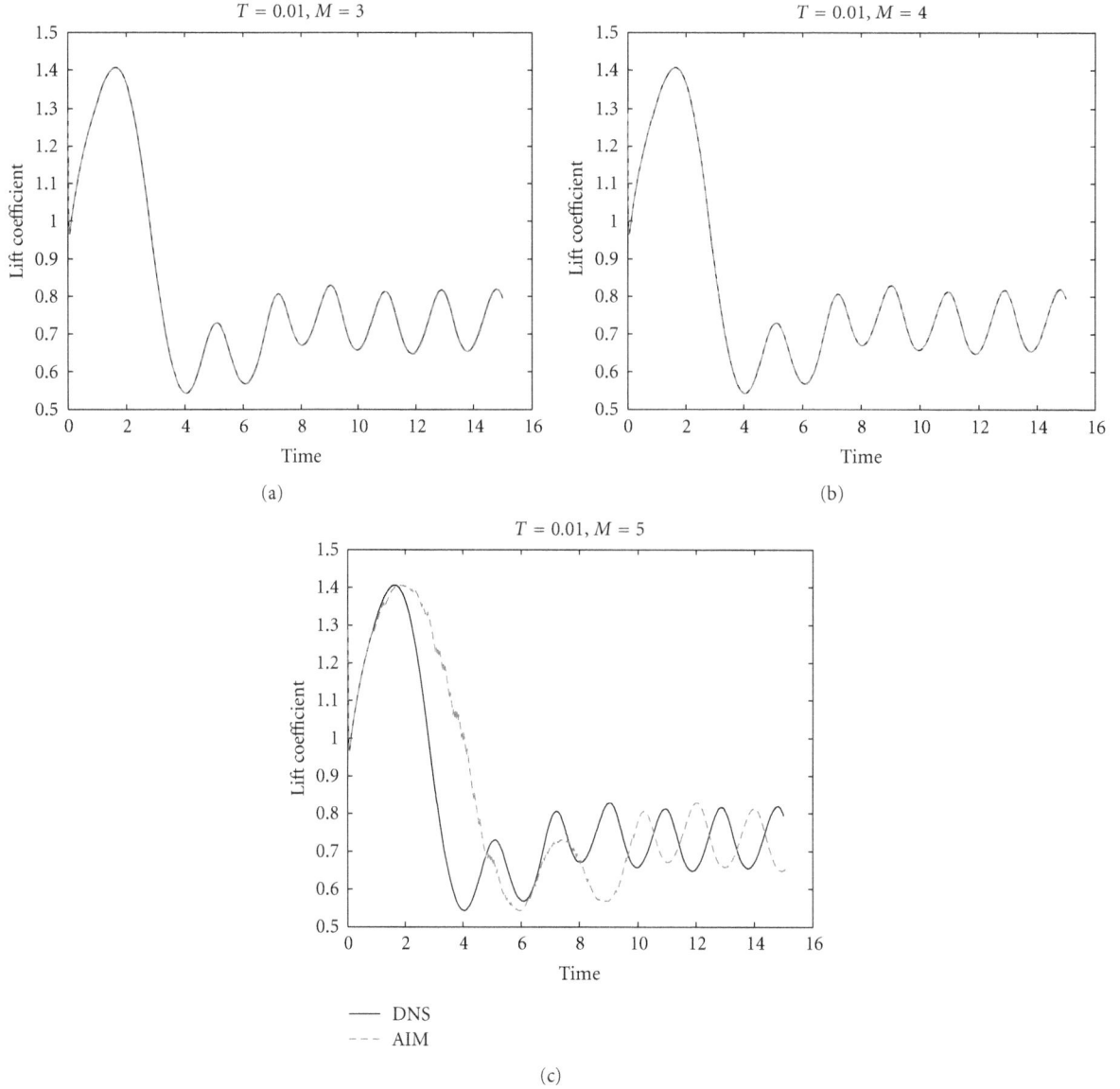

FIGURE 5: Transient dynamics: Re = 400, lift coefficient for $T = 0.01$ with various number of M. From top to bottom: $M = 3, 4,$ and 5.

these processes by two operators: a *restriction operator* $\mathcal{R}$ and a *lifting operator* $\mathcal{L}$ such that

$$\mathbf{a}(t) = \mathcal{R}\mathbf{u}(t, \mathbf{x}) \equiv \{\langle \mathbf{u}(t, \mathbf{x}), \phi_k(\mathbf{x}) \rangle, t \in A, \forall k\}, \quad (6)$$

$$\mathbf{u}(t, \mathbf{x}) = \mathcal{L}\mathbf{a}(t) \equiv \sum_{k=0}^{\infty} a_k(t)\phi_k(\mathbf{x}). \quad (7)$$

In a discrete computation, we can approximate (7) by K-term of POD expansion. The representation can be expressed as

$$\mathbf{u}_K(t, \mathbf{x}) = \sum_{k=1}^{K} a_k(t)\phi_k(\mathbf{x}), \quad (8)$$

and the *truncated restriction* and *truncated lifting* operators are defined as $\mathcal{R}_K$ and $\mathcal{L}_K$, respectively. The convergence of the K-term POD expansion is the form

$$\|\mathbf{u} - \mathbf{u}_K\| \longrightarrow K^{-\gamma}, \quad \text{as } K \longrightarrow \infty, \quad (9)$$

where convergence rate, $\gamma > 0$, is large enough.

In general, we can write the evolution of the POD coefficient $\mathbf{a}(t)$ by

$$\frac{d\mathbf{a}}{dt} = \mathbf{g}(\mathbf{a}(t)), \quad (10)$$

where the explicit form of $\mathbf{g}$ remains unknown. Thus, the derivative of POD coefficients must be approximated, rather than explicitly evaluated, to march forward in time. It has been noted that one can find an explicit form of $\mathbf{g}$ by projecting the governing PDEs onto the POD modes [6, 14–18]. It is a traditional Galerkin projection approach that is

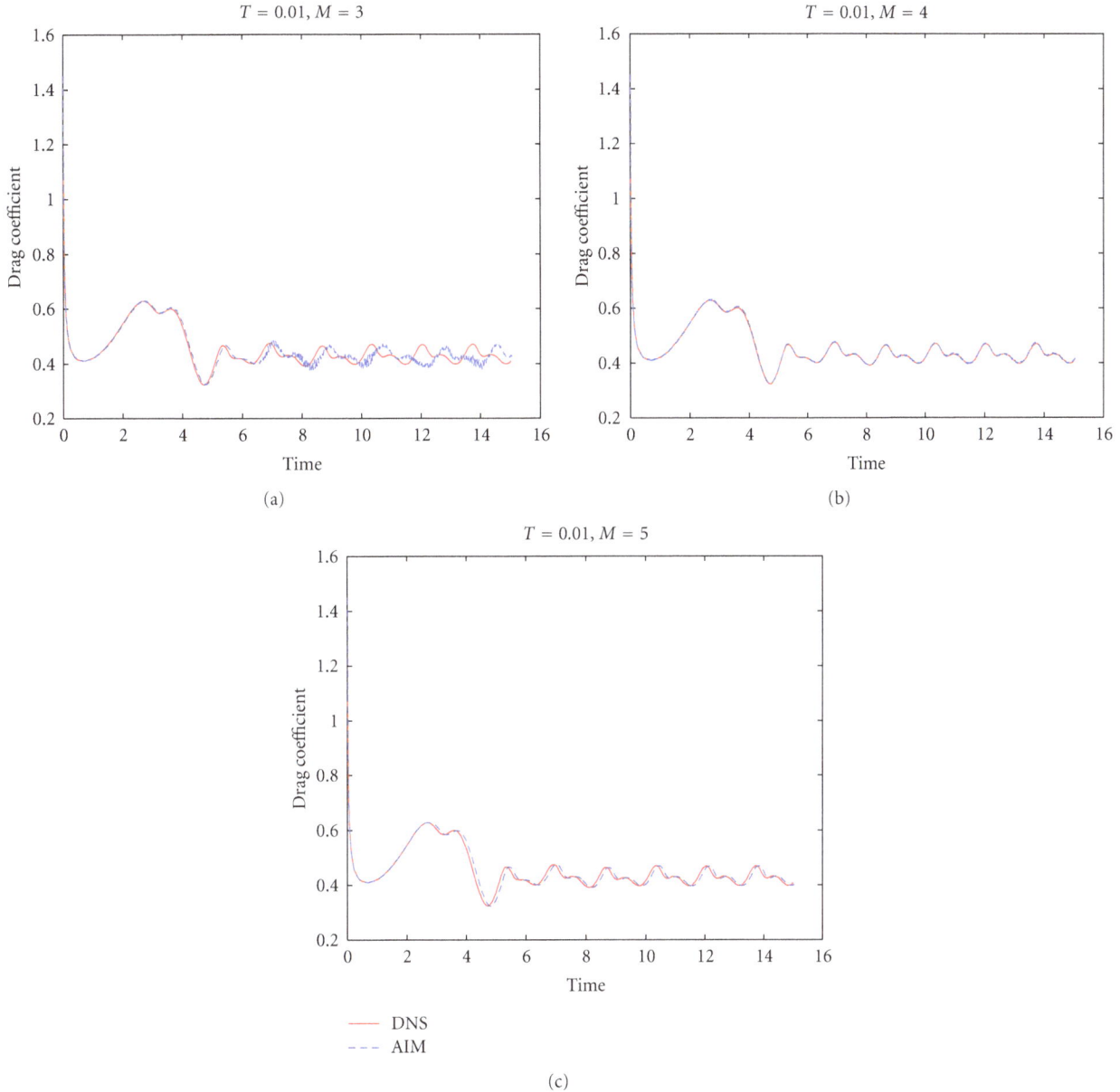

Figure 6: Transient dynamics: Re = 800, drag coefficient for $T = 0.01$ with various number of M. From top to bottom: $M = 3, 4$, and 5.

different from our present method. We will provide more details of restriction and lifting processes for solving the incompressible flows in the next subsections.

2.3.1. Restriction and Lifting. We employ the *snapshot method* [4] to extract the set of POD bases, $\{\phi_k(\mathbf{x})\}$, from the ensemble of previously saved solutions. In our study, these solutions or snapshots of flow fields at each time step are obtained from the full DNS in a small finite time interval. We refer to this computation as a fine-scale computation. Then we project this set of snapshots to the POD bases and compute forward in time on the slow manifold space. We call this step a projective step that enables us to perform relatively large time step on the manifold space when

comparing with the time step of the fine scale computation. Hence, the projective step is sometimes called a coarse-scale computation.

In the fine-scale time interval, the solutions or snapshots $\mathbf{u}(t_i, \mathbf{x})$ at time t_i are obtained by solving numerically the incompressible Navier-Stokes equations through the spectral/hp element method, where $t^n \leq t_i \leq t_c^n$, $i = 1, \ldots, n_f$, where t_c and n_f denote the final time value and the number of time steps of fine-scale computations, respectively. From (5), the POD bases are then determined discretely by

$$\phi_k(\mathbf{x}) = \sum_{i=1}^{n_f} a_k(t_i) \mathbf{u}(t_i, \mathbf{x}) \, dt, \quad \forall k, \qquad (11)$$

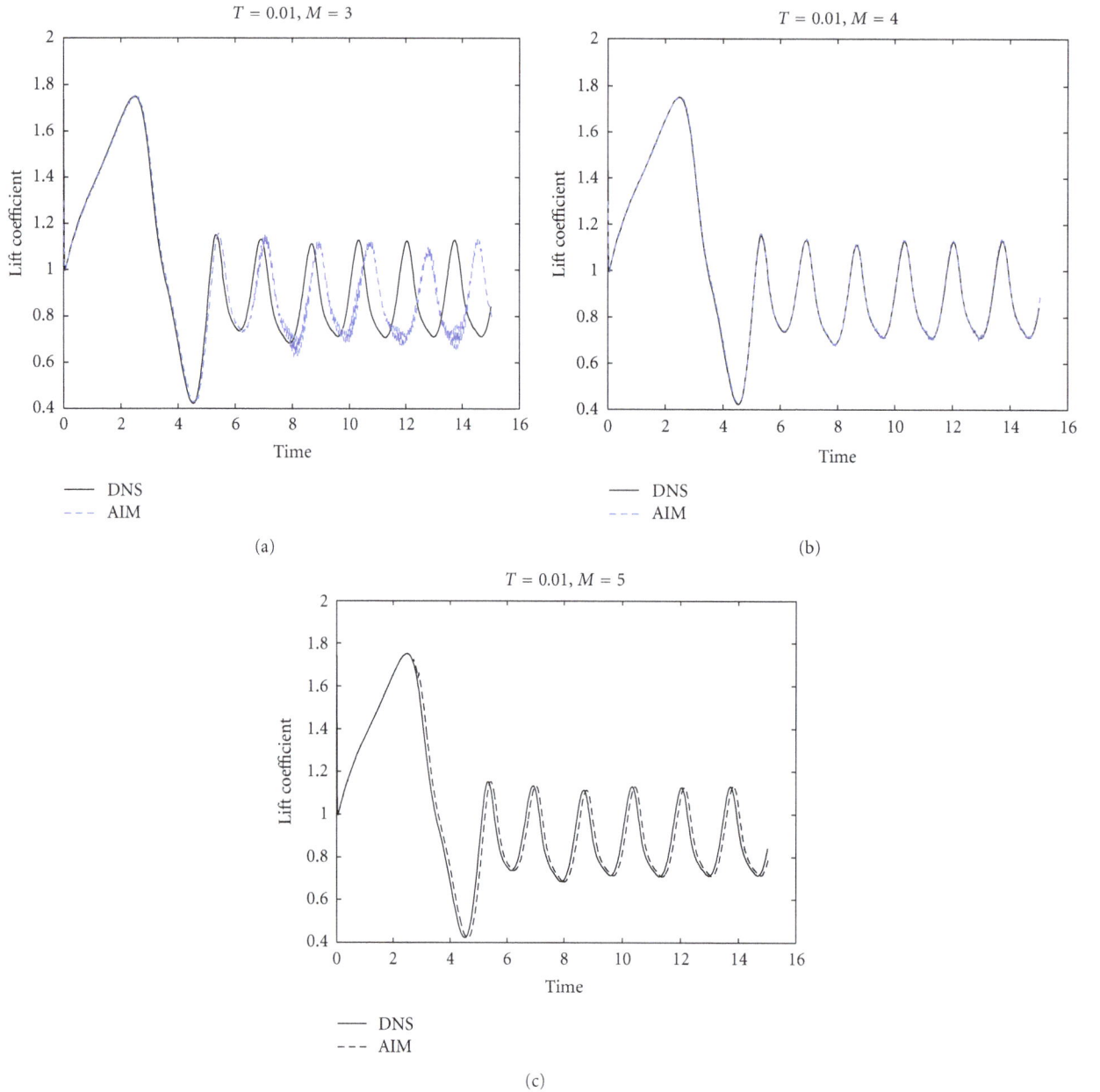

FIGURE 7: Transient Dynamics: Re = 800, lift coefficient for $T = 0.01$ with various number of M. From top to bottom: $M = 3, 4$, and 5.

where $\{a_k\}$ are obtained by solving the correlation matrix (3). Once the POD basis functions are determined from (11), we can restrict any solution $\mathbf{u}(t, \mathbf{x})$ for any given t to obtain the corresponding POD coefficients a_k by (6). The derivative of the POD coefficients can then be approximated and used to march forward in time by the projective integration technique (coarse-scale). Details are summarized in the next subsection. After the projective step is completed, the lifting procedure is required to return the computations back to a fine-scale resolution. It is the reverse process of the restriction, that is, for a given set of computed POD coefficients at time t, we can reconstruct the corresponding solution by using (7).

2.3.2. *Projective Integration.* The projective integration procedure is described as follows.

(i) Approximate the RHS of (10) at $t = t_c^n$ using

$$\mathbf{g}(t_c^n) = \sum_{j=0}^{n_e} \alpha_j \mathbf{a}(t_j) = \frac{d\mathbf{a}}{dt}(t_c^n) + O\left(\delta t^{J_f}\right), \quad (12)$$

where $1 \le n_e \le n_f$, $t_j = t_c^n + j\delta t$, and J_f denotes the order of approximation. Here $\{\alpha_j\}_{j=0}^{n_e}$ is a set of consistent coefficients such that $\sum \alpha_j f(t_j) = df/dt(t_c^n) + O(\delta t^{J_f})$.

(ii) Once the RHS of the typical reduced-order model (10) is estimated numerically, one can use any standard ODE solver to integrate the numerical solution in time. For

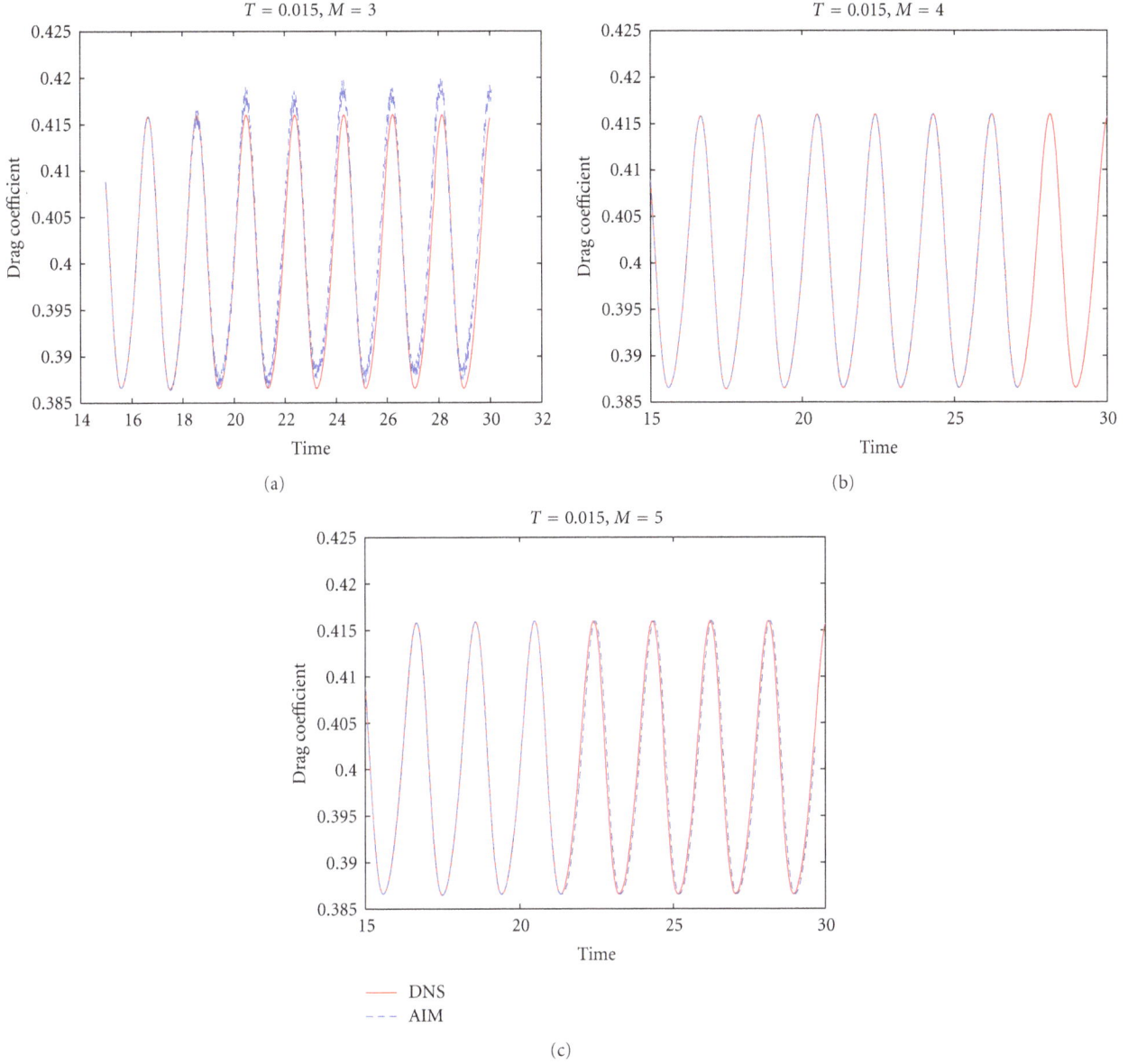

FIGURE 8: Periodic case: Re = 400. Computed drag coefficient for $T = 0.015$ with various number of M. From top to bottom: $M = 3, 4$, and 5.

instance, given a coarse time step $\Delta t_c \equiv n_c \delta t$, where $n_c \geq 1$, such that $t^{n+1} = t_c^n + \Delta t_c = t^n + (n_f + n_c)\delta t$, the single step forward Euler method takes the form

$$\mathbf{a}(t^{n+1}) = \mathbf{a}(t_c^n) + \Delta t_c \cdot \mathbf{g}(t_c^n) + O(\Delta t_c^2). \tag{13}$$

It should be noted that other higher-order explicit integration schemes (and possibly implicit ones) can be used to achieve better accuracy and/or stability properties. For instance, one can use the following scheme:

$$\mathbf{a}(t^{n+1}) = \mathbf{a}(t_c^n) + \sum_{k=1}^{J_c} \frac{(\Delta t_c)^k}{k!} \frac{\partial^{(k-1)}}{\partial t^{k-1}} \mathbf{g}(t_c^n) + O\left(\Delta t_c^{J_c+1}\right). \tag{14}$$

The higher order temporal derivatives of $\mathbf{g}(t)$ are approximated in a way similar to (12). It can be seen that (14) is

a high-order single-step method. In this work, we perform the projective step using forward Euler method. This scheme provides smaller oscillations than using higher order scheme such as (14).

2.4. Equation-Free POD-Assisted Projective Integration.
The overall steps of equation-free POD-assisted projective integration are summarized in this section.

Generally, one global time step of this method is composed of two types of time scales which are fine-scale and coarse-scale time steps, respectively. We start calculations by using fine-scale integrator with n_f time steps and then perform coarse-scale integration for n_c steps. The fine-scale integrator can be performed by any accurate methods whereas the coarse-scale integrator is chosen to be the

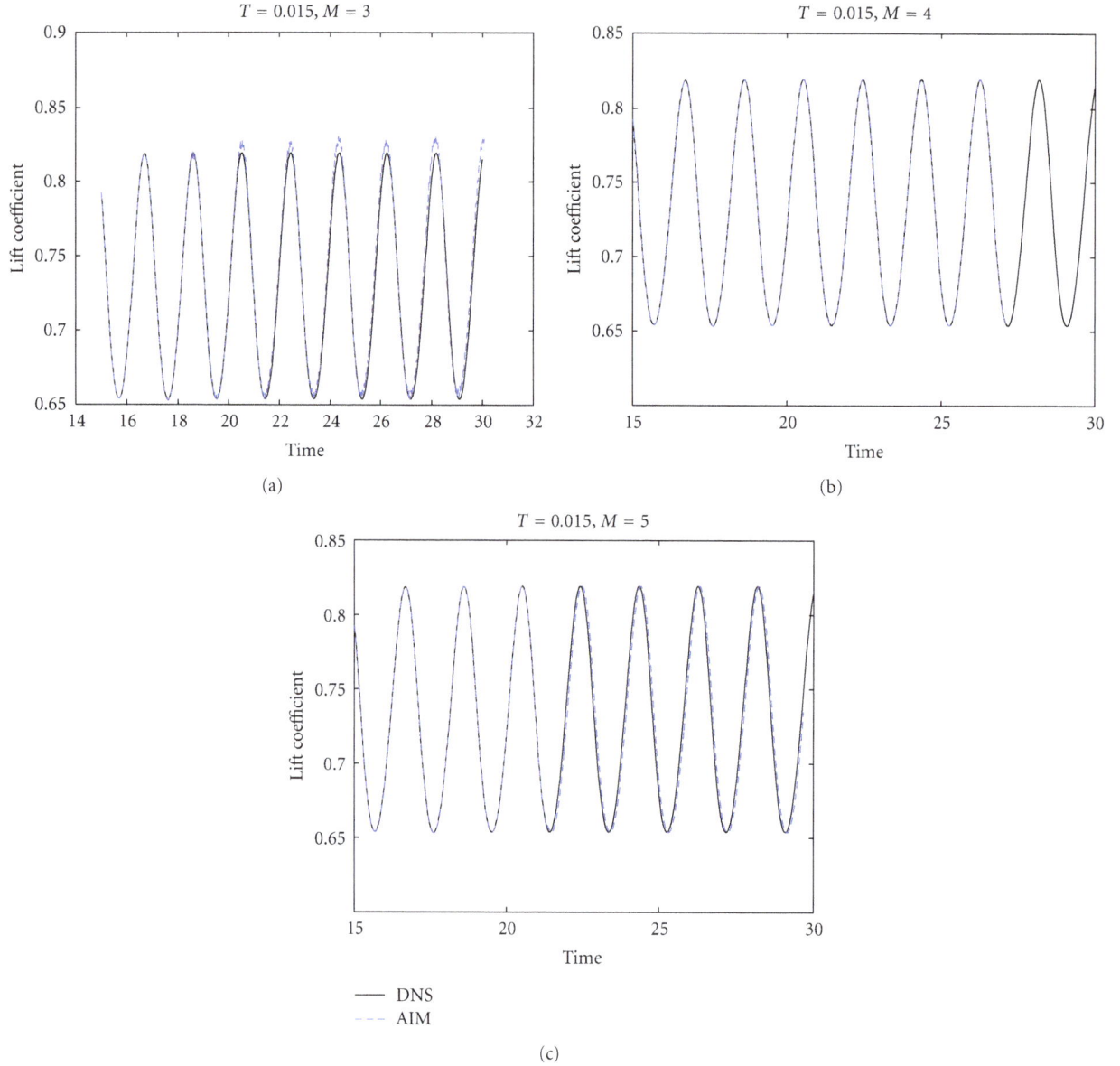

FIGURE 9: Periodic case: Re = 400. Computed lift coefficient for $T = 0.015$ with various number of M. From top to bottom: $M = 3, 4$, and 5.

forward Euler's method in this study. A single step of the equation-free POD-assisted projective integration method, from $t = t^n$ to $t = t^{n+1}$, consists of the following main steps.

(i) *Fine-scale computation:* we perform this step by the DNS. The DNS provides numerical solutions at each time step δt, and so at $t_c^n = t^n + \Delta t_f$, where $\Delta t_f = n_f \delta t$, and n_f is the number of fine-scale time steps.

(ii) *Restriction:* derive the POD coefficients using the previously saved solutions. Here, we use the method of snapshot for deriving POD modes and their corresponding coefficients. We also estimate the time derivatives of each POD mode coefficient at this stage.

(iii) *Projective integration:* time integration using the Euler method is performed on the POD hyperspace with

time step T. This step is called the coarse-scale computation because we can use a relatively large T when comparing with Δt_f. Thus the efficiency of the method depends directly on how large coarse-time step T can be.

(iv) *Lifting:* the projected solution at time $t^{n+1} = \Delta t_f + T$ is lifted from POD hyperspace to the physical domain.

This completes the single time step of the equation-free POD-assisted projective integration method. Finally, we repeat the whole processes until the final time is reached.

3. Numerical Results

In this section, we investigate the accuracy and efficiency of the projective integration method for solving incompressible

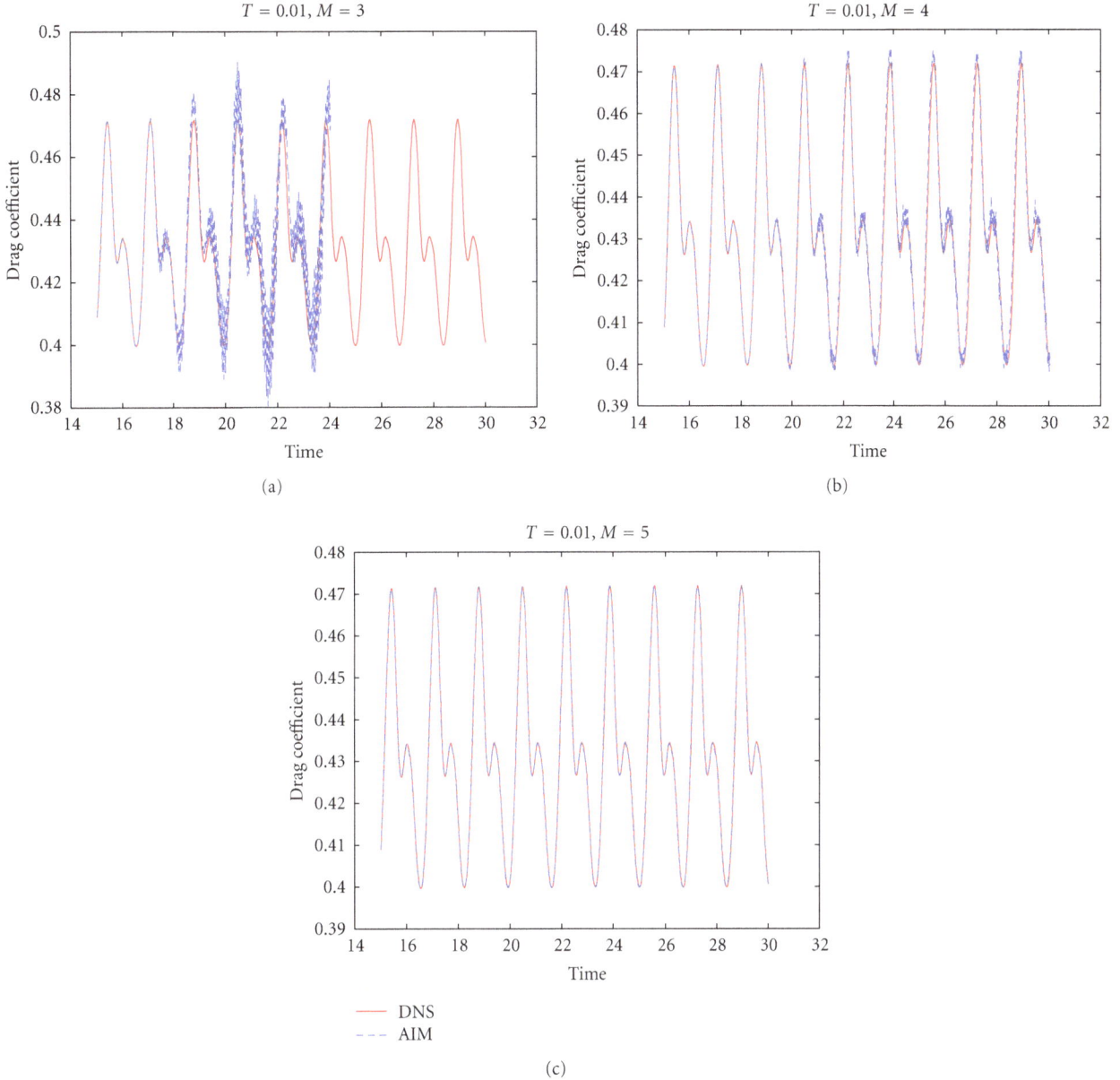

FIGURE 10: Periodic case: Re = 800. Computed drag coefficient for $T = 0.01$ with various number of M. From top to bottom: $M = 3, 4$, and 5.

flows. The purpose is to explore the representations of online POD modes. So, we restrict our investigations to two cases of Reynolds number, Re = 400 and 800. For each case, we separate flow pattern into two types: transient and quasi-periodic flows. We have checked the accuracy of our DNS results and found that they are in good agreement with the experimental results reported by Williamson et al. [19]. Thus the DNS results in our simulations can be regarded as the exact solutions which will be used to validate the accuracy of the numerical results obtained from the equation-free POD-assisted projective integration method.

For all simulations, there are two time scales: fine time scale, δt, and coarse time scale, T. The total time step of one projective step is $n_f \delta t + T$, where n_f is the number of

fine-scale steps. We have set five snapshots for each projective step, thus the maximum number of POD modes in one projective time step is five as well. This number is enough to capture the dynamics of solutions because flow dynamics change rather slightly in one projective step. It is different from the offline method where the number of POD modes used to describe the entire flow dynamics must be much greater.

We calculate time dependent solutions from $t = 0$ to 30. The unsteady flows can be classified into two cases: transient case when $0 \leq t < 15$, and periodic case with Karman vortex street when $15 \leq t \leq 30$. Note that periodic flow in both cases of Re occurs before $t = 15$, but we classify the flow pattern by this value to make sure that the flow is entirely

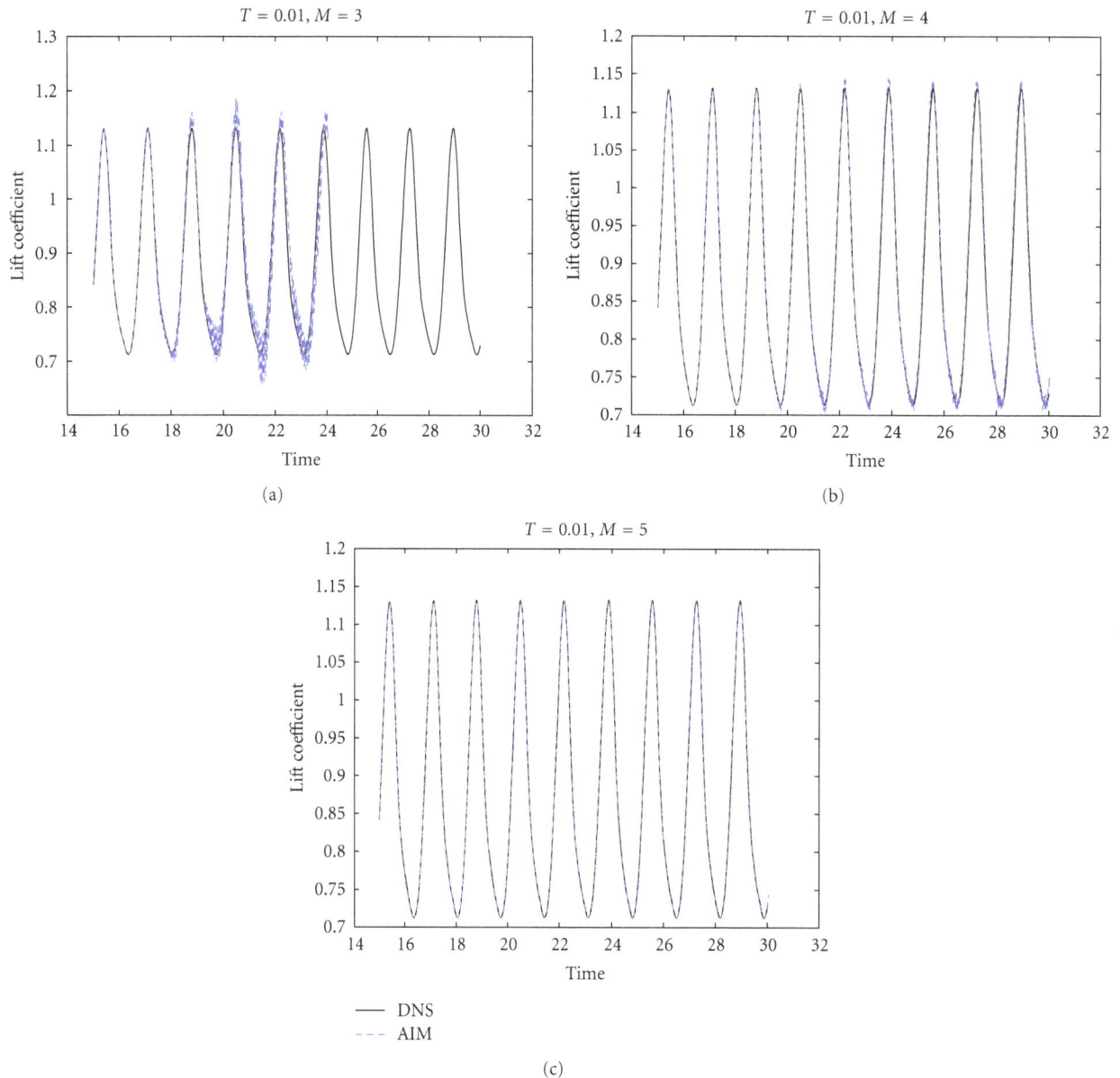

FIGURE 11: Periodic case: Re = 800. Computed lift coefficient for $T = 0.01$ with various number of M. From top to bottom: $M = 3, 4$, and 5.

periodic. Vorticity plots of DNS results when $t = 13$ for Re = 400 and 800 are shown in Figures 2 and 3, respectively. Efficiency of the numerical scheme for both transient and periodic flows will be shown in the next sections.

3.1. Transient Dynamics. In this section, we apply the equation-free POD-assisted projective integration method to numerically solve unsteady flow problems on $0 \leq t < 15$. We set $\delta t = 0.0005$. The accuracy and efficiency of the present method can be observed by varying the values of coarse time step and the number of POD modes, M.

Figures 4 and 5 show drag and lift coefficients for Re = 400 using various numbers of POD modes in the simulations. It can be seen that flows become periodic when $t \approx 8$. The label of AIM (approximated inertial manifold)

in these figures refers to the results obtained by the projective integration method. We have set coarse time step as $T = 0.01$. This time step is very large (twenty times as much) when comparing with the fine-scale. To extract POD bases, we have used five snapshots from DNS where each snapshot is collected at 0.01 time step. So, we run on fine-scale system 0.05 time steps. The last two snapshots are used in the projective step by the forward Euler method. Thus, we have run on the fine-scale system by the DNS with 0.05 time steps and on the slow manifold (coarse-scale) with 0.01 time steps.

Next, we investigate the appropriate number of POD modes by trial and error. It is found that solutions by using $M = 3$ or 4 are in good agreement with the DNS results. However, some phase shifts in both drag and lift coefficients appear for $M = 5$. Thus, the appropriate number of POD modes should be 3 or 4. Adding just one more

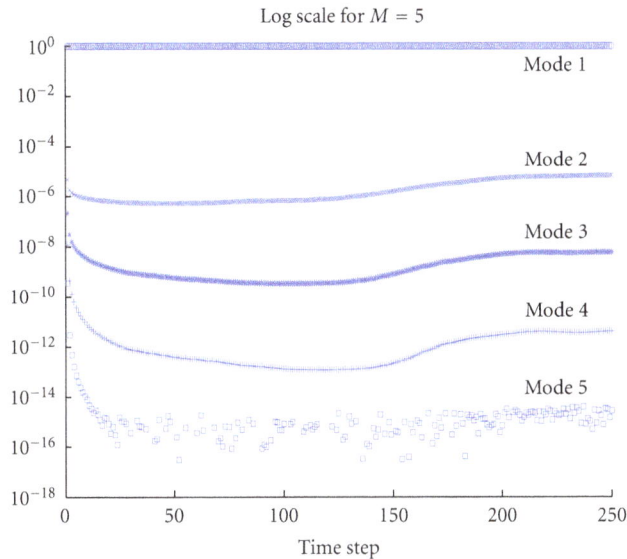

FIGURE 12: Normalized energy versus convective time (projective time step) in transient case, Re = 400.

POD mode in the restriction step results in an incorrect phase-shift pattern. These results motivate us to further investigate the representations of each POD mode to get better understanding of how to choose the most appropriate number of POD modes. Some observations will be shown in Section 4.

The cases of smaller coarse time step, $T = 0.005$ and 0.007, are also investigated. Numerical results are in good agreement with the DNS results for $M = 3$ and 4. The profiles of drag and lift coefficients are the same as those for the case of $T = 0.01$. But the results of using $M = 5$ and $T = 0.005$ still produce some phase shifts. This numerical error comes from the POD expansion part. The numerical results are clearly incorrect when we use larger coarse-scale step, $T = 0.02$. The numerical solutions diverge very rapidly for $M = 4$ and 5. When we set $T = 0.015$, the numerical method still provides accurate solutions for $M = 4$. Thus, the highest efficiency for simulating this problem occurs when we set $0.01 < T < 0.015$ and $M = 4$.

Figures 6 and 7 show drag and lift coefficients for Re = 800 with various number of POD modes. In this case, these coefficients are higher than the values in the case of Re = 400. Flows become periodic when $t \approx 6$. We have set coarse time step to $T = 0.01$. Accurate numerical results are obtained by using only $M = 4$. Using not enough POD modes, for example $M = 3$, produces some oscillations. On the other hand, increasing just one more POD mode in solution expansion has affected the phase-shift pattern. We have checked for the cases of smaller T. It is found that $M = 4$ provides accurate results. But as we set $T = 0.015$, numerical solution diverges rapidly for any number of POD modes constructed. This shows the effects of truncation error in the forward Euler method that becomes a dominant factor causing the instability of the numerical scheme. For Re = 800, we conclude that the projective integration method has the highest efficiency when we set $0.01 < T < 0.0125$ and $M = 4$.

3.2. Periodic Dynamics. In this section, periodic flows when $15 < t < 30$ are investigated. The main propose is to explore the efficiency of the online Galerkin-free POD-assisted projective integration method in order to capture the periodic pattern of vorticity downstream. For Re = 400, computed drag coefficients when $T = 0.015$ are shown in Figure 8. The projective integration results (AIM) are in good agreement with the DNS results for $M = 4$ and 5. Using not enough POD modes, for example $M = 3$, results directly to the stability of numerical solutions. Similar results of computed lift coefficients are shown in Figure 9. Only using $M = 4$ and 5 provides accurate results. These observations are relatively different from the transient case where increment of M results in some phase-shift patterns. This suggests computations in the case of periodic flow stay within some range of the attractor, in constrast to the case of transient flow.

For Re = 800, computed drag and lift coefficients when $T = 0.01$ are shown in Figures 10 and 11. Only using $M = 5$, provides accurate results for both drag and lift coefficients. The numerical scheme is more stable as M increased. But, solutions diverge rapidly as we increase time step to $T = 0.015$. Increment of M cannot improve the stability of the numerical scheme. Truncation error from coarse-scale time step is now a dominant factor destroying the stability.

4. Interplay of POD Modes

In this section, the representations of each individual POD mode in the online POD-assisted projective integration method are investigated. Because it is an online process, it implies that we must investigate the POD modes during time integration. This approach is different from the traditional reduced-order model by the Galerkin projection. This shows one advantage of the present scheme allowing one to gain more understanding of the dynamical representations for each POD mode.

In our problem, the normalized energy for each POD mode is shown in Figure 12. These normalized values are calculated from the set of eigenvalues extracted from the restriction stage of one projective time step. Log-scale plot shows dynamically the POD energy in time. The first POD mode has the maximum energy of order $O(1)$ whereas the fifth POD mode has the minimum energy of $O(10^{-15})$. The first POD mode captures the most energy in every projective step. Higher POD modes have functions for improving solution accurracy. The energy plot shows clearly that, in energy sense, the combination of the first four POD modes is enough to build up the online model. The energy of the fifth POD mode is very small. Also, this POD mode produces some oscillations in the simulations. It is shown by phase-shift appearance of the drag and lift coefficients in Figures 4 and 5.

In order to understand the representations for each POD mode in time, the interplay among the first five POD modes is investigated. We proceed as follows. We compute solutions using only $M = 1$, for example, only the first POD mode included in the projective integration. Next, we calculate

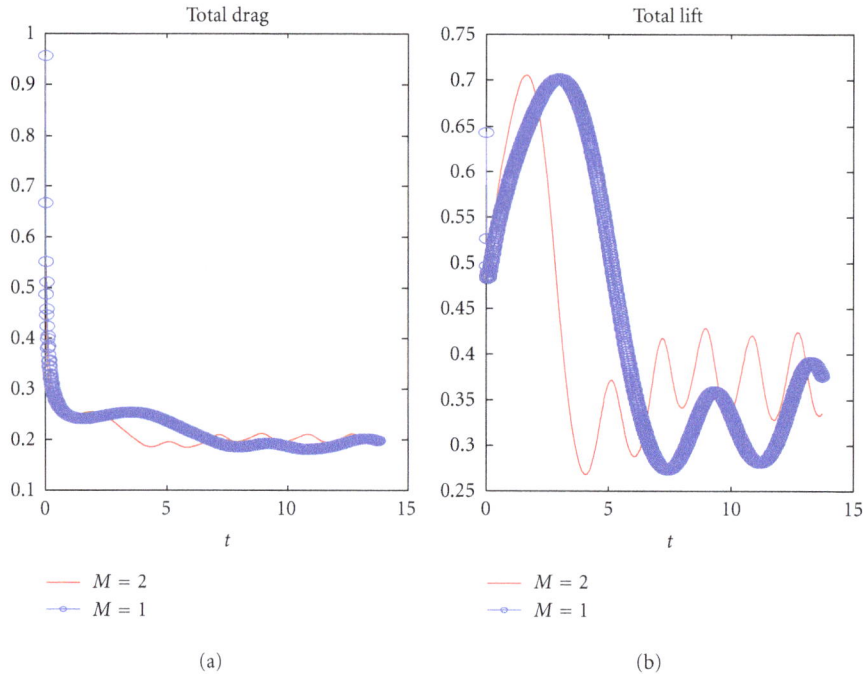

(a)

(b)

FIGURE 13: Drag and lift coefficients for Re = 400 when using $M = 1$ and $M = 2$.

solutions using M = 2. The difference between the first and second calculations is regarded as the interplay of the first two POD modes. Similar processes can be performed for investigating the interplay of other higher POD modes. We have set $T = 0.005$ in all calculations to minimize numerical errors. Observed results can be summarized as follows.

4.1. The First and Second POD Modes. Drag and lift calculations for the case of projective integration with $M = 1$ and $M = 2$ are shown in Figure 13. Comparing with the DNS results in Figures 4 and 5 shows that using only the first POD mode produces incorrect result. It can be seen that the second POD mode assists the first POD mode by recovering the frequency and amplitude of drag and lift forces. The differences of individual components in the lift and drag coefficients (i.e., pressure and viscous components) are also shown. The different values are relatively large especially at the early stage of the transient case $(0 < t < 5)$.

Apart from investigating the overall variables like the lift and drag coefficients, we also investigate the local differences in the primary results obtained from the projective integration method with $M = 1$ and $M = 2$. Plots of time series for u, v, and p at various probe numbers, #1, #2, and #3 (see Figure 1) are presented in Figure 14. The second POD mode assists the first POD mode by increasing frequency but representing the same order of amplitude similar to that observed in the lift and drag coefficients. The representations of the second POD mode are similar in all of these three probes.

4.2. The Third POD Mode. We now focus on the third POD mode. Comparing the results in Figure 15 with those in

Figure 13, we see that inclusion of the third mode brings both phase and amplitude of the lift and drag coefficients in good agreement with DNS results. We see that the third POD mode does provide a significant contribution to the solution in terms of further correcting phase and amplitude. The differences in drag and lift coefficients between $M = 2$ and $M = 3$ are smaller than those values in the cases of $M = 1$ and $M = 2$. The first three POD modes are very important in the online method because they can capture all basic flow behaviors.

4.3. The Fourth and the Fifth POD Modes. We perform the same investigations on higher POD modes. It is found that adding one more POD mode which is the fourth POD mode into the solution expansion of $M = 3$ does not provide any significant improvement. The total drag and lift by applying $M = 3$ and $M = 4$ are shown in Figure 16. Sets 1 and 2 refer to using $M = 3$ and 4, respectively. Because all major characteristics of flow solution are largely captured by the first three POD modes, including too many POD modes in the projective method does not guarantee the improvement in solution accuracy. This can be seen in Figure 17. The inconsistency can be explained by the fact that the approximations of the higher mode derivatives are inaccurate. They exhibit small-amplitude high-frequency oscillatory bahavior, which results in the phase-shift pattern seen in the solution when we include the fifth POD mode.

To reveal the representations of each POD mode by the online method, we also show some transient simulations when using various POD modes. Results are presented in Figures 18–21. The vorticity plots in the case of Re = 400 at time $t = 1.9, 6.3, 10.7,$ and 12.9 are shown in Figure 18.

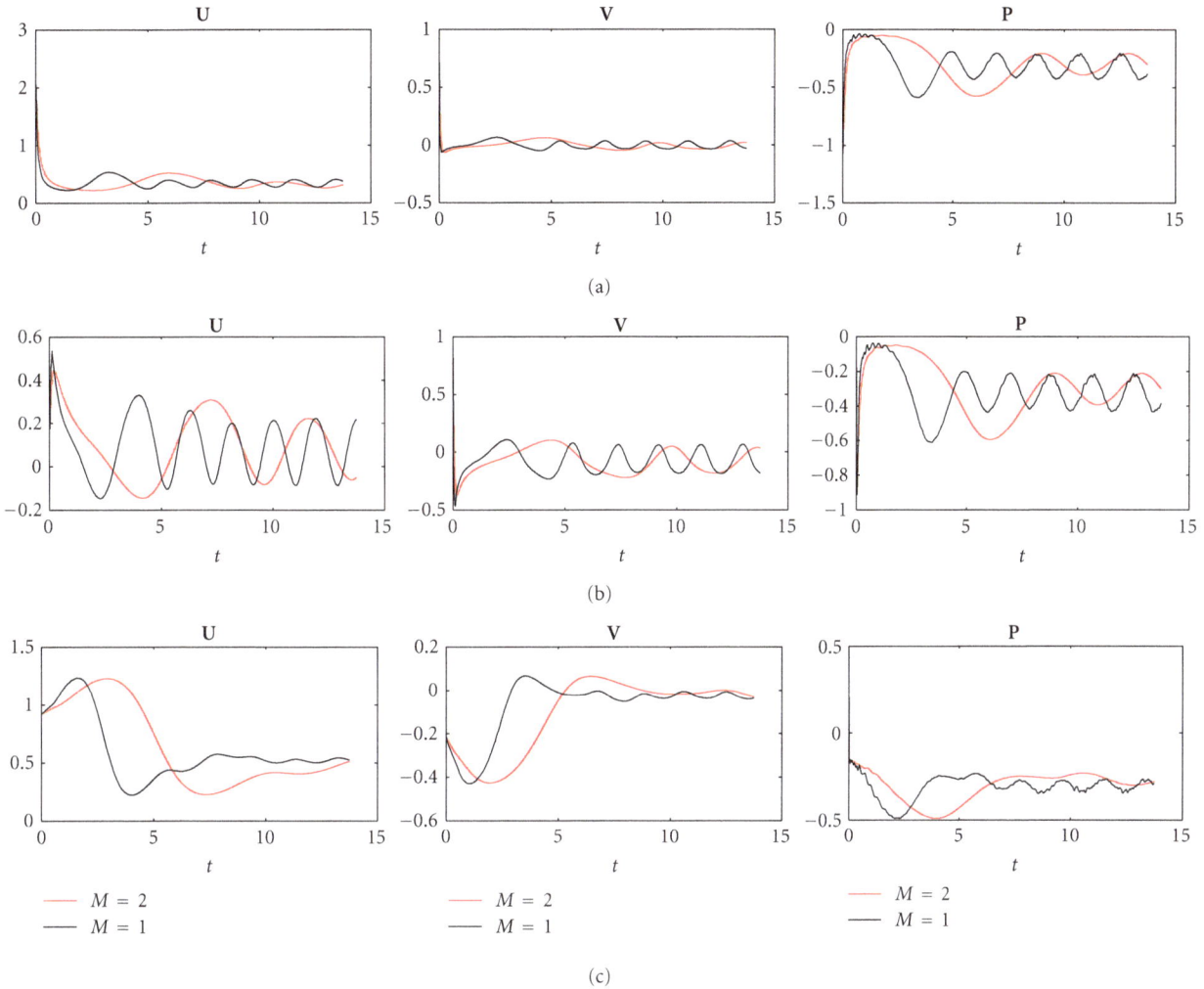

FIGURE 14: Time series plot of the primary variables. From top to bottom: Probes #1, #2, and #3.

It can be seen that the first POD mode can capture only basic flow behaviors, but its representation is inaccurate. It represents as the mean mode due to its highest contained energy (see Figure 12). Using the first two POD modes is shown in Figure 19. The second POD mode assists the first POD mode by improving the strength and frequency of vorticity field downstream. When comparing with the DNS results in Figure 2, flow patterns are nearly the same. There are some small differences in vorticity magnitude. However, if we increase the number of POD modes to $M = 3$, the results from the online projective method are in good agreement to those results from the DNS (see Figures 20 and 2 at approximately $t = 13$). The third POD mode can correct the magnitude of vorticity field. Hence, using the combination of the first three POD modes is enough in our simulations. The results by the online method when using the first four POD modes are shown in Figure 21. The results are the same as those by the first three POD modes.

The appropriate number of extracted POD modes should be 3 or 4 in our simulations. However, the results by the online method are not accurate when using the combination

of the first five POD modes (see Figures 4 and 5). It is due to the fact that the fifth POD mode has very high frequency with even smaller amplitude. Including this high POD mode in the projective method produces some noise. These observations also show the difference of reduced model structure between the online and offline methods. For the online method, the first few POD modes are enough to represent flow behaviors whereas a large number of POD modes is needed in the offline method.

5. Conclusions

The POD-assisted projective integration method based on the Galerkin-free approach is presented in this paper. We apply the approach to simulate the two-dimensional incompressible fluid flows past the NACA0012 airfoil. The present approach is referred to as the online method. POD bases used in solution expansions are extracted during time integration from the data set of direct numerical simulation. The present approach is different from the offline method in which refered POD based are extracted from a priori known

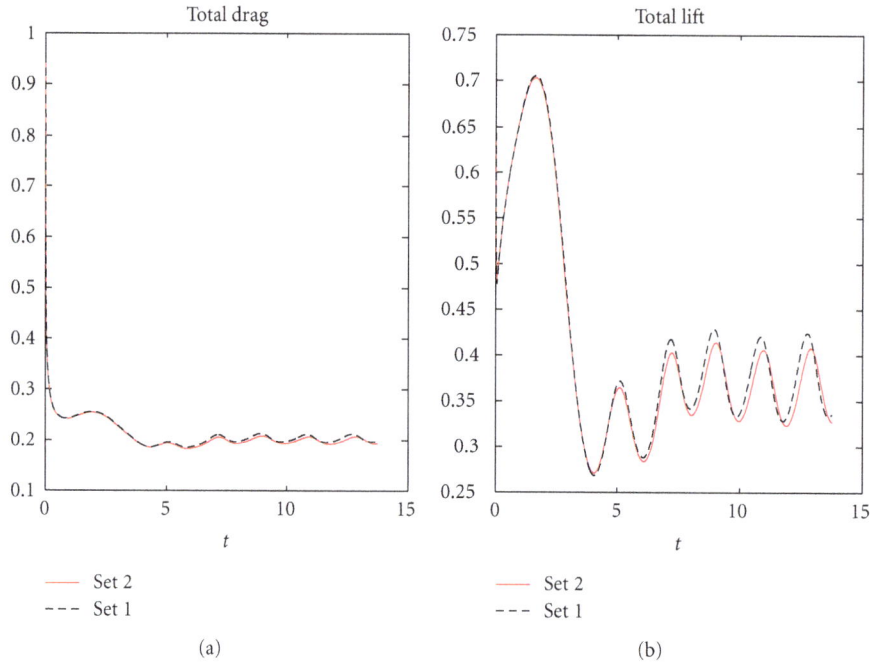

FIGURE 15: Drag and lift coefficients when using $M = 2$ and $M = 3$.

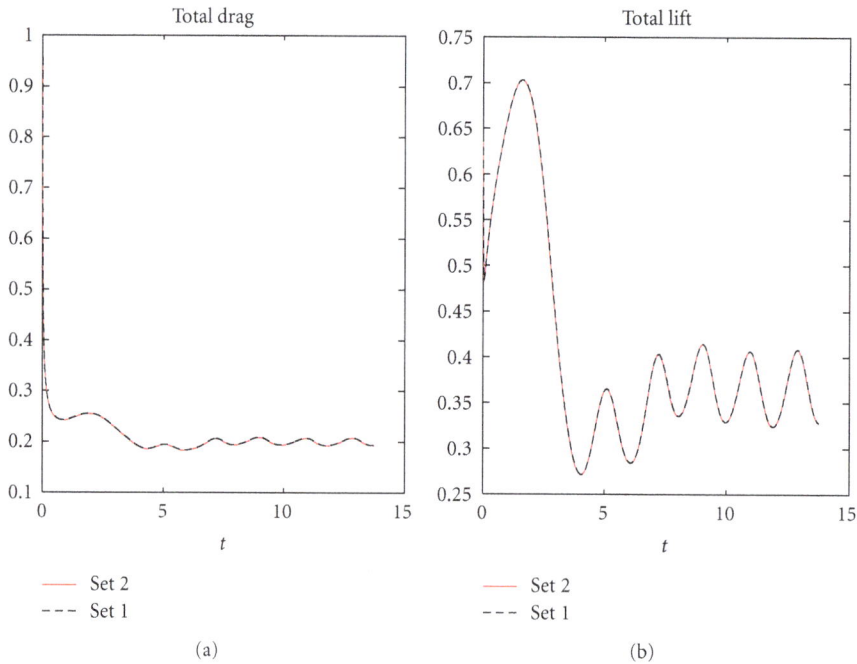

FIGURE 16: Total drag and total lift when using $M = 3$ and $M = 4$.

data set in the whole time integration process, not during time integration process. We have applied the online method for simulating the incompressible flows in two cases of Re, 400 and 800, for time $0 < t < 30$. There are two time scales which are fine and coarse scales. Fine (δt) and coarse (T) scales are the time scales for the DNS and the projective integration, respectively. We have set $\delta t = 0.0005$ in all simulations. It is observed that the maximum values of T are 0.015 and 0.0125 for Re $= 400$ and 800, respectively.

This shows the efficiency of the present method that can use very large jump in the projective step when comparing with the DNS time step. The number of POD modes used in the projective method is also an important issue. We found that using only the first few POD modes in the online method is enough for representing flow behavior. We have also revealed the representations of each dominant POD mode. The first mode has the highest energy. The basic flow behavior can be represented by using this mode but

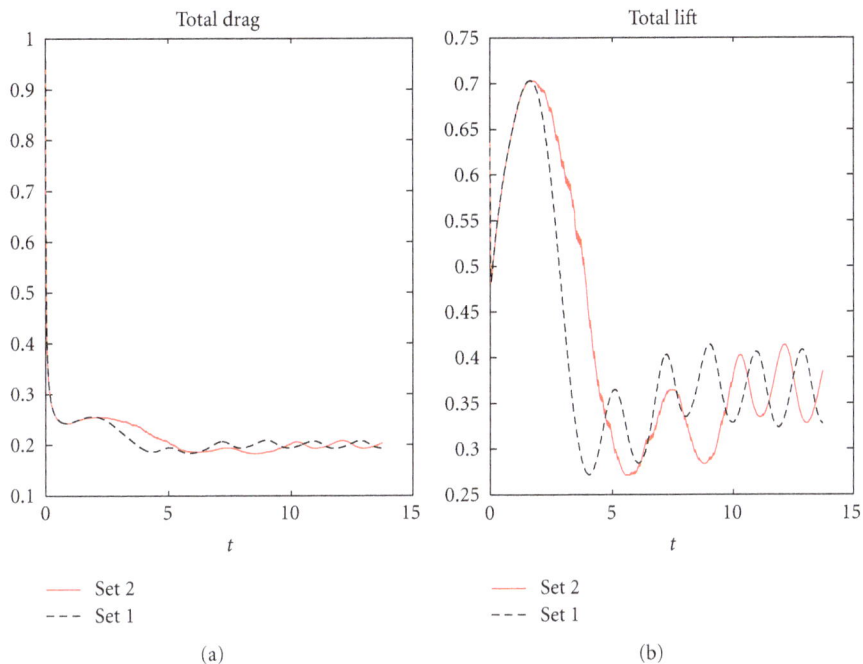

FIGURE 17: Total drag and total lift when using $M = 4$ and $M = 5$.

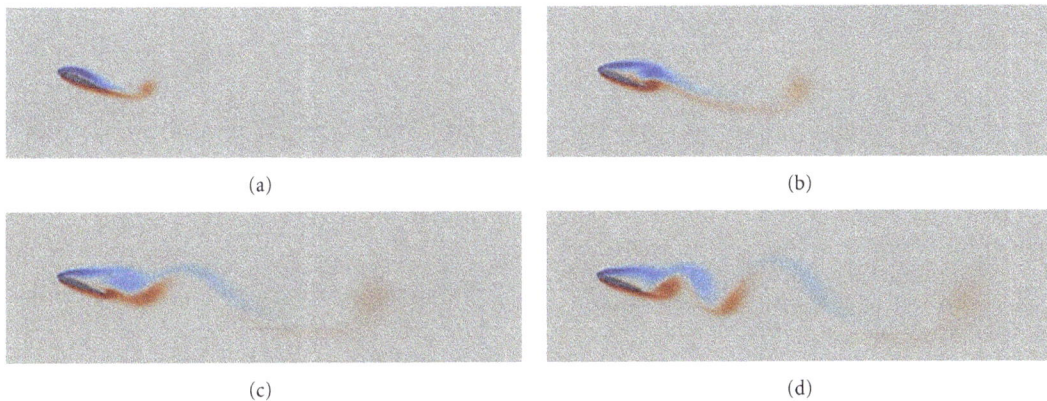

FIGURE 18: Vorticity plots of Re $= 400$ when $M = 1$ for $t = 1.9$ (a), 6.3 (b), 10.7 (c), and 12.9 (d).

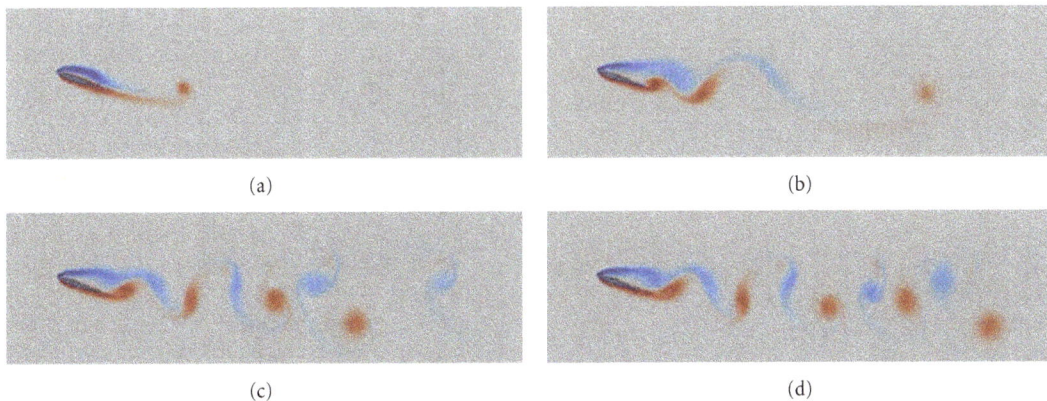

FIGURE 19: Vorticity plots of Re$= 400$ when $M = 2$ for $t = 1.9$ (a), 6.3 (b), 10.7 (c), and 12.9 (d).

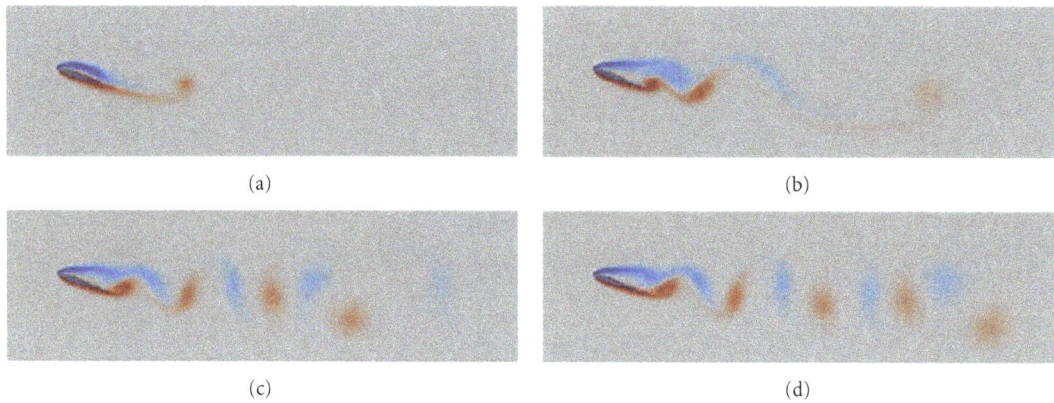

FIGURE 20: Vorticity plots of Re = 400 when $M = 3$ for $t = 1.9$ (a), 6.3 (b), 10.7 (c), and 12.9 (d).

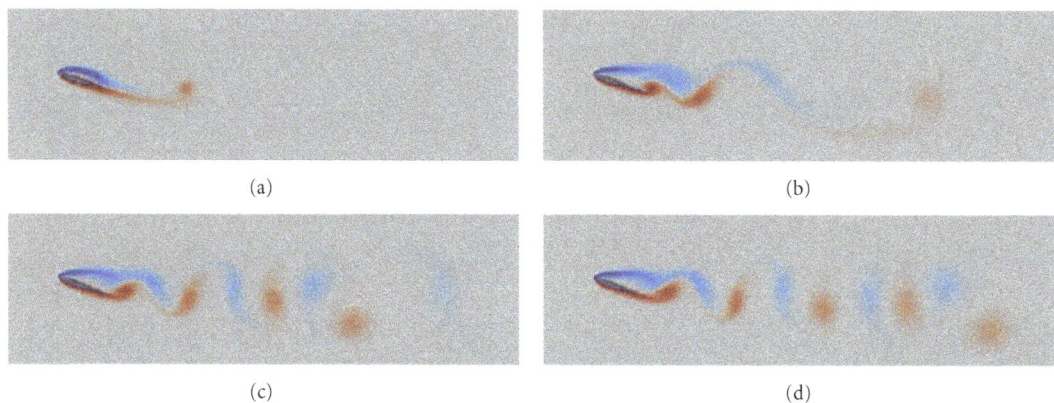

FIGURE 21: Vorticity plots of Re = 400 when $M = 4$ for $t = 1.9$ (a), 6.3 (b), 10.7 (c), and 12.9 (d).

the overall results are inaccurate. Including the second and third POD modes can improve solution accuracy because the higher POD modes contain high energy and degree of fluctuations. The derivative approximations for temporal mode are less accurate as well. Thus, including higher POD mode can produce some incorrect results in simulations. Although, there are many cases of Re that have not been investigated. Hopefully our observations may provide some insights of how to choose discretized parameters and the number of POD modes employed in other simulations.

Extensions of this work would be to study and compare the efficiency of the online and offline approaches, as well as the relationship of POD modes. Our preliminary studies show that the first POD mode in the online method is almost the same as the first POD mode in the offline method. However, it is not the case for the second POD mode in the online method. It is represented by a combination of higher offline POD modes.

Acknowledgment

The authors would like to thank Professor A. L. Pardhanani at the Earlham College for some useful comments and proofreading throughout this paper.

References

[1] I. G. Kevrekidis, C. W. Gear, and G. Hummer, "Equation-free: the computer-aided analysis of complex multiscale systems," *AIChE Journal*, vol. 50, no. 7, pp. 1346–1355, 2004.

[2] S. Sirisup, G. E. Karniadakis, D. Xiu, and I. G. Kevrekidis, "Equation-free/Galerkin-free POD-assisted computation of incompressible flows," *Journal of Computational Physics*, vol. 207, no. 2, pp. 568–587, 2005.

[3] D. Xiu, I. G. Kevrekidis, and R. Ghanem, "An equation-free, multiscale approach to uncertainty quantification," *Computing in Science and Engineering*, vol. 7, no. 3, pp. 16–23, 2005.

[4] L. Sirovich, "Turbulence and the dynamics of coherent structures, parts I, II and III," *Quarterly of Applied Mathematics*, vol. 45, pp. 561–590, 1987.

[5] C. Foias, M. S. Jolly, I. G. Kevrekidis, and E. S. Titi, "Dissipativity of numerical schemes," *Nonlinearity*, vol. 4, no. 3, pp. 591–613, 1991.

[6] S. Sirisup and G. E. Karniadakis, "A spectral viscosity method for correcting the long-term behavior of POD models," *Journal of Computational Physics*, vol. 194, no. 1, pp. 92–116, 2004.

[7] O. P. Le Maître and L. Mathelin, "Equation-free model reduction for complex dynamical systems," *International Journal for Numerical Methods in Fluids*, vol. 63, no. 2, pp. 163–184, 2010.

[8] Y. Hoarau, M. Braza, Y. Ventikos, and D. Faghani, "First stages of the transition to turbulence and control in the

incompressible detached flow around a NACA0012 wing," *International Journal of Heat and Fluid Flow*, vol. 27, no. 5, pp. 878–886, 2006.

[9] N. Okong'o and D. D. Knight, "Implicit unstructured Navier-Stokes simulation of leading edge separation over a pitching airfoil," *Applied Numerical Mathematics*, vol. 27, no. 3, pp. 269–308, 1998.

[10] S. G. Siegel, J. Seidel, C. Fagley, D. M. Luchtenburg, K. Cohen, and T. McLaughlin, "Low-dimensional modelling of a transient cylinder wake using double proper orthogonal decomposition," *Journal of Fluid Mechanics*, vol. 610, pp. 1–42, 2008.

[11] J. Seidel, S. Siegel, K. Cohen, and T. McLaughlin, "POD based separation control on the NACA0015 airfoil," in *Proceedings of the 43rd AIAA Aerospace Sciences Meeting and Exhibit*, AIAA-2005-0297, pp. 12059–12069, January 2005.

[12] G. E. Karniadakis and S. J. Sherwin, *Spectral/hp Element Methods for CFD*, Oxford University Press, 1999.

[13] G. Berkooz, P. Holmes, and J. L. Lumley, "The proper orthogonal decomposition in the analysis of turbulent flows," *Annual Review of Fluid Mechanics*, vol. 25, no. 1, pp. 539–575, 1993.

[14] W. Cazemier, R. W. Verstappen, and A. E. Veldman, "Proper orthogonal decomposition and low-dimensional models for driven cavity flows," *Physics of Fluids*, vol. 10, no. 7, pp. 1685–1699, 1998.

[15] A. E. Deane, I. G. Kevrekidis, G. E. Karniadakis, and S. A. Orszag, "Low-dimensional models for complex geometry flows: application to grooved channels and circular cylinders," *Physics of Fluids A*, vol. 3, no. 10, pp. 2337–2354, 1991.

[16] X. Ma, G. S. Karamanos, and G. E. Karniadakis, "Dynamics and low-dimensionality of a turbulent near wake," *Journal of Fluid Mechanics*, vol. 410, pp. 29–65, 2000.

[17] X. Ma and G. E. Karniadakis, "A low-dimensional model for simulating three-dimensional cylinder flow," *Journal of Fluid Mechanics*, vol. 458, pp. 181–190, 2002.

[18] B. R. Noack, K. Afanasiev, M. Morzyński, G. Tadmor, and F. Thiele, "A hierarchy of low-dimensional models for the transient and post-transient cylinder wake," *Journal of Fluid Mechanics*, no. 497, pp. 335–363, 2003.

[19] C. Williamson, R. Govardhson, and A. Prasad, "Experiments on low reynolds number NACA0012 aerofoils," Tech. Rep., Cornell University, 1994.

12

Robust Adaptive Tracking Control of a Class of Robot Manipulators with Model Uncertainties

G. Solís-Perales[1] and R. Peón-Escalante[2]

[1] *Departamento de Electrónica, CUCEI Universidad de Guadalajara, Avenida Revolución No. 1500, 44430 Guadalajara, JAL, Mexico*
[2] *Facultad de Ingeniería, Universidad Autónoma de Yucatán, Avenida Industrias no Contaminantes, Apdo. Postal 150 Cordemex, Mérida, Yucatán, Mexico*

Correspondence should be addressed to G. Solís-Perales, gualberto.solis@cucei.udg.mx

Academic Editor: Ahmed Rachid

A robust tracking controller for robot manipulators measuring only the angular positions and considering model uncertainties is presented. It is considered that the model is uncertain; that is, the system parameters, nonlinear terms, external perturbations, and the friction effects in each robot joint are considered unknown. The controller is composed by two parts, a linearizing-like control feedback and a high-gain estimator. The main idea is to lump the uncertain terms into a new state which represents the dynamics of the uncertainties. This new state is then estimated in order to be compensated. In this way the resulting controller is robust. A numerical example for a RR robot manipulator is provided, in order to corroborate the results.

1. Introduction

Control of manipulators is a classical control problem [1–4], due to the innumerable applications, for instance, in manufacturing processes, biomedical engineering, and aeronautical. In the tracking problem, it is known that the tracking error should be maintained as small as possible, even in the presence of parameter variations, external perturbations, modeling errors, and even more the presence of faults in the system. These aspects produce, a source of uncertainties and then the controller should be robust against these phenomena.

Tracking and control of manipulators have been studied using several approaches. For instance, in [5] authors presented an approach for controlling a robot using only joint position measurements. In such approach an observer is proposed in order to estimate the joint velocities, but uncertainties in the model are not considered. A fuzzy scheme was presented in [6]; however, the methodology requires the knowledge of a nominal values for the inertia matrix. An alternative approach was presented in [7], where the problem of estimation of time-varying parameters was

obtained via the multiestimation, which consists in the integration of an adaptive controller to a multi-model-based technique. An application of control of a dual cooperative manipulator based on singularly perturbed formulation and a position/force controller is designed [8], which does not require the velocity measures, as an observer to estimate the velocity and reduce the number of sensors.

The present contribution consists in considering uncertain the parameters of the manipulator and the effects of the friction; moreover, it is not necessary to estimate or to adapt to any parameter. In this sense, the controller is considered robust against parameter variations, nonmodeled dynamics (friction dynamics), and external disturbances. We illustrate that the robust control can reject external perturbations, attenuate parameter variation effects, and compensate nonmodeled dynamics. It is known that the friction is a physical phenomenon and many models exist for its study. Thus, the unknown of the friction model is a challenging problem and, from the control view point, compensating this phenomenon is determinant in the performance of the system. Thus in the present contribution a controller which compensate friction and modeling errors is described. In this sense, the controller is capable to compensate modeling errors, friction, parameter variations, and external perturbations using only the measure of the angular position of the robots.

The controller is constructed departing from the nonlinear control theory. First, a linearizing control law is determined, then all the uncertain terms are lumped into a new state, which is unknown. The linearizing-like controller requires the knowledge of such an unknown state. To tackle this problem, we use a high-gain state estimator to obtain an estimated value for this uncertain term [9]. This value is used by the controller which now can compensate all the uncertainties.

The paper is organized as follows: in Section 2, the dynamic model for a robot manipulator is presented, Section 3 contains the main contribution on nonlinear robust control, Section 4 presents the simulation results, and finally, in Section 5, some conclusions are given.

2. Dynamic Model of the Manipulator

Let us consider the well-known standard equation describing the dynamics of a n-DOF rigid joint robot manipulator with friction

$$\mathbf{M}(q)\ddot{q} + \mathbf{V}(q,\dot{q}) + \mathbf{G}(q) + \mathbf{B}(\dot{q}) = \mathbf{T}(t) - \mathbf{T}_d(t), \quad (1)$$

where $\mathbf{M}(q) \in \mathbb{R}^{n \times n}$ is a positive definite inertia matrix, $\mathbf{V}(q,\dot{q}) \in \mathbb{R}^{n \times 1}$ the Coriolis and centrifugal forces, $\mathbf{G}(q) \in \mathbb{R}^{n \times 1}$ is the vector for the gravitational forces and $\mathbf{B}(\dot{q})$ is a vector field of the friction terms, $\mathbf{T}, \mathbf{T}_d \in \mathbb{R}^{n \times 1}$ stand for the control and disturbance torques, respectively, and $q(t) \in \mathbb{R}^n$ is the state vector for the joint angular positions of the manipulator. In this contribution we consider the friction terms as follows:

$$
\begin{aligned}
B_i(\dot{q}_i) = {} & B_{v,i}\dot{q}_i + B_{f_1,i}\left(1 - \frac{2}{1 + e^{2\omega_1 \dot{q}_i}}\right) \\
& + B_{f_2,i}\left(1 - \frac{2}{1 + e^{2\omega_2 \dot{q}_i}}\right),
\end{aligned}
\quad (2)
$$

where B_v, i stands for the viscous friction; the second and third terms model the Coulomb and Stribeck friction effects in the ith robot link; for details see [10, 11] and references therein.

3. Nonlinear Control of Robot Manipulators

Let us consider (1) and defining $x_1 = q = [q_1, q_2, \ldots, q_n]^T$ as the n-vector of the joint positions and $x_2 = \dot{q} = \dot{x}_1 = [\dot{q}_1, \dot{q}_2, \ldots, \dot{q}_n]^T$ as the n-vector of angular velocities; thus one can write the system as follows:

$$\dot{x}_1 = x_2,$$

$$\dot{x}_2 = \mathbf{M}(x_1)^{-1}(\mathbf{T} - \mathbf{T}_d - \mathbf{V}(x_1, x_2) - \mathbf{G}(x_1) - \mathbf{B}(x_2)), \quad (3)$$

$$y = x_1,$$

where the output y is the vector of angular positions, without losing generality, for a manipulator with n joints, the affine system can be written as follows:

$$\dot{\chi} = \Psi(\chi) + \sum_{j=1}^{n} \Gamma_j(\chi)\tau_j$$

$$y_i = h_i(\chi), \quad (4)$$

where $\chi \in \mathbb{R}^{2n}$ is the state vector of positions and velocities of the manipulator given by $\chi = [q_1, q_2, \ldots, q_n, \dot{q}_1, \ldots, \dot{q}_n]^T$, $\Psi(\chi){:}\mathbb{R}^{2n} \to \mathbb{R}^{2n}$ is a sufficiently smooth vector field describing the robot dynamics, $\Gamma_j(\chi) \in \mathbb{R}^{2n}$ is the input vector given by $\Gamma_j(\chi) = [0_{n \times 1}, \mathfrak{m}_{1,j}(\chi), \ldots, \mathfrak{m}_{n,j}(\chi)]^T$ where $\mathfrak{m}_{i,j}$ are the entries of the inverse inertia matrix, and τ_j is the control torque in the jth joint. It is important to stress that the friction terms are included into the dynamics $\Psi(\chi)$. System (4) consists of a system with n inputs and n outputs; therefore the multiple-input and multiple-output nonlinear control tools are used [12]. The underlying idea of the nonlinear feedback control is to find functions τ_j, such that the desired dynamical behavior is induced for any initial condition $\chi_0 = \chi(0)$ in an attraction basin $U_0 \subset \mathbb{R}^{2n}$. To begin with, we depart from the relative degree condition for a MIMO system [12].

Definition 1. The multiple input and multiple output affine system (4) has relative degree $(\rho_1, \rho_2, \ldots, \rho_n)$ at the point χ^0 if

(i) $L_{\Gamma_j}L_\Psi^k h_i(\chi) = 0$, for all $1 \le j, i \le n, k < \rho_i - 1$, and for all χ in the neighborhood of χ^0,

(ii) the $n \times n$ matrix

$$
\mathcal{A}_\chi = \begin{pmatrix}
L_{\Gamma_1}L_\Psi^{\rho_1 - 1}h_1(\chi) & \cdots & L_{\Gamma_n}L_\Psi^{\rho_1 - 1}h_1(\chi) \\
L_{\Gamma_1}L_\Psi^{\rho_2 - 1}h_2(\chi) & \cdots & L_{\Gamma_n}L_\Psi^{\rho_2 - 1}h_2(\chi) \\
\cdots & \cdots & \cdots \\
L_{\Gamma_1}L_\Psi^{\rho_n - 1}h_n(\chi) & \cdots & L_{\Gamma_n}L_\Psi^{\rho_n - 1}h_n(\chi)
\end{pmatrix}
\quad (5)
$$

is nonsingular at $\chi = \chi^0$.

In the fully actuated manipulator, the relative-degree matrix $\mathcal{A}_\chi$ is square, and it is given by the positive definite matrix $M(x_1)^{-1}$. Thus, due to the invertibility of the matrix $\mathcal{A}_\chi$, a diffeomorphic transformation can be determined based on the Lie derivatives, of the outputs along the vector fields. Without losing generality, one can propose the transformation $z = \Phi(\chi) = [\chi_1, \chi_{n+1}, \chi_2, \chi_{n+2}, \ldots, \chi_n, \chi_{2n}]^T$, $\Phi : \mathbb{R}^{2n} \to \mathbb{R}^{2n}$, for $z \in \Omega \subset U$, such that the affine form (4) takes a linearizable canonical form. In terms of the Lie derivatives the transformation is given as $z_{2k-1} = h_k(\chi) = \chi_k$ and $z_{2k} = L_\Psi h_k(\chi) = \chi_{n+k}, k = 1, 2, \ldots, n$. Therefore, the system in the new-state variables is described by n set of equations as follows:

$$\dot{z}_{2k-1} = z_{2k},$$

$$\dot{z}_{2k} = \zeta_k(z) + \sum_{j=1}^{n} \vartheta_{k,j}(z)\tau_j, \qquad (6)$$

$$y_k = z_{2k-1}, \quad k = 1, 2, \ldots, n.$$

This system corresponds to the kth joint, and it is fully linearizable via feedback, and $\zeta_k(z) = L_\Psi^2 h_k(\Phi(z)^{-1})$, $\vartheta_{i,j}(z) = L_{\Gamma_j} h_k(\Phi(z)^{-1})$. These functions represent the dynamics of the system and the way as the control is acting onto the system, respectively. Note that the friction term, the Coriolis, centrifugal forces the gravity effect, and all possible non modeled dynamics are in function $\zeta_k(z)$ and the inertia terms are in function $\vartheta_{k,j}(z)$. From system (6), the vector of control torques is calculated as follows:

$$T = \mathcal{A}_\chi^{-1}(\chi)\left\{-\Im(z) + \mathbf{q}_r^{(2)} + \mathbf{K}_1(\mathbf{Z}_1 - \mathbf{q}_r) + \mathbf{K}_2\left(\mathbf{Z}_2 - \mathbf{q}_r^{(1)}\right)\right\}, \qquad (7)$$

where $\Im(z)$ is the vector of functions $\zeta_k(z) = L_\Psi^2 h_k(\Phi(z)^{-1})$, $\mathbf{q}_r$ is the reference angular position vector, $\mathbf{K}_1$ and $\mathbf{K}_2$ are control gain matrices, and $\mathbf{Z}_1$ and $\mathbf{Z}_2$ are the vector of transformed angular positions and velocities, respectively. This controller is called the perfect controller since it requires the perfect knowledge of the dynamics $\Im(z)$, the angular velocities $\mathbf{Z}_2$, and the matrix $\mathcal{A}_\chi^{-1}(\chi)$ in order to track the reference trajectory. In this sense such a controller is not realistic and cannot be implemented. Thus, we assume that only the angular positions are available for feedback and that $\text{sign}[\hat{\vartheta}_{k,j}(z)] = \text{sign}[\vartheta_{k,j}(z)]$ at any $z \in U^0 \subset \mathbb{R}^{2n}$ of z^0 is known. Note that it is not necessary to know a nominal value of the inertia matrix.

Thus, system (6) can be written in an extended form defining $\delta_k(z) = \vartheta_{k,j}(z) - \hat{\vartheta}_{k,j}(z)$, $\Theta_k(z,u) = \zeta_k(z) + \delta_k(z)\tau_j$, and $\eta_k = \Theta_k(z,\tau_j)$. Thus system (6) takes the form

$$\dot{z}_{2k-1} = z_{2k}$$

$$\dot{z}_{2k} = \eta_k + \sum_{j=i}^{n} \hat{\vartheta}_{k,j}(z)\tau_j, \qquad (8)$$

$$\dot{\eta}_k = \Xi_k\left(z, \eta_k, \tau_j, \dot{\tau}_j\right)$$

$$y_k = z_{2k-1}, \quad k = 1, 2, \ldots, n,$$

where the function $\Xi_k(\cdot) = \Sigma_{k=1}^{n}((\partial\Theta_k(\cdot)/\partial z_{2k-1})z_{2k} + (\partial\Theta_k(\cdot)/\partial z_{2k})(\eta_k + \Sigma_{j=1}^{n}\hat{\vartheta}_{k,j}(z)\tau_j) + \delta_k(z)\dot{\tau}_j)$. The new state η_k provides the dynamics of the uncertain terms, such as the friction and the uncertain parameter values [9].

System (8) is dynamically equivalent to system (6) and comprises the lumping state; however, such a state is again unknown. Although the state η_k is unknown, there are only two unknown variables, the new-state η_k and the state z_{2k} which corresponds to the angular velocity.

At this point the controller is given by

$$T = \mathcal{A}_\chi^{-1}(\chi)\left\{-\mathbf{H} + \mathbf{q}_r^{(2)} + \mathbf{K}_1(\mathbf{Z}_1 - \mathbf{q}_r) + \mathbf{K}_2\left(\mathbf{Z}_2 - \mathbf{q}_r^{(1)}\right)\right\}, \qquad (9)$$

where $\mathbf{H}$ is the vector with entries η_k and it cannot be implemented since the states η_k and z_{2k} are not available for feedback. To tackle this problem we use a state estimator which provides an estimated value of η_k and z_{2k} [9]. The state estimator for the kth link is given by

$$\dot{\hat{z}}_{2k-1} = \hat{z}_{2k} + \lambda_k \kappa_{1,k}(z_{2k-1} - \hat{z}_{2k-1})$$

$$\dot{\hat{z}}_{2k} = \hat{\eta}_k + \sum_{j=i}^{n} \hat{\vartheta}_{k,j}(\hat{z})\tau_j + \lambda_k^2 \kappa_{2,k}(z_{2k-1} - \hat{z}_{2k-1}), \qquad (10)$$

$$\dot{\hat{\eta}}_k = \lambda_k^3 \kappa_{3,k}(z_{2k-i} - \hat{z}_{2k-1}),$$

$$y_{e,k} = \hat{z}_{2k-1}, \quad k = 1, 2, \ldots, n,$$

With these estimates the control command can be given by

$$T = \mathcal{A}_e^{-1}\left\{-\hat{\mathbf{H}} + \mathbf{q}_r^{(2)} + \mathbf{K}_1(\mathbf{Z}_1 - \mathbf{q}_r) + \mathbf{K}_2\left(\hat{\mathbf{Z}}_2 - \mathbf{q}_r^{(1)}\right)\right\}, \qquad (11)$$

where $\mathcal{A}_e$ is an estimated matrix where its elements satisfy the condition $\text{sign}[\hat{\vartheta}_{k,j}(z)] = \text{sign}[\vartheta_{k,j}(z)]$. This controller only requires the value of the angular position vector $\mathbf{Z}_1$, the time derivatives of the reference signals $y_{r,k}$, and the estimated values of the angular velocity. The estimation parameters λ_k and $\kappa_{i,k}$ are the high gain parameters and the state estimator gains, respectively, which are chosen in such way that the close-loop system be stable.

Proposition 2. *Let $e_k \in \mathbb{R}^3$ be an estimation error vector whose components are defined as $e_{1,k} = \lambda_k(z_{2k-1} - \hat{z}_{2k-1})$, $e_{2,k} = \lambda_k^2(z_{2k-1} - \hat{z}_{2k-1})$, and $e_{3,k} = \eta_k - \hat{\eta}_k$. For sufficiently large λ_k, the dynamics of the estimation error decays globally exponentially to zero if $\text{sign}(\hat{\vartheta}_{k,j}(z)) = \text{sign}(\vartheta_{k,j}(z))$.*

Proof. Combining systems (8) and (10), the dynamics of the estimation error can be written as follows:

$$\dot{e}_k = \lambda_k \mathbf{D}e_k + \Delta\left(z, \eta_k, \tau_{e,j}\right), \qquad (12)$$

where $\Delta(z, \eta_k, \tau_{e,j}) = [0, 0, \Xi(z, \eta_k, \tau_{e,j})]^T$ and the matrix $\mathbf{D}$ is

$$\mathbf{D} = \begin{pmatrix} -\kappa_{1,k} & 1 & 0 \\ -\kappa_{2,k} & 0 & 1 \\ -\kappa_{3,k} & 0 & 0 \end{pmatrix}, \qquad (13)$$

since $\text{sign}(\hat{\vartheta}_{k,j}(z)) = \text{sign}(\vartheta_{k,j}(z))$, thus $\mathbf{D}$ is obviously Hurwitz; thus there exists a positive definite and symmetric matrix $\mathbf{P}$ such that $\mathbf{PD} + \mathbf{D}^T\mathbf{P} = -\mathbf{Q}$, with $\mathbf{Q}$ a positive definite and symmetric matrix. Choosing $V(e_k) = e_k^T\mathbf{P}e_k$ as the Lyapunov-like function candidate, one has

$$\dot{V}(e_k) = \dot{e}_k^T\mathbf{P}e_k + e_k^T\mathbf{P}\dot{e}_k = \lambda_k e_k^T\left(\mathbf{PD} + \mathbf{D}^T\mathbf{P}\right)e_k$$

$$+ 2e_k^T\mathbf{P}\Delta\left(z, \eta_k, \tau_{e,j}\right)$$

$$\dot{V}(e_k) \leq -\lambda_k\sigma_{\min}(\mathbf{Q})\|e_k\|^2 + 2\sigma_{\max}(\mathbf{P})\|e_k\|\left\|\Delta\left(z, \eta_k, \tau_{e,j}\right)\right\|,$$

$$\dot{V}(e_k) \leq -\lambda_k\sigma_{\min}(\mathbf{Q})\|e_k\|^2 + 2\sigma_{\max}(\mathbf{P})\|e_k\|r,$$

$$(14)$$

where $\sigma_{\min}(\mathbf{Q})$ and $\sigma_{\max}(\mathbf{P})$ are the minimum and maximum eigenvalues of the matrices $\mathbf{Q}$ and $\mathbf{P}$, respectively, and the function $\Delta(z, \eta_k, \tau_{e,j})$ satisfies $\|\Delta(z, \eta_k, \tau_{e,j})\| \leq r$ for some $r > 0$. Therefore, the error e_k decays, and it can be seen that it is ultimately bounded. Also note that as λ_k increases, e_k will decrease, which also decreases the exponential estimation error bound; therefore, λ_k should be made as large as possible, and this achieves the proof.

It is important to note that the proposition is valid for $k = 1, 2, \ldots, n$, therefore it is possible to obtain an estimated value for the unknown states. The previous proposition states that there exists a Lyapunov function which depends on the parameter λ_k and provides the global exponential convergence of the estimation error dynamics.

Proposition 3. *System* (1) *asymptotically tracks a prescribed sufficiently smooth reference under the feedback* (11) *via the stabilization of the extended system* (8).

Proof. Substituting the robust feedback (9) into (8) and considering the estimation error system (12), the closed-loop dynamics is given by

$$\dot{z}_{2k-1} = z_{2k},$$

$$\dot{z}_{2k} = \eta_k + \sum_{j=i}^{n}\hat{\vartheta}_{k,j}(z)\tau_j,$$

$$(15)$$

$$\dot{\eta}_k = \Xi_k\left(z, \eta_k, \tau_j, \dot{\tau}_j\right),$$

$$\dot{e}_k = \lambda_k\mathbf{D}e_k + \Delta\left(z, \eta_k, \tau_j\right).$$

Since $\eta_k = \Theta_k(z, \tau_j)$ and $\tau_j = (1/\hat{\vartheta}_j(z))(-\hat{\eta}_k + \nu_k)$, it follows that $\eta_k = \mathfrak{Z}_k(z, \eta_k, e_k, \tau_j, t)$ (which can be computed from the first integral of $\dot{\eta}_k = \Xi_k(z, \eta_k, \tau_j, \dot{\tau}_j)$; i.e, $\eta_k = \int \Xi_k(z, \eta_k, e_k, \tau_j, \sigma)d\sigma$). Then, according to the contraction mapping theorem the state η_k can be expressed globally and uniquely as a function of the coordinates (z, e_k). Now, note that since the matrix $\mathbf{D}$ is Hurwitz by construction and the nonlinear function $\Delta(z, \eta_k, \tau_j)$ is bounded, the estimation error system (12) is asymptotically stable. In this sense given a compact set of initial conditions $\Omega_k \subset \mathbb{R}^3$ containing the origin, there exists an upper bound $\tau_{j,\max}$ such that $\tau_j \leq \tau_{j,\max}$ and a high-gain estimator parameter λ_k such that Ω_k is contained into the attraction basin. Hence, the closed-loop

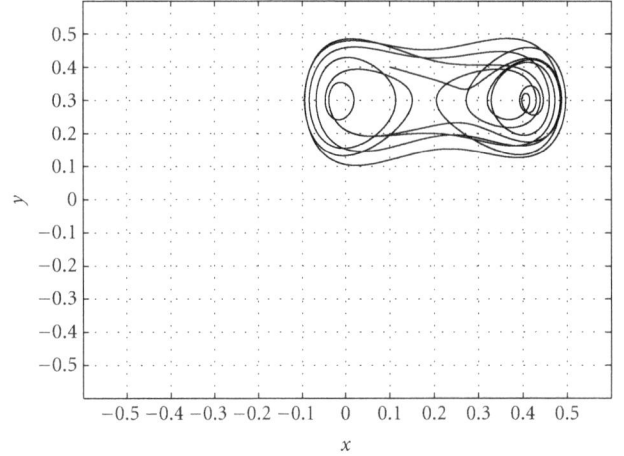

FIGURE 1: Reference irregular (chaotic) trajectory used to be followed by the RR manipulator.

system is semiglobally practically stable; that is, $(e_k, \eta_k) \rightarrow (0, 0)$, and this achieves the proof.

The previous proposition is a simple strategy for robust tracking of robot manipulators, despite the uncertain knowledge of the models.

4. Numerical Simulations

To corroborate the performance of the robust controller in the effects of parameter variations, modeling errors, and external perturbations, we propose to apply the robust controller to a RR manipulator (rotational-rotational, $n = 2$, $q = [q_1, q_2]^T$). We propose a smooth chaotic trajectory as a reference, which is obtained from the solutions of the Duffing chaotic system

$$\dot{y}_1 = y_2,$$

$$\dot{y}_2 = y_1 + 0.25y_2 + y_3, \qquad (16)$$

$$Y_R = \mathcal{C}y.$$

This system provides irregular, unpredictable but bounded smooth trajectories in the cartesian coordinates of the manipulator; therefore using the inverse kinematics transformation the reference trajectories in the articular space are given by the functions $q_r = [\varphi_1(Y_R)\varphi_2(Y_R)]^T$. In this case we used these trajectories in order to illustrate that the controller makes the manipulator tracks an irregular trajectory. The reference trajectories are illustrated in Figure 1.

For the RR configuration and from (3), we have

$$M(q) = \begin{bmatrix} P_1 + P_2 + 2P_3\cos q_2 & P_2 + P_3\cos q_2 \\ P_2 + P_3\cos q_2 & P_2 \end{bmatrix},$$

$$V(q, \dot{q}) = \begin{bmatrix} -2P_3\sin q_2\dot{q}_1\dot{q}_2 - P_3\sin q_2\dot{q}_2^2 \\ P_3\sin q_2\dot{q}_2^2 \end{bmatrix}, \qquad (17)$$

$$G(q) = \begin{bmatrix} g_1\cos q_1 + g_2\cos(q_1 + q_2) \\ g_2\cos(q_1 + q_2) \end{bmatrix},$$

where $P_i, g_j, i = 1, 2, 3, j = 1, 2$ are constant parameters given by $P_1 = m_1 lcm_1^2 + m_2 l_1^2 + Izz_1$, $P_2 = m_2 lcm_2^2 + Izz_2$, $P_3 = m_2 l_1 lcm_2$, $g_1 = (m_1 lcm_1 + m_2 l_1)g$, and $g_2 = m_2 lcm_2 g$ and are considered unknown. Now, performing the change of variable $x_1(t) = [q_1, q_2]^T = [\chi_1, \chi_2]^T$ and $x_2(t) = [\dot{q}_1, \dot{q}_2]^T = [\chi_3, \chi_4]^T$, the system in affine form is written as follows:

$$\dot{\chi}_1 = \chi_3,$$

$$\dot{\chi}_2 = \chi_4,$$

$$\dot{\chi}_3 = f_1(\chi) + M_{11}\tau_1 + M_{12}\tau_2, \qquad (18)$$

$$\dot{\chi}_4 = f_2(\chi) + M_{21}\tau_1 + M_{22}\tau_2,$$

$$Y = \mathcal{C}\chi,$$

where $f_1(\chi) = -\mathfrak{m}_{11}(V_1(\chi) + G_1(\chi) + B_1(\chi)) - \mathfrak{m}_{1,2}(V_2(\chi) + G_2(\chi) + B_2(\chi))$ and $f_2(\chi) = -\mathfrak{m}_{21}(V_1(\chi) + G_1(\chi) + B_1(\chi)) - \mathfrak{m}_{22}(V_2(\chi) + G_2(\chi) + B_2(\chi))$ are the nonlinear terms (considered uncertain) of the robot manipulator and the output is given by the matrix $\mathcal{C} = [I_{2\times2}, 0_{2\times2}]^T$. Notice that, in this nonlinear function appear the friction terms (2), $\mathfrak{m}_{k,j}$ stand for the elements of the inverse inertia matrix, $V_1(\chi) = -2P_3 \sin \chi_2 \chi_3 \chi_4 - P_3 \sin \chi_2 \chi_4^2$, $V_2(\chi) = P_3 \sin \chi_2 \chi_4^2$, $G_1(\chi) = g_1 \cos \chi_1 + g_2 \cos(\chi_1 + \chi_2)$, and $G_2(\chi) = g_2 \cos(\chi_1 + \chi_2)$. Therefore, calculating the relative degree matrix (see Definition 1) for this system we have

$$\mathcal{A}_\chi = \begin{pmatrix} L_{\Gamma_1} L_\Psi h_1(\chi) & L_{\Gamma_2} L_\Psi h_1(\chi) \\ L_{\Gamma_1} L_\Psi h_2(\chi) & L_{\Gamma_2} L_\Psi h_2(\chi) \end{pmatrix}$$

$$= \begin{pmatrix} \mathfrak{m}_{11} & \mathfrak{m}_{12} \\ \mathfrak{m}_{21} & \mathfrak{m}_{22} \end{pmatrix} = M(\chi)^{-1}. \qquad (19)$$

Note that the relative degree matrix is invertible due to the positive definiteness of the inertia matrix; however, it is considered uncertain and a nominal model is not required. The parameter values used for this example but considered unknown in the controller were for link 1: $l_1 = 0.35$ m, $lcm_1 = 0.175$ m, $Izz_1 = 0.0064$ Kg m^2, $m_1 = 0.479$ Kg. For link 2: $l_2 = 0.30$ m, $lcm_2 = 0.145$ m, $Izz_2 = 0.004$ Kg m^2, $m_2 = 0.341$ Kg. Thus for $k = 2$ the transformed extended system is written as follows:

$$\dot{z}_1 = z_2,$$

$$\dot{z}_2 = \eta_1 + \hat{\vartheta}_{1,1}(z)\tau_1 + \hat{\vartheta}_{1,2}(z)\tau_2,$$

$$\dot{\eta}_1 = \Xi_1(z, \eta_1, \tau_k, \dot{\tau}_k),$$

$$\dot{z}_3 = z_4, \qquad (20)$$

$$\dot{z}_4 = \eta_1 + \hat{\vartheta}_{2,1}(z)\tau_1 + \hat{\vartheta}_{2,2}(z)\tau_2,$$

$$\dot{\eta}_2 = \Xi_2(z, \eta_2, \tau_k, \dot{\tau}_k),$$

$$y_1 = z_1,$$

$$y_2 = z_3;$$

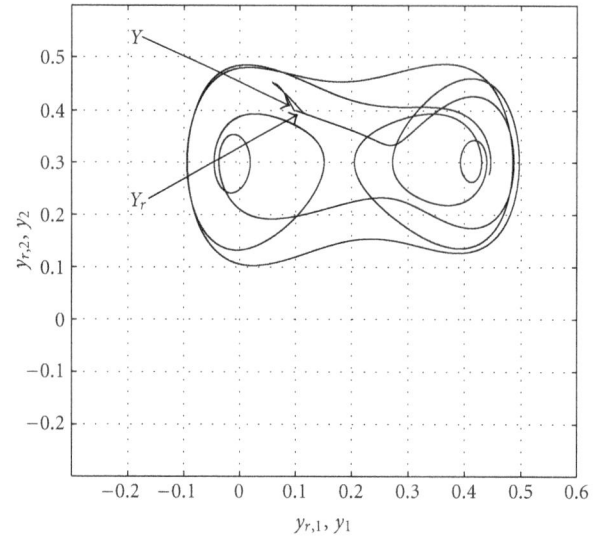

FIGURE 2: Tracking of the reference trajectory, using the robust approach.

since we assume that only the measured states are available for feedback and by Propositions 2 and 3, we can design the state estimator as follows:

$$\dot{\hat{z}}_1 = \hat{z}_2 + \lambda_1 \kappa_{1,1}(z_1 - \hat{z}_1),$$

$$\dot{\hat{z}}_2 = \hat{\eta}_1 + \hat{\vartheta}_{1,1}(\hat{z})\tau_1 + \hat{\vartheta}_{1,2}(\hat{z})\tau_2 + \lambda_1^2 \kappa_{2,1}(z_1 - \hat{z}_1),$$

$$\dot{\hat{\eta}}_1 = \lambda_1^3 \kappa_{3,1}(z_1 - \hat{z}_1),$$

$$\dot{\hat{z}}_3 = \hat{z}_4 + \lambda_2 \kappa_{1,2}(z_2 - \hat{z}_2), \qquad (21)$$

$$\dot{\hat{z}}_4 = \hat{\eta}_2 + \hat{\vartheta}_{2,1}(\hat{z})\tau_1 + \hat{\vartheta}_{2,2}(\hat{z})\tau_2 + \lambda_2^2 \kappa_{2,2}(z_2 - \hat{z}_2),$$

$$\dot{\hat{\eta}}_2 = \lambda_2^3 \kappa_{3,2}(z_2 - \hat{z}_2);$$

thus, the required robust control vector is given by

$$\begin{bmatrix} \tau_1 \\ \tau_2 \end{bmatrix} = A_e^{-1} \begin{bmatrix} -\hat{\eta}_1 + q_{r,1}^{(2)} + \nu_1 \\ -\hat{\eta}_2 + q_{r,2}^{(2)} + \nu_2 \end{bmatrix}, \qquad (22)$$

where ν_1 and ν_2 are the new control inputs and are given by

$$\begin{bmatrix} \nu_1 \\ \nu_2 \end{bmatrix} = \begin{bmatrix} K_{1,1}(z_1 - q_{r,1}) + K_{2,1}(\hat{z}_2 - q_{r,1}^{(1)}) \\ K_{1,2}(z_3 - q_{r,2}) + K_{2,2}(\hat{z}_4 - q_{r,2}^{(1)}) \end{bmatrix}. \qquad (23)$$

Therefore, with these controllers the RR robot manipulator tracks the trajectory given by the Duffing equation. The tracking of the reference trajectory is illustrated in Figure 2, where Y_r and Y are the reference and the robot manipulator trajectories, respectively, and Y is obtained by the kinematics transformation. Note that the perturbation torques do not affect the performance. That is, the perturbations are rejected by the controller. The error in the robot workspace is illustrated in Figure 3, where it remains close to the origin even in the presence of the uncertainties.

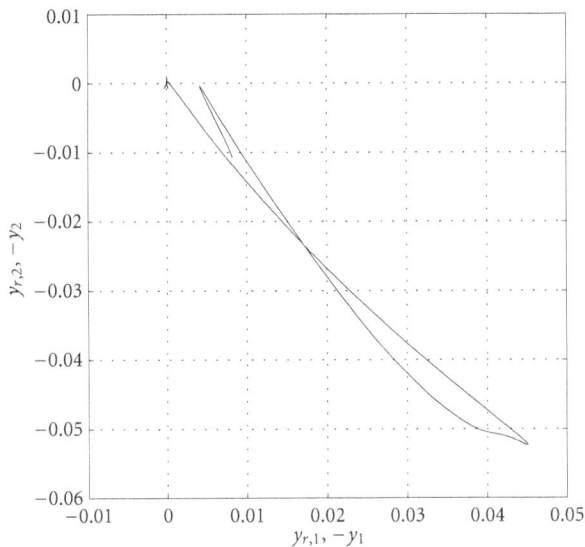

FIGURE 3: Tracking error in the robot workspace.

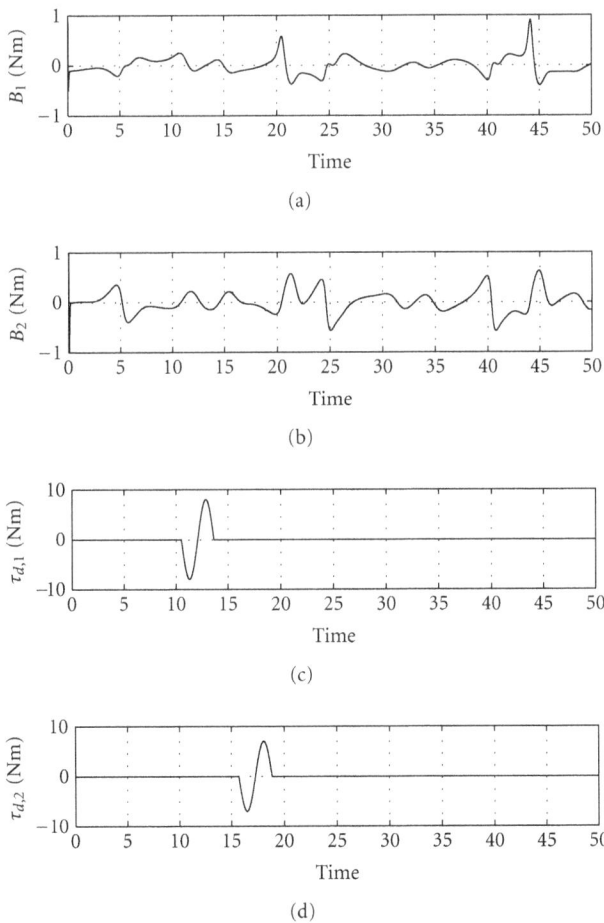

FIGURE 4: Friction dynamics in each robot joint and external torque perturbations.

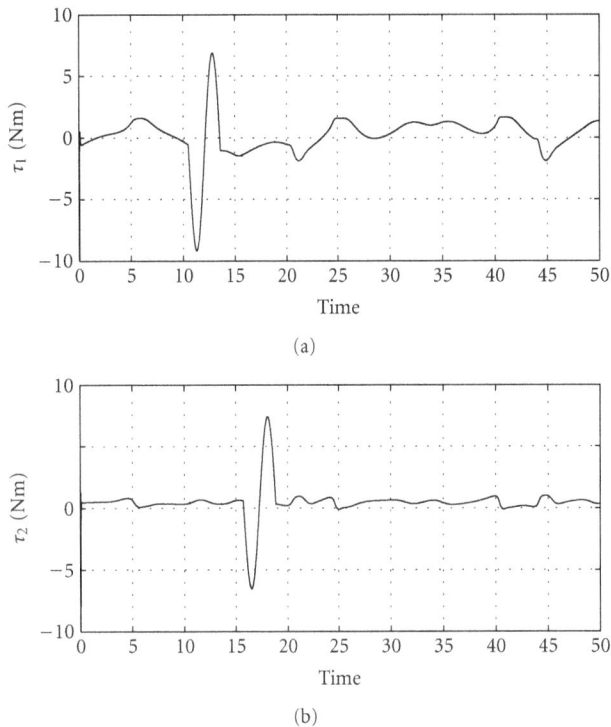

FIGURE 5: Control torques provided by the robust controller.

In Figure 4 the dynamics of the friction and perturbations in joints are illustrated. The friction was obtained with $B_v = 0.45$, $B_{f_1} = 0.1$, $B_{f_2} = 0.27$, $\omega_1 = 0.005$, and $\omega_2 = 0.0025$ for the first link and $B_v = 0.45$, $B_{f_1} = 1$, $B_{f_2} = 2.7$, $\omega_1 = 0.005$, and $\omega_2 = 0.0025$ for the second. The perturbation torques are given by a sinusoidal function of amplitude 8 Nm and 7 Nm for the first and second robot elements. Note that this friction parameters and the specific functional for the friction are completely unknown. Therefore, it can be considered as time varying the parameters and the friction parameter, without retuning the controller, that is, the tracking of the trajectory, is sustained.

Finally, in Figure 5 the resulting control torques for the robot are illustrated. Note that the perturbation torques are compensated by the controllers; however, the controllers do not have any knowledge about them.

5. Conclusions

In this work we present a robust tracking control for the compensation of modeling errors, parameter variations, and external perturbation for a class of robot manipulators. The robust scheme is such that it only requires to measure the angular positions. In this sense, the controller can compensate the uncertainty in parameter values, uncertain knowledge of the model, and external perturbations via an augmented state. This approach does not require to estimate or adapt any parameter; it is only needed to know the sign of some parameters but not a nominal value. In this sense the control strategy is considered robust; in the case of the friction it is not required of the precise function to be compensated. The key feature is to lump the uncertain terms into a new state, which provides the dynamics of the uncertainties via a state estimator, such that instead of estimate every parameter it is only required to estimate the dynamic of the uncertainties. This robust control approach can be applied to other robot configurations provided that the relative degree matrix be invertible.

References

[1] J. J. E. Slotine, "The robust control of robot manipulators," *International Journal of Robotics Research*, vol. 4, no. 2, pp. 49–61, 1985.

[2] J. J. Craig, P. Hsu, and S. S. Sastry, "Adaptive control of mechanical manipulators," *International Journal of Robotics Research*, vol. 6, no. 2, pp. 16–28, 1987.

[3] N. Sadegh and R. Horowitz, "Stability and robustness analysis of a class of adaptive controllers for robotic manipulators," *International Journal of Robotics Research*, vol. 9, no. 3, pp. 74–92, 1990.

[4] H. G. Sage, M. F. de Mathelin, and E. Ostertag, "Robust control of robot manipulators: a survey," *International Journal of Control*, vol. 72, no. 16, pp. 1498–1522, 1999.

[5] S. Nicosia and P. Tomei, "Robot control by using only joint position measurements," *IEEE Transactions on Automatic Control*, vol. 35, no. 9, pp. 1058–1061, 1990.

[6] E. Kim, "Output feedback tracking control of robot manipulators with model uncertainty via adaptive fuzzy logic," *IEEE Transactions on Fuzzy Systems*, vol. 12, no. 3, pp. 368–378, 2004.

[7] A. Ibeas, M. de la Sen, and S. Alonso-Quesada, "Stable multi-estimation model for single-input single-output discrete adaptive control systems," *International Journal of Systems Science*, vol. 35, no. 8, pp. 479–501, 2004.

[8] H. Bolandi and A.F. Ehyaei, "Position/force control of a dual cooperative manipulator system based on a singularly perturbe dynamic model," *International Journal of Robotics and Automation*, vol. 27, pp. 76–91, 2012.

[9] R. Femat, J. Alvarez-Ramírez, and G. Fernández-Anaya, "Adaptive synchronization of high-order chaotic systems: a feedback with low-order parametrization," *Physica D*, vol. 139, no. 3-4, pp. 231–246, 2000.

[10] H. Olsson, K. J. Åström, C. Canudas de Wit, M. Gäfvert, and P. Lischinsky, "Friction models and friction compensation," *European Journal of Control*, vol. 9, pp. 629–636, 2001.

[11] R. H. A. Hensen, G. Z. Angelis, M. J. G. van de Molengraft, A. G. de Jager, and J. J. Kok, "Grey-box modeling of friction: an experimental case-study," *European Journal of Control*, vol. 6, no. 3, pp. 258–267, 2000.

[12] A. Isidori, *Noninear Control Systems*, Springer, 1995.

13

Average Bandwidth Allocation Model of WFQ

Tomáš Balogh and Martin Medvecký

Institute of Telecommunications, Faculty of Electrical Engineering and Information Technology, Slovak University of Technology,
Ilkovicova 3, 812 19 Bratislava, Slovakia

Correspondence should be addressed to Tomáš Balogh, tomas.balogh@stuba.sk

Academic Editor: Agostino Bruzzone

We present a new iterative method for the calculation of average bandwidth assignment to traffic flows using a WFQ scheduler in IP based NGN networks. The bandwidth assignment calculation is based on the link speed, assigned weights, arrival rate, and average packet length or input rate of the traffic flows. We prove the model outcome with examples and simulation results using NS2 simulator.

1. Introduction

The current trends in telecommunication infrastructure with packet oriented networks bring up the question of supporting Quality of Service (QoS). Methods, that are able to assign priorities to flows or packets and then service them differently according to their needs in network nodes, were proposed for the demands of QoS support. Queue Scheduling Discipline (QSD) algorithms are responsible for choosing packets to output from queues. They are designed to divide the output capacity fairly and optimally. Algorithms that are able to make this decision according to priorities are the basic component of modern QoS supporting networks [1].

For an optimal configuration of these algorithms we need to calculate or simulate the result of our setting to expect the impact on QoS. The network nodes can be modeled using Markovian models [2].

Most of the existing WFQ bandwidth allocation models do not consider variable utilization of queues or bandwidth redistribution of unassigned link capacity. For this reason we proposed our iterative mathematical model for bandwidth allocation of WFQ. The model can be used for the analysis of the impact of weight settings, analyzing the stability of the system and modeling of delay and queue length of traffic classes.

The next sections of the paper are structured as follows. At first the WFQ algorithm is presented followed by a short presentation of common used bandwidth constraint models. The third section of the paper describes the proposed model for average bandwidth allocation of WFQ followed by examples of WFQ bandwidth allocation and simulation results proving the proposed model.

2. Bandwidth Allocation

There are many scheduling algorithms and several bandwidth allocation models proposed for bandwidth allocation estimation. We focused on WFQ and bandwidth allocation models proposed for MPLS traffic engineering.

2.1. Weighted Fair Queuing. WFQ was introduced in 1989 by Demers et al. and Zhang [3, 4]. The algorithm provides fair output bandwidth sharing according to assigned weights. The decision which packet should be read from the packet queue and sent next is done by calculating a virtual finish-time. The scheduler assigns the finish time to each packet as it arrives in the queue. The time corresponds with the time, in which the packet would be completely sent bit by bit from each queue as in the Generalized Processor Sharing (GPS) algorithm. The number of bits calculated in one turn corresponds with the assigned weights. The packet with the smallest finish time is chosen for output. WFQ guarantees that each traffic class gets a portion of the output bandwidth and shares it proportional to the assigned weights.

2.2. Bandwidth Constraint Models. One of the goals of Diff-Serv or MPLS traffic engineering is to guarantee bandwidth reservations for different service classes. For these goals two functions are defined [5]:

    (i) class-type (CT) is a group of traffic flows, based on QoS settings, sharing the same bandwidth reservation;

    (ii) bandwidth constraint (BC) is a part of the output bandwidth that a CT can use.

For the mapping between BCs and CTs the maximum allocation model (MAM), max allocation with reservation (MAR), and Russian dolls model (RDM) are defined.

Maximum Allocation Model. The MAM model [6] maps one BC to one CT. The whole bandwidth is strictly divided and no sharing between CTs is allowed.

Max Allocation with Reservation. MAR [7] is similar to MAM in that a maximum bandwidth is allocated to each CT. However, through the use of bandwidth reservation and protection mechanisms, CTs are allowed to exceed their bandwidth allocations under conditions of no congestion but revert to their allocated bandwidths when overload and congestion occurs [6].

Russian Dolls Model. The RDM model is more effective in bandwidth sharing. It assigns BCs to groups of CTs. For example CT7 with the highest QoS requirements gets its own BC7. The CT6 with lower QoS requirements shares its BC6 with CT7, and so forth. In extreme cases the lower priorities get less bandwidth as they need or even starve [8].

3. WFQ Bandwidth Allocation Model

In general, WFQ and some other scheduling algorithms like WRR, WF^2Q+, and so forth allocate bandwidth differently as the models described in Section 2.2. The available bandwidth is divided between service classes or waiting queues according to assigned weights. The sharing of unused bandwidth is allowed and is divided between the other queues again according to assigned weights.

The proposed model is a part of the research of modelling of traffic parameters of NGN networks, is a modification a presented model for bandwidth allocation of the WRR algorithm, and will be further used for delay and queue length modeling of these algorithms.

3.1. Definitions and Notations. We assume a network node with P priority classes or waiting queues. Each queue i has a weight w_i assigned. Packets enter the queue with an arrival rate λ_i and mean packet size L_i. The product of these two variables represents the input bandwidth of the priority:

$$I_i = \lambda_i L_i. \tag{1}$$

The total available output bandwidth T will be divided between the priority classes and each of them will get B_i.

For the bandwidth calculation an iterative method will be used. The kth iteration of B_i will be noted as $B_{i,k}$.

3.2. Model Proposal. To describe the bandwidth allocation of WFQ, we have to analyze all possible situations that can occur. We will use an iterative method for the analysis.

Let us take a look at the possible situations that can appear in the first step of bandwidth allocation. The WFQ algorithm works at the principle that a number of bits represented by the weight value are sent at once to a virtual output. The bits are then reorganized to the original packets and the packet which is completely transmitted in this way is dequeued as the first. This assures an exact bandwidth allocation between queues according to assigned weights. The distribution of the available bandwidth can be written as follows:

$$T \frac{w_i}{\sum_{j=1}^{P} w_j}. \tag{2}$$

After the bandwidth is divided between the queues according to (2), there are 3 possible situations.

    (i) The first possibility is that each queue gets and uses the bandwidth calculated in (2). No additional sharing of unused bandwidth will happen. This will happen if

$$T \frac{w_i}{\sum_{j=1}^{P} w_j} \leq \lambda_i L_i = I_i, \quad i = 1, 2, \ldots, P. \tag{3}$$

    (ii) The second option is that each queue is satisfied with the assigned bandwidth. In this case:

$$T \frac{w_i}{\sum_{j=1}^{P} w_j} \geq \lambda_i L_i = I_i, \quad i = 1, 2, \ldots, P. \tag{4}$$

In these two cases, the bandwidth assignment is finished in the first iteration step. No unused bandwidth needs to be divided between other queues. A queue gets the bandwidth which it needs (1) or the proportion of bandwidth based on the WFQ rules (2):

$$B_{i,1} = \min\left(I_i, T \frac{w_i}{\sum_{j=1}^{P} w_j}\right). \tag{5}$$

This (5) represents also our first iteration step.

If the conditions (3) or (4) are not met, we have to calculate the bandwidth assignment in the next iteration steps. This means some queues need more bandwidth than has been assigned using (2), some others use only the bandwidth calculated in (1), and the rest of the bandwidth is unused and can be shared. We will reassign the unused bandwidth only between the queues whose requirements are not satisfied. The queues that do not need more bandwidth can be identified as follows:

$$I_i - B_{i,k-1} = 0. \tag{6}$$

If the queues bandwidth requirements are met, the result of (6) will be zero. On the other hand a positive number indicates that the queue needs more bandwidth. This will help us to identify the queues with enough bandwidth or with bandwidth shortage.

The reallocation of the unused capacity will be done only between the queues whose bandwidth requirements are not satisfied until all capacity is divided or all queue requirements met and can take $P - 1$ steps in the worst case. The next iterative step can be written as follows:

$$
\begin{aligned}
B_{i,k} = \Bigg(& I_i, B_{i,k-1} + \left(T - \sum_{j=1}^{P} B_{j,k-1} \right) \\
& \times \frac{w_i}{\sum_{j=1}^{P} w_j \min \left(I_j - B_{j,k-1}, 1 \right)} \Bigg).
\end{aligned} \tag{7}
$$

Equation (7) will be used for calculation of all other iterations from $k = 2$ to $k = P$. The calculation has to stop after all bandwidth requirements of the queues are met otherwise it leads to division by zero. The conditions for the termination of the calculation are as follows.

  (i) The whole output bandwidth is already distributed between the queues:

$$
T = \sum_{i=1}^{P} B_{i,k}, \tag{8}
$$

  (ii) or all the requirements of the queues are satisfied:

$$
B_{i,k} = I_i = \lambda_i L_i, \quad i = 1, 2, \ldots, P. \tag{9}
$$

These conditions are also met if in the next iteration no redistribution of bandwidth occurs:

$$
B_{i,k} = B_{i,k-1}, \quad i = 1, 2, \ldots, P. \tag{10}
$$

4. Analysis of Different Behavior Variants

Let us demonstrate the performance of our model in the comparison with WFQ on some examples. In these examples we will assume 4 priority classes. We will show 4 different behaviors. The first example presents the situation, where all traffic classes get the required bandwidth. The second one shows the case in which the bottleneck link has less capacity than is needed and the distribution is done according to packet size and weights. The third example shows us the worst case in which redistribution of bandwidth occurs and the calculation of bandwidth takes P iterations. The last example demonstrates also bandwidth reallocation but the reallocation process will stop after less than P iterations.

Example 1. In this example we will assume a 100 Mbps output link. The first class represents a VoIP flow with high traffic. The mean packet size is set to 100 B what equals to $L_1 = 800$ bits. The packets enter the system with a mean inter packet interval 10 ms, which represents an arrival rate of $\lambda_1 = 100$. The second class represents a video conference

with $L_2 = 8000$ bits and $\lambda_2 = 10$. The third class represents video streaming with the same parameters. The last fourth class transports data with lowest priority settings. The traffic parameters are $L_4 = 12000$ bits and $\lambda_4 = 1$.

The input bandwidths calculated using (1) are $I_1 = 80$ kbps, $I_2 = 80$ kbps, $I_3 = 80$ kbps, and $I_4 = 12$ kbps.

The weights are set in the following way: $w_1 = 4$, $w_2 = 2$, $w_3 = 2$, and $w_4 = 1$. The bandwidth allocated to the queues according to (2) is 40 Mbps, 30 Mbps, 20 Mbps, and 10 Mbps, which is more than all the queues need. In this case the iterations are stopped after the first step (5) and the bandwidth used by the queues is the lower value of this equation I_i. We stopped the iterations according to the condition defined in (9).

Example 2. This example uses the same traffic settings as in Example 1. The only difference is that the output link capacity is set to 50 kbps.

The bandwidth allocation calculated using (2) is 20 kbps, 15 kbps, 10 kbps, and 5 kbps. This represents the whole 50 kbps output capacity. None of the traffic classes has enough capacity to redistribute and the bandwidth allocation is done again in the first iteration.

Example 3. In this example we will show the worst case in which the bandwidth allocation stops after the maximal P steps. We will use the same packet size $L_i = 375$ B in all queues. The weights are again set as follows: $w_1 = 4$, $w_2 = 2$, $w_3 = 2$, and $w_4 = 1$. There are different arrival rates that are set to modify the required bandwidth and present the reallocation. The arrival rates are set to $\lambda_1 = 1000$, $\lambda_2 = 1041.667$, $\lambda_3 = 1000$, and $\lambda_4 = 416.667$. The output link capacity is set to 10 Mbps.

This settings result into the following bandwidth requirements calculated using (1): 3 Mbps, 3.125 Mbps, 3 Mbps, and 1.25 Mbps, where the sum of these bandwidths is higher than the output capacity. All the bandwidth calculations are also visible in Table 1.

In the first iteration the bandwidth allocated using (2) is 4 Mbps, 3 Mbps, 2 Mbps, and 1 Mbps. The first traffic class can use only 3 Mbps of the assigned capacity and the remaining 1 Mbps is divided between the remaining 3 classes. This result corresponds with the proposed model (5).

In the second iteration the result of (6) is equal to zero for the first flow which means that the remaining capacity will be divided between classes 2, 3, and 4. The remaining 1 Mbps is divided again according to the weights in the following way: 0.5 Mbps, 0.333 Mbps, and 0.167 Mbps and added to the already assigned bandwidth.

In the 3rd iteration the remaining capacity of 0.375 Mbps is divided between classes 3 and 4 by the ratio of 2 : 1 due to the assigned weights. This capacity is added to the previously assigned and results into 3 Mbps, 3.125 Mbps, 2.583 Mbps, and 1.292 Mbps.

In the 4th and last reallocation of bandwidth the unused capacity 0.042 Mbps of class 4 is reassigned to the last unsatisfied class 3 and fully used. The resulting allocation of bandwidth is as follows: 3 Mbps, 3.125 Mbps, 2.625 Mbps, and 1.25 Mbps.

TABLE 1: Allocation of bandwidth in Example 3.

Traffic class	Required (Mpbs)	Bandwidth						
		1st iteration		2nd iteration		3rd iteration		4th iteration
		Allocated (Mpbs)	Unallocated (Mpbs)	Allocated (Mpbs)	Unallocated (Mpbs)	Allocated (Mpbs)	Unallocated (Mpbs)	Allocated (Mpbs)
Traffic class 1	3	4	1	3		3		3
Traffic class 2	3.125	3		3.5	0.375	3.125		3.125
Traffic class 3	3	2		2.333		2.583		2.625
Traffic class 4	1.25	1		1.167		1.292	0.045	1.25

All these results correspond with the proposed models (5) and (7).

Example 4. This example describes the bandwidth allocation, where the calculation has to be stopped after the conditions in (8) or (9) are met.

The weight and packet size settings are the same as in the previous Example 3. The output bandwidth is set again to 10 Mbps. The arrival rates are set to 1000, 1000, 750, and 500 pps (packets per second). These settings lead to bandwidth requirements of 3, 3, 2.25, and 1.5 Mbps as a result of (1).

In the first iteration using (2) we allocate 4, 3, 2, and 1 Mbps to the queues. In this case the first queue has 1 Mbps remaining for reallocation and the second queue is already satisfied with the allocated bandwidth.

The second iteration reassigns the 1 Mbps divided using the ratio 2 : 1 to queues 3 and 4. The bandwidth assigned to them is 2.667 Mbps and 1.333 Mbps, but queue 3 needs only 2.25 Mbps output capacity and the remaining part of the capacity can be reassigned to the last unsatisfied queue 4.

In the third iteration we assign 3 Mbps to the first queue, 2 Mbps to the second queue, 2.25 Mbps to the third queue, and 1.75 Mbps to the last queue. The fourth queue needs only 1.5 Mbps and this means that bandwidth requirements of all queues are met. The iterations have to stop at this moment according to (9) otherwise the model would lead to dividing by zero.

We can change the arrival rate of the fourth queue to 750 pps and raise the bandwidth requirements 2.25 Mbps. In this case in the 3rd iteration the bandwidth allocations are 3, 3, 2.25, and 1.75 Mbps. This means that the whole output capacity is divided to the queues (8) and we can stop the iteration. Otherwise each next step would lead to same results.

5. Simulations

To proove the results of our mathematical model we used simulations in the NS2 simulation software [9] (version 2.29) with DiffServ4NS patch [10].

For the simulations a simple network model with 4 transmitting nodes (1–4) and four receiving nodes (6–9) was used. The transmitting and receiving nodes are interconnected with one link between nodes 0 and 5. The node 0 uses WFQ to schedule packets on this bottleneck link

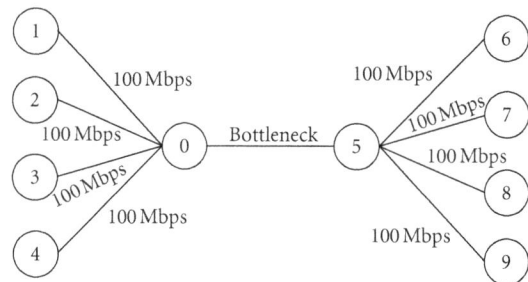

FIGURE 1: Simulation model.

where the mentioned bandwidths are set. All other links have a capacity of 100 Mbps. The model is shown in Figure 1. The queues at node 0 have enough capacity so no packet loss will occur.

We used two types of traffic sources. The first one generates packets only with one packet size and constant packet interval. These settings are easier to simulate and represent a $D/D/1/\infty$ Markovian model.

The second traffic source type represents an $M/M/1/\infty$ model. There is a lack of possibility to generate traffics with different packet sizes in NS2 simulator. For this reason the M/M/1 source is modeled using an ON/OFF source where each node generates one packet with a random size (exponential distribution) and the interval for the next packet transmission is a random time (again a random number with exponential distribution).

An example of input data generated at one node with the mean packet size 375 B and arrival rate 1000 pps is shown in Figures 2 and 3. The red line represents the number of packets generated corresponding with the exponential probability calculated for these settings and the blue bars represent the histogram of packets generated in the simulation that lasted 100 s.

We made many simulations under different parameter settings. The presented results correspond with described examples or present different extreme settings. The results of simulations of M/M/1 and D/D/1 models and the results of our proposed model are shown in Table 2.

We measured the bandwidth after achieving "steady state." The measurement started after 20 s of simulation when the bandwidth was stable and queues filled up with waiting packets [11].

TABLE 2: Simulation results compared with mathematical model results.

Simulation variant (#)	(1)	(2)	(3)	(4)	(5)	(6)	(7)	(8)	(9)
Mean packet size (B)	100, 1000, 1000,1500	100, 1000, 1000, 1500	375, 375, 375, 375	375, 375, 375, 375	375, 375, 375, 375	1000, 100, 10, 1	1000, 1000, 1000, 1000	1000, 100, 1000, 100	1000, 1000, 1000, 1000
Mean arrival rate (pps)	100, 10, 10, 1	100, 10, 10, 1	1000, 1041.67, 1000, 416.67	1000, 1000, 750, 500	1000, 1000, 750, 750	100, 100, 100, 100	1, 10, 100, 1000	1000, 100, 1000, 100	100, 100, 100, 100
Input bandwidth (Mbps)	0.08, 0.08, 0.08, 0.012	0.08, 0.08, 0.08, 0.012	3, 3.125, 3, 1.25	3, 3, 2.25, 1.5	3, 3, 2.25, 2.25	0.8, 0.08, 0.008, 0.0008	0.008, 0.08, 0.8, 8	8, 0.08, 8, 0.08	0.8, 0.8, 0.8, 0.8
Weight settings	4, 3, 2, 1	4, 3, 2, 1	4, 3, 2, 1	4, 3, 2, 1	4, 3, 2, 1	1, 1, 1, 1	4, 3, 2, 1	4, 3, 2, 1	40, 3, 2, 1
Link capacity (Mbps)	100	0.05	10	10	10	0.5	4	8	1.6
D/D/1 simulation results (Mbps)	0.079, 0.079, 0.079, 0.012	0.019, 0.015, 0.01, 0.005	3.00, 3.15, 2.59, 1.248	2.99, 2.99, 2.249, 1.50	2.99, 3.00, 2.25, 1.749	0.41, 0.08, 0.0079, 0.0008	0.0078, 0.08, 0.8, 3.111	5.226, 0.079, 2.6133, 0.079	0.79, 0.4, 0.267, 0.133
M/M/1 simulation results (Mbps)	0.0804 ± 0.47%, 0.0801 ± 1.34%, 0.0802 ± 1.26%, 0.0122 ± 1.69%	0.0200 ± 0.06%, 0.0150 ± 0.10%, 0.010 ± 0.12%, 0.0050 ± 0.36%	3.0041 ± 0.13%, 3.1296 ± 0.14%, 2.6134 ± 0.30%, 1.2529 ± 0.27%	3.0061 ± 0.43%, 3.0022 ± 0.23%, 2.2501 ± 0.32%, 1.5007 ± 0.33%	3.0054 ± 0.34%, 3.0030 ± 0.13%, 2.2564 ± 0.21%, 1.7352 ± 0.66%	0.4095 ± 0.21%, 0.0808 ± 1.07%, 0.0085 ± 1.01%, 0.0013 ± 0.77%	0.0080 ± 2.73%, 0.0799 ± 0.97%, 0.7991 ± 0.67%, 3.1129 ± 0.17%	5.2261 ± 0.01%, 0.0804 ± 0.42%, 2.6131 ± 0.07%, 0.0805 ± 0.57%	0.8015 ± 0.56%, 0.3993 ± 0.57%, 0.2662 ± 0.57%, 0.1331 ± 0.56%
Model results (Mbps)	0.08, 0.08, 0.08, 0.012	0.02, 0.015, 0.01, 0.005	3, 3.125, 2.625, 1.25	3, 3, 2.25, 1.5	3, 3, 2.25, 1.75	0.411, 0.08, 0.008, 0.0008	0.008, 0.08, 0.8, 3.112	5.2266, 0.08, 2.6133, 0.08	0.8, 0.4, 0.266, 0.133

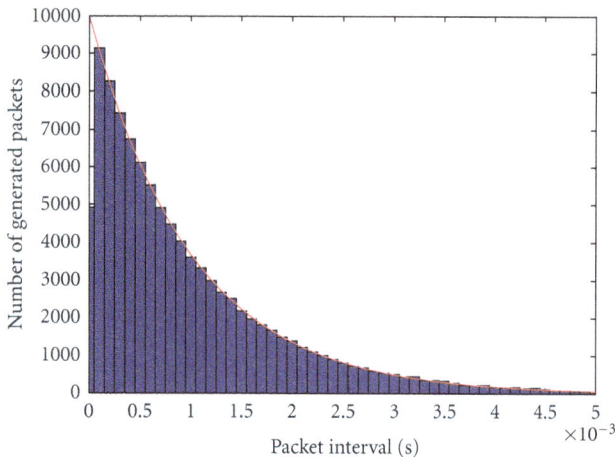

Figure 2: Exponential probability distribution of arrival rate with mean value 1000 pps.

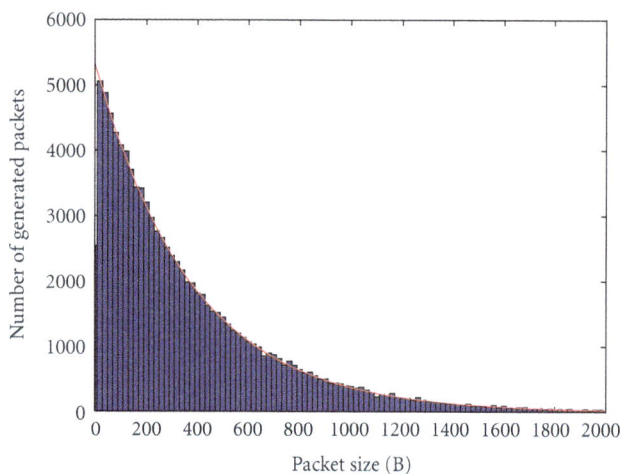

Figure 3: Exponential probability distribution of packet sizes with mean value 375 B.

The results of the mathematical model mostly correspond with the simulation results. The results of the D/D/1 simulation model are more exact due to the exact setting of packet size. The small inaccuracy can be caused by measurement errors, where the bandwidth calculation is stopped closely to an arrival of a packet when small arrival rates are set. Due to the deterministic parameter settings there is no difference between more runs of simulations and no result variance occurs.

The presented results for the M/M/1 simulations are an average value calculated from 10 simulation runs and the standard deviation of the runs is also provided. The simulation runs for most parameter settings lasted 200 s. In cases an extreme low arrival rate was set we extended the simulation duration up to 1000 s. We provided also simulations with WF²Q+ [12] scheduler instead of WFQ. The simulation results correspondent with the presented results and proved that this model is applicable also for other WFQ based schedulers that use packet size for the dequeue order decision.

6. Conclusion

We presented a new iterative bandwidth allocation model for WFQ in IP based NGN networks. The proposed model uses the weight settings of the WFQ scheduler and average input bandwidth of different flows for the bandwidth calculation. The variable utilization of different queues and packet redistribution is considered in the calculations. The proposed model allows to easily predict the impacts of the scheduler, traffic shapers, and input traffics on QoS of the transported data.

The functionality of the model was presented on five different examples and confirmed by simulations in the NS2 simulator for both D/D/1 and M/M/1 input traffics.

The proposed iterative bandwidth allocation model was tested with WF²Q+ scheduler with the same simulation results. Therefore we can say that proposed model is also applicable on other WFQ based schedulers.

The results of this bandwidth allocation model will be used in further research of delay and packet loss modeling using Markovian queue models.

Acknowledgments

This work is a part of research activities conducted at Slovak University of Technology Bratislava, Faculty of Electrical Engineering and Information Technology, Institute of Telecommunications, within the scope of the project "Support of Center of Excellence for SMART Technologies, Systems and Services II., ITMS 26240120029, cofunded by the ERDF."

References

[1] T. Mišuth and I. Baroňá, "On queueing systems application in IP networks," in *EE ČAsopis Pre Elektrotechniku A Energetiku*, vol. 16, pp. 91–94, 2010.

[2] T. Mišuth, E. Chromý, and M. Kavacký, "Prediction of traffic in the contact centers," in *Proceedings of the 6th International Conference on Electrical and Electronics Engineering (ELECO'09)*, pp. 111–114, Bursa, Turkey, November 2009.

[3] A. Demers, S. Keshav, and S. Shenker, "Analysis and simulation of a fair queuing algorithm," in *Proceedings of the ACM Computer Communication Review (SIGCOMM '89)*, pp. 3–12, 1989.

[4] L. Zhang, "Virtual clock: a new traffic control algorithm for packet switching networks," in *Proceedings of the ACM Transactions on Computer Systems*, vol. 9, no. 2, pp. 101–124, 1990.

[5] K. Molnar and M. Vlcek, "Evaluation of bandwidth constraint models for MPLS networks," *Electronics*, vol. 1, pp. 172–175, 2009.

[6] RFC, 4125, *Maximum Allocation Bandwidth Constraints Model for DiffServ-Aware MPLS Traffic Engeneering*, The Internet Society, 2005.

[7] RFC, 4126, *Max Allocation with Reservation Bandwidth Constraints Model for Diffserv-Aware MPLS Traffic Engineering & Performance Comparisons*, The Internet Society, 2005.

[8] RFC, 4127, *Russian Dolls Bandwidth Constraints Model for DiffServ-Aware MPLS Traffic Engeneering*, The Internet Society, 2005.

[9] The Network Simulator-ns-2, http://www.isi.edu/nsnam/ns/.

[10] Andreozzi, "Sergio. DiffServ4NS," http://sergioandreozzi .com/research/network/diffserv4ns/.

[11] K. Pawlikowski, H. D. J. Jeong, and J.S. R. Lee, "On credibility of simulation studies of telecommunication networks," *IEEE Communications Magazine*, vol. 40, no. 1, pp. 132–139, 2002.

[12] J. C. R. Bennet and H. Zhang, "Hierarchical packet fair queuing algorithms," in *Proceedings of the IEEE/ACM Transactions on Networking*, pp. 675–689, 1997.

14

Parallel Mesh Adaptive Techniques Illustrated with Complex Compressible Flow Simulations

Pénélope Leyland,[1] Angelo Casagrande,[1] and Yannick Savoy[2]

[1] EPFL STI GR-SCI-IAG, Station 9, 1015 Lausanne, Switzerland
[2] APCO Technologies, Chemin de Champex 10, CH-1860 Aigle, Switzerland

Correspondence should be addressed to Angelo Casagrande, angelo.casagrande@epfl.ch

Academic Editor: Antonio Munjiza

The aim of this paper is to discuss efficient adaptive parallel solution techniques on unstructured 2D and 3D meshes. We concentrate on the aspect of parallel a posteriori mesh adaptation. One of the main advantages of unstructured grids is their capability to adapt dynamically by localised refinement and derefinement during the calculation to enhance the solution accuracy and to optimise the computational time. Grid adaption also involves optimisation of the grid quality, which will be described here for both structural and geometrical optimisation.

1. Introduction

The accuracy of a numerical simulation is strongly dependant on the distribution of grid points in the computational domain. For this reason grid generation remains a topical task in CFD applications. Prior knowledge of the flow solution is usually required for a grid to be efficient, that is, matching the features in the flow field with appropriate grid resolution. This, however, may not be available, requiring human intervention in analysing the results of an initial solution, going back to the preprocessing stage, and taking an educated guess at how the mesh should be modified. Alternatively, a generally fine grid over most parts of the domain is generated to obtain a relatively good solution. Both of the above cases however, require excessive time, effort, and computational resources.

Let us consider the case with manual intervention by the user. This step can be automated by adaptation, whereby the flow solution is analysed automatically, following some predefined criteria, and the grid resolution adjusted to the problem. The use of such techniques allows for computationally precise distribution of grid points (rather than eye precision) and for extremely reduced user intervention,

thus addressing the time and effort issues. It also resolves problems related to computational time and costs, as the adapted grid can have fewer overall points, with similar resolution in areas of interest, than an unadapted fine mesh.

Grid enrichment (h-refinement) is used here; that is, the density of grid points is increased in regions in order to minimise the space discretisation error [1].

In this method the mesh topology is drastically changed, as nodes are added and removed in order to capture flow features and at the same time reduce the computational load in areas where the solution is sufficiently smooth. Therefore it is particularly suitable for unstructured grids, where the structure can undergo significant changes.

The criteria for refinement and derefinement can be based on solution-based criteria and/or error estimation criteria. Grid enrichment may be further divided into two main streams, grid remeshing and grid subdivision. We will be using the second method, with the grid being divided into smaller elements where necessary. New nodes are added to edges that are identified for refinement, and in turn the cells are divided. Therefore, it is easy to see how the use of unstructured grids can be particularly beneficial. The advantage of this method is its speed and efficiency.

Drawbacks of this technique are the complex data structure and most often, the lack of information of the underlying geometry on the bounding surfaces.

This technique can be approached in two different ways, with a hierarchical framework which saves parent-child relationship between cells at every step and a non-hierarchical approach which discards the history of the original mesh during the filiation of successive grids. The method adopted here is completely nonhierarchical [2], since higher quality meshes can be achieved. This is due to the greater flexibility gained from the omission of the original macromesh, which in turn allows the use of high performance structural optimisation algorithms. With the use of efficient (de)refinement techniques, this approach is well suited for transient problems or for producing coarse grids to be used in multigrid algorithms. In fact, the resulting grids are almost equivalent to those obtained by remeshing, with much less computational time required.

The paper is organised as follows. Grid adaption is treated in detail in Section 2. Numerical results are shown in Section 3. Parallel performance aspects are discussed in Section 4. Finally, Section 5 outlines the conclusions.

2. Parallel Grid Adaptation Techniques

The algorithm outlined in [3] allows the solution of the problem, resulting from a space discretisation on a given grid, say $\mathcal{T}_h^{(0)}$. Once a preliminary solution of $\mathcal{T}_h^{(0)}$ has been obtained, one or several steps of grid adaptation cycles are considered in order to improve the solution accuracy and to optimise the computational resources. This means that, after convergence on $\mathcal{T}_h^{(0)}$, the grid is adapted one or several times. At adaptation cycle i, the solution $\mathbf{U}^{(i)}$ corresponding to the solution of the pseudotransient problem on the grid $\mathcal{T}_h^{(i-1)}$ is projected on the grid $\mathcal{T}_h^{(i)}$ and then used as a starting solution for the problem

$$S\frac{d\mathbf{U}}{dt} + \mathbf{R}(\mathbf{U}) = \mathbf{0}, \qquad (1)$$

discretised on $\mathcal{T}_h^{(i)}$, with S a nonsingular (lumped) mass matrix.

Note that different space discretisation techniques can be used at different grid adaptation cycles. In general, we start with first-order schemes on nonadapted grids, then we turn to second-order schemes on adapted grids.

The goal of grid adaptation is to increase the accuracy of the solution process by locally enforcing the h-adaptivity using smaller discretisation elements. This process tends to uniformly equidistribute the local error η_h throughout the grid, $\mathcal{T}^h$.

The first step in a grid adaptation algorithm is therefore to locally evaluate criteria corresponding to the solution error estimate and mark out the zones to be modified in order to minimise globally the error. The criteria used should be as close as possible to the error estimations of the underlying discretisation scheme, taking as adaptation criteria functions of the current solution field (local error). There are several derivations of the adaptation criteria. One

way is to evaluate the error based on the form of the original equations, a priori, which is a challenging task for nonlinear complex systems such as the hyperbolic-elliptic system of the Euler equations. Another is to evaluate an optimisation procedure based on derivatives. For the grid adaptation procedures developed here, a strategy based on an a-posteriori error estimate where the computed residual of the solution is used to define the error [4, 5] has proved to be robust and precise for inviscid flows and for tracking discontinuities.

Strictly speaking, all the error estimation-based criteria require a complete formulation of the underlying discretisation scheme of the non-linear system that is used to model the physical problem. For finite element discretisations, there are several rules for assuring admissibility, conformality, and regularity of the geometrical properties of the grid elements [6]. Also, the various forms of the adaptation criteria can be exactly deduced from the discretised system which assures accuracy and stability. This is applicable for model problems and even the incompressible Navier-Stokes equations. For numerical schemes for hyperbolic problems, and even more so for the compressible Euler or Navier-Stokes systems, the exact formulation of the discretisation schemes is still incomplete, especially for equivalent finite volume-type schemes, where the error estimation can be obtained by duality arguments on model problems. For the multidimensional upwind schemes used in the present work, this is still an open issue. In [7], some advances are made in this direction. A special mention must be made on recent works of discontinuous Galerkin methods, which allow deep and complex mathematical background and hence render "exact" error estimates [8, 9]. However, these techniques are not the concern here. The choice of an adaptation criterion starts with the study of the partial derivative operators of the underlying equations and hence reflects the physical phenomena. It is hence logical that physical criteria enter into the criterion. The criteria used here are mostly based on physical quantities evaluated on the solution u_h and the residual R_h. Another concern is the regularity of the grid, which also is a function of the physics coming from the equations. Indeed, anisotropic grid adaptation techniques, for example, for boundary layer adaptation, are based on working on the regularity of the grid within a certain metric coming from the equation system and detecting the dominating directionality [10–12]. Here, isotropic grid adaptation is required as the fundamental properties of the system are of wave nature. Regularity and grid optimisation strategies are developed using techniques based on error redistribution and spring analogy techniques.

Adaptation requires in all cases a local error estimate per grid cell, $\eta(T_k)$, where $\mathcal{T}_h = \bigcup_k T_k$, ponderated by some tolerance levels:

$$\max_{T_k} \frac{\eta(T_k)}{\tilde{\eta}(T_k)} = \delta. \qquad (2)$$

Here, $\tilde{\eta}(T_k)$ can be the average on the neighbours of T_k. If the ratio $\eta(T_k)/\tilde{\eta}(T_k) > \delta$, then T_k is to be refined.

In this work local estimates are all based on a posteriori criteria, which require a solution on the starting grid. Then, grid refinement and coarsening operations are performed, taking as adaptation criteria functions of the current solution field (local error) and geometrical properties of the current grid (optimisation).

These criteria are used to perform both grid refinement and derefinement operations at first. Then, an optimisation step follows, based on geometrical properties of the current grid, followed by repartition, reordering, and renumbering. These phases are now detailed.

2.1. Adaptation Criteria. The physical adaptation criteria adopted in this work are based on flow quantities such as density, Mach number, pressure, and entropy, as are also error estimators, but differ in the simplicity of their construction. In fact these use directly the physical quantities mentioned. A first method is to take the difference between the values at the nodes of a segment and use its absolute value as an indicator for the adaptation process. Although this may seem as a very crude way of identifying flow features, it is very effectively applied to the grid enrichment method mentioned earlier. Another method employed is the undivided gradient along an edge [13]:

$$\varepsilon = \frac{\partial u}{\partial x} h, \tag{3}$$

which discretely can be written as

$$\varepsilon = \Delta u. \tag{4}$$

From this it can be clearly seen that the value of ε, which should approximate the error, decreases as the mesh size h becomes smaller.

Various modifications to this method have been developed, such as inclusion of local mesh length scale:

$$\varepsilon = \Delta u \cdot \Delta x. \tag{5}$$

This leads to a more effective refinement criterion [14], as the simple form of (4) remains approximately constant in the vicinity of shock waves, due to the steepened shock wave profile as the mesh is refined and the jumps remaining relatively constant. The drawback is a heavier weight of larger cells than smaller ones because of the additional length scale, even in regions of smooth flow, leading to global refinement. Although these criteria have been successfully employed, they are not optimal. This is due to the tendency of excessively refining the mesh.

In fact the so-called "physical" adaptation criteria correspond to exact mathematical error estimators. In [5], it is shown that for the linear advection diffusion problems, the evaluation of the jump of a characteristic variable across an edge is equivalent to the evaluation of the discrete H^1 norm of the solution. The relation between physical and mathematical criteria is therefore very close.

Two adaptation criteria used herein are based on physical criteria. Let us consider a solution field U that has been evaluated over the entire mesh for the selected criterion function [15]. A low- or high-pass filter is then applied on the solution field, which we will then denote as $\hat{U}$. For each

segment, the function $f = f(\hat{U})$ is computed, where $f(\hat{U})$ is one of the following:

(i) the difference of $\hat{U}$ between the vertices of the segment

$$\left| \hat{U}_a - \hat{U}_b \right|. \tag{6}$$

(ii) the gradient of $\hat{U}$ between the vertices of the segment

$$\frac{\left| \hat{U}_a - \hat{U}_b \right|}{l_{ab}}, \tag{7}$$

where a and b are the end nodes of the segment and l_{ab} is the segment length. For the 2D case only, the choice also includes

(i) the upwind flux of $\hat{U}$ through the segment,

(ii) the downwind flux of $\hat{U}$ through the segment.

Further control on the field is obtained through the use of a filter F that removes part of the segments from the field. This can be applied in two ways, as an offset value, or as a cut-off value (Figure 1). In the first case each node of the mesh is examined and the following is applied:

$$\hat{U} = \max \left(0, \hat{U} - F \right). \tag{8}$$

In the second case, the above changes to

$$\hat{U} = \min \left(F, \hat{U} \right). \tag{9}$$

The refinement criterion is then built with $\overline{f}$, the mean value of f, over the grid. The segment i is then marked, for splitting in the case of refinement or for remaining in the case of the derefinement step, if

$$f_i \geq \mathcal{F}\overline{f}, \tag{10}$$

where $\mathcal{F}$ is a factor used to set the criterion as a function of the mean value $\overline{f}$ (Figure 2). In other words $\mathcal{F}$ is a coefficient that multiplies the mean value of the segment field, and all segments with higher values are marked.

2.2. Mesh Refinement and Coarsening. The adaptation procedures developed here apply for general shaped elements in 2 or 3 dimensions. They are based on the concept of an element built up as an agglomerate of a cell and its neighbours, called a shell. The filters work on the shells, as the structural optimisation procedures that are described below. The first step is the local refinement and coarsening step, which requires evaluation of the criteria and successive marking out of the interior properties of the shells.

Initially, the algorithm tests all the segments of the existing grid and decides whether a new node has to be created on each segment or not. Usually, a low-pass and a high-pass filter are applied to the solution fields. These filter operations render an error estimation in an appropriate norm, and also detect discontinuities. By filtering the

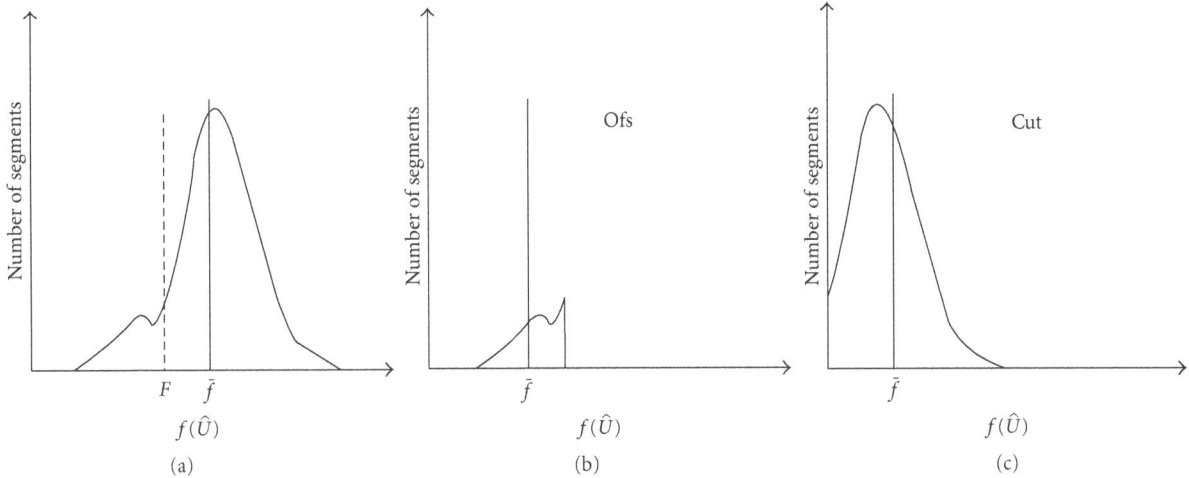

FIGURE 1: Graphical representation of the filter F and its effect.

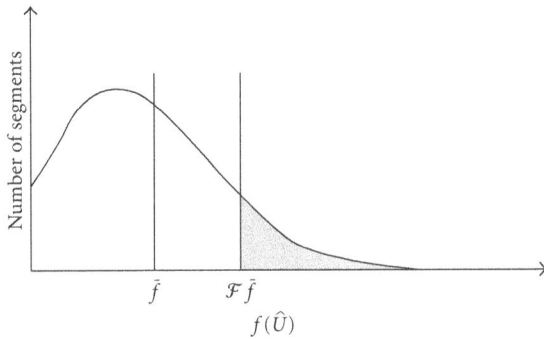

FIGURE 2: Graphical representation of factor $\mathcal{F}$ and $\overline{f}$ values used to limit segment marking.

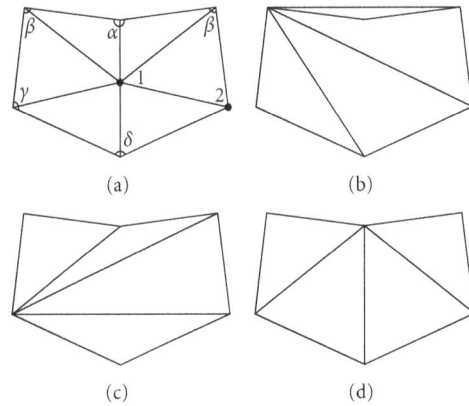

FIGURE 3: Segment collapsing with shell control: (a) initial shell, (b) shell inversion collapse, (c) collapse with element distortion, and (d) best collapse available.

gradient of the pressure or the Mach number, for instance, a local densification and stretching of the grid is applied. Also criteria based on the original geometry of the grid are used to optimise the grid structure. These criteria will depend on factors such as absolute segment length, related to neighbouring cells, and so forth.

Then, in order to minimise the number of nodes and elements and to improve the geometrical quality of the grid, a grid coarsening algorithm is employed. The procedure is essentially as the one outlined for grid refinement. As a result, a set of nodes to be deleted is found. These nodes are removed using an edge collapsing procedure (see Figure 3).

The low- or high-pass filter which is applied on the solution field u_h becomes a certain function $\hat{u}_h$. This can be, for example, the difference of $\hat{u}_h$ between the two vertexes of the segment. Let f_i be the value of $\hat{u}_h$ on the segment i. The refinement is performed on the segment i if $f_i \geq \mathcal{F}\overline{f}$, or it is kept unchanged otherwise. The factor $\mathcal{F}$ is used to set the criterion as a function of the mean value $\overline{f}$. This filtering process acts as a cutoff to the *a posteriori* error, by limiting its domain of influence to only a bandwidth of values.

This segment collapsing method used herein, developed by Savoy and Leyland [15], consists in building a shell around the node marked for removal with its surrounding elements. Let us consider the shell created around node 1 in Figure 3(a). Vertex 2 will not be considered as it is also marked, whilst at all other vertices the inner curvature angle will be calculated. The next step consists in collapsing one of the inner segments in order to delete the centre node. The choice will fall onto the segment that connects the centre node to the neighbouring node with the greater angle associated to it, in this case α. Note that with this technique the risk of cell inversion (Figure 3(b)) and element distortion (Figure 3(c)) is minimised, resulting in improved mesh quality. However, shell volume conservation is also checked, in order to prevent accidental element inversion from happening.

The procedure works in both two and three dimensions and is very efficient in the first case. Efficiency is somewhat reduced in the 3D case due to a large number of constraints imposed during the marking, especially when avoiding

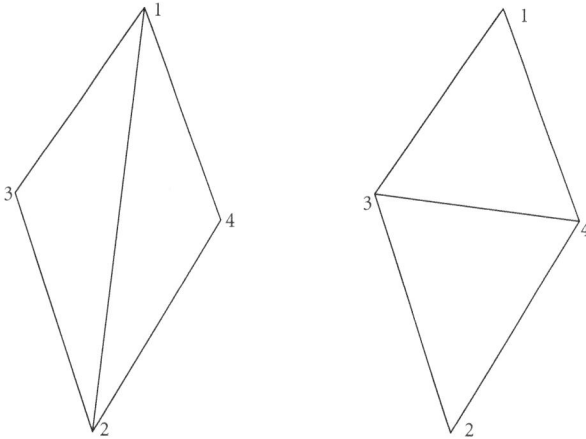

FIGURE 4: Two-dimensional edge swapping.

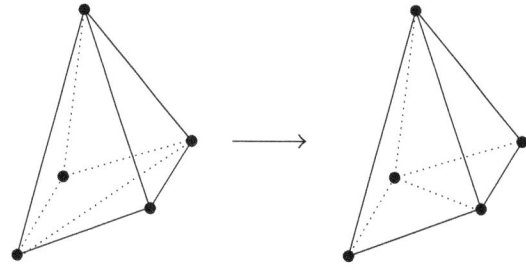

FIGURE 5: Three-dimensional face swapping.

element inversion, which in turn does not allow to remove a large amount of nodes.

2.3. Optimisation Techniques. Mesh quality and precision of the underlying discretisation are highly dependent on the shape of elements and shells just described. Therefore an equilibrium state would be desirable in the cells. This is achieved by equilateral triangles in 2D and equilateral-type tetrahedra in 3D. However, the mesh obtained after the refinement and coarsening steps will be far from this desired equilibrium state. This is due to the different local node density and strong variations between element sizes and nodes angles. The number of node neighbours may also differ dramatically between vertices. In order to overcome these problems arising from the previous steps, the mesh must be optimised. This is done in several ways that may be grouped into two major strategies:

(1) structural optimization:

    (a) diagonal swapping,
    (b) edge collapsing,

(2) geometrical Optimisation:

    (a) spring analogy,
    (b) boundary smoothing,
    (c) inverted elements.

2.3.1. Structural Optimisation. In this step the mesh is analysed and modified in function of the number of node neighbours $\mathcal{N}_i$. Following the Delaunay criterion [16], where the optimal element should be equilateral, $\mathcal{N}_{opt}$ is then related to the number of equilateral elements needed to fill the area around the node. In the two-dimensional case, $\mathcal{N}_{opt} = 6$ and can be easily calculated by considering $\pi/3$, in the Euclidean metric, as the optimal node angle. For the three-dimensional case, the spherical angle of the tetrahedron at each vertex is considered and the number of neighbouring elements calculated. The Euler-Descartes

relation is then used to find the number of neighbouring nodes, leading to $13 < \mathcal{N}_{opt} < 14$ (for further details see [7, 15]).

Diagonal Swapping. This consists in swapping the internal edge of two neighbouring triangles, as shown in Figure 4, for the two-dimensional case.

The procedure is carried out to reduce the number of node neighbours $\mathcal{N}_i$ when this is greater than $\mathcal{N}_{opt}$. This is done by checking $\mathcal{N}_i$ on all vertices implicated in the operation. In particular the swapping is performed if the following conditions are satisfied:

$$\mathcal{N}_3 + \mathcal{N}_4 + 2 < \mathcal{N}_1 + \mathcal{N}_2 \tag{11}$$

or

$$\mathcal{N}_3 + \mathcal{N}_4 + 2$$
$$\max(\mathcal{N}_3, \mathcal{N}_4) + 1 < \max(\mathcal{N}_1, \mathcal{N}_2). \tag{12}$$

The three-dimensional case requires more effort and attention, as the swap implies a face swapping, leading to complete remeshing of the shell built with the elements surrounding the deleted segment. The volume conservation must also be checked in order to avoid cell inversions during the shell remeshing. An example of a face swap is shown in Figure 5.

Edge Collapsing. This intervention is done when $\mathcal{N}_i < \mathcal{N}_{opt}$, and although the method is similar to the one shown in Section 2.2 the scope is completely different. As for the swapping, the collapsing criteria are applied to the segments. Let $\mathcal{N}_1$ and $\mathcal{N}_2$ be the node neighbour numbers for the two vertices of the given segment, with $\mathcal{N}_1 \leq \mathcal{N}_2$. The collapsing is done by deleting the node which corresponds to $\mathcal{N}_1$. The collapsing is performed if

$$\mathcal{N}_2' \leq \mathcal{N}_2 \quad \text{or} \quad \mathcal{N}_2' \leq \mathcal{N}_{opt}, \tag{13}$$

where $\mathcal{N}_2'$ is the node neighbours' number resulting from the collapsing. It can be deduced from $\mathcal{N}_1$, $\mathcal{N}_2$, and $\mathcal{M}$ the number of cells surrounding the segment:

$$\mathcal{N}_2' = \mathcal{N}_1 + \mathcal{N}_2 - \mathcal{M} - 2. \tag{14}$$

These criteria are valid in both two and three dimensions. An example of edge collapsing in 2D is shown in Figure 6.

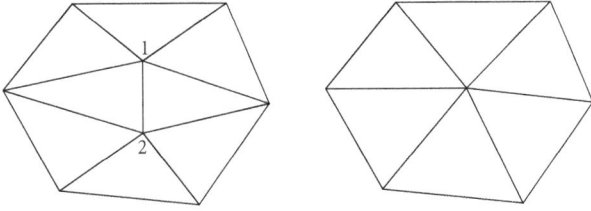

FIGURE 6: Edge collapsing in 2D.

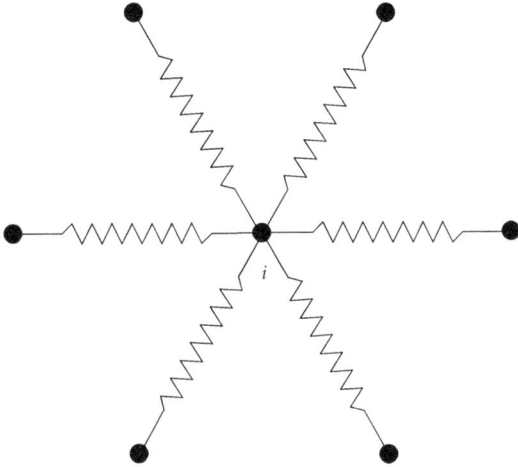

FIGURE 7: Spring analogy: springs replacing segments.

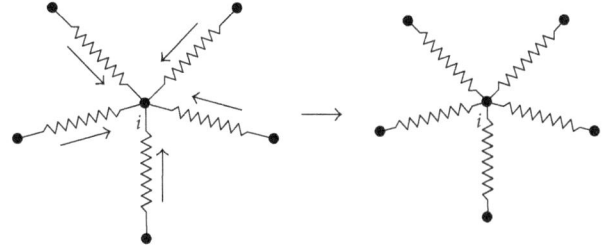

FIGURE 8: Springs' movement based on node neighbours: springs contracting.

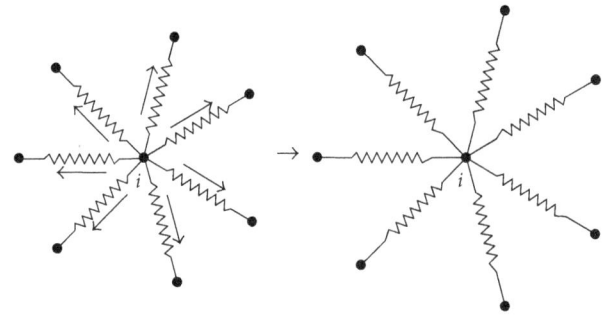

FIGURE 9: Springs' movement based on node neighbours: springs expanding.

2.3.2. Geometrical Optimisation. The goal of this step is to modify the mesh without changes to the global data structure. This is achieved primarily by means of node displacement, based on spring analogy. However, other techniques must be applied to ensure a better handling of the node displacement. Node neighbours' number, for example, will be employed again for adjusting the spring stiffness. Particular care will be given to nodes lying on the bounding geometry and avoiding element inversion.

Spring Analogy. This technique has been heavily developed for moving mesh algorithms [17–19], for instance. We have adapted these concepts, to the present strategy of parallel mesh adaptation. Here each segment in the mesh is replaced by an elastic spring (Figure 7). The objective is then to minimise the deforming energy of the overall elastic system. This will result in the force **F** at node i obtained using Hooke's law:

$$\mathbf{F}_i = \sum_{j \in k(i)} \alpha_{ij} \left(\mathbf{x}_j - \mathbf{x}_i \right) = \mathbf{0}, \qquad (15)$$

where $k(i)$ represents the set of node neighbours of vertex i, with size $\mathcal{N}_i$, and α_{ij} denotes the spring stiffness of the segment joining node i with neighbour j. Hence the equilibrium position at coordinates $\mathbf{x}_i$ can be expressed as

$$\mathbf{x}_i = \frac{\sum_{j \in k(i)} \alpha_{ij} \mathbf{x}_j}{\sum_{j \in k(i)} \alpha_{ij}}, \qquad (16)$$

which can be resolved using a Jacobi iterative scheme.

Spring Stiffness. Node neighbours' number is once again very useful for mesh optimisation. In fact, if spring stiffness α_{ij} were to be set to one in order to produce equilateral elements, the following would occur:

(i) if $\mathcal{N}_i < \mathcal{N}_{\text{opt}}$, $k(i)$ move towards i (Figure 8);

(ii) if $\mathcal{N}_i > \mathcal{N}_{\text{opt}}$, $k(i)$ move away from i (Figure 9).

To partially avoid this problem, the following weight function can be used to determine the spring stiffness:

$$\alpha_{ij} = \alpha_j = \max \left[1, \mathcal{N}_{\text{opt}} + \mathcal{A} \left(\mathcal{N}_i - \mathcal{N}_{\text{opt}} \right) \right]. \qquad (17)$$

This relates the spring stiffness to $\mathcal{N}_j$, the number of neighbours for the node $j \in k_i$. It also introduces the smoothing lineal factor $\mathcal{A}$, which is set manually.

Boundary Nodes. Nodes lying on the geometric boundaries have to be moved with caution (if moved at all). Whatever the method used for positioning the node on the underlying geometry, a sufficient node density must be guaranteed within critical regions where the boundary curvature is large. This can be achieved by maintaining boundary nodes with a new spring joining the reference point $\widetilde{\mathbf{x}}$ and the new position $\mathbf{x}^{n+1}$. The stiffness β_i of this new spring is then chosen as a function of the local maximum boundary curvature. The resulting force is then calculated as

$$\mathbf{F}_i = \sum_{j \in k_\Gamma(i)} \alpha_j \left(\mathbf{x}_j - \mathbf{x}_i \right) + \beta_i (\widetilde{\mathbf{x}}_i - \mathbf{x}_i) = \mathbf{0}, \qquad (18)$$

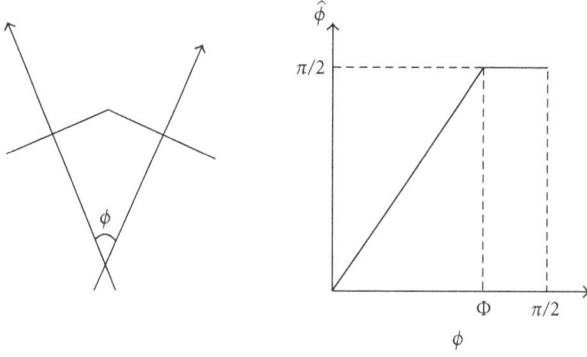

FIGURE 10: Curvature angle and filter.

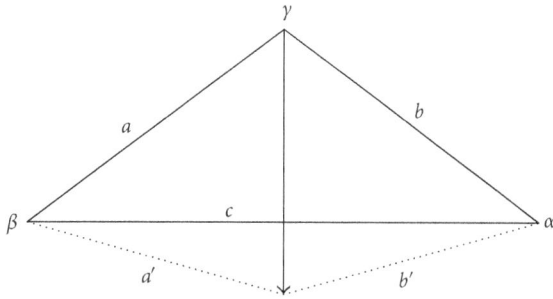

FIGURE 11: Inverted elements' *snap-through*.

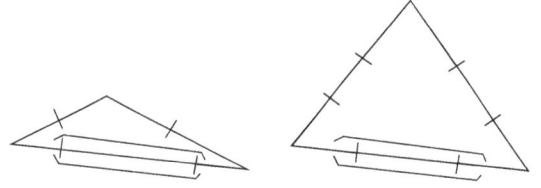

FIGURE 12: Cell inversion *stops*.

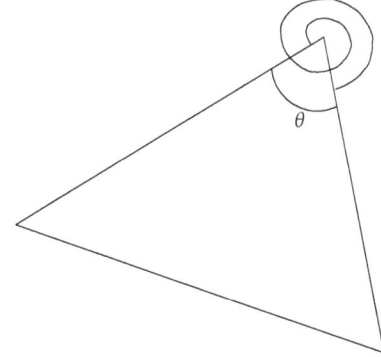

FIGURE 13: Torsional spring.

where $k_\Gamma(i)$ represents the subset of $k(i)$ which contains all the node neighbours located on the boundary. The following formulation may then be obtained substituting $\tilde{\mathbf{x}}$ by $\mathbf{x}^n$:

$$\mathbf{x}_i^{n+1} = \frac{\sum_{j \in k_\Gamma(i)} \alpha_j \mathbf{x}_j^n + \beta_i \mathbf{x}_i^n}{\sum_{j \in k_\Gamma(i)} \alpha_j + \beta_i}. \qquad (19)$$

The stiffness of the new spring is then defined as a function of the curvature angle ϕ. This angle is first filtered such that the node displacement is restricted, especially when it exceeds a given value Φ (Figure 10):

$$\beta_i = \mathcal{B}\left(\frac{1}{\cos^2\hat{\phi}} - 1\right) \quad \text{with } \mathcal{B} \geq 0, \qquad (20)$$

where $\mathcal{B}$ is a user-defined boundary stiffness factor.

Inverted Elements or Torsional Springs. This is a major issue, [18], which needs to be controlled thoroughly, as it causes loss in overall volume mesh conservation. It may occur when a vertex crosses over the opposite face of the element, which inverts the cell volume. This phenomenon, called *snap-through*, is shown in Figure 11 and is prone to happening on the boundary when this moves. The configuration shown has a low energy as the springs a and b rotate. To remedy this, the segment spring analogy is used together with initially rigid mesh boundaries, then semitorsional springs are placed in the corner between adjacent edges, that is, the stiffness of segment c is divided by the angle between segments a and b. As the sum of the angles is equal to π, the stiffness is

approximately unchanged if the triangle is equilateral. For deformed elements instead, the vertex angles that are closer to 0 or π become rigid.

Cell inversion may also occur inside the heart of the mesh. A method to avoid this can be devised by setting critical cells rigid, with segment springs working in only one direction, rendering a relaxation of the elements. The vertex movement is then made free if it increases the element quality, which means that the introduced segment springs work like *stops* (Figure 12).

To determine the stiffness of the segment spring when it acts as a *stop*, the angular deformation energy of the cells is computed. A torsion spring is set at the opposite angle of each cell surrounding the segment (Figure 13):

$$\mathbf{C} = \mathcal{C}\left(\frac{1}{\sin^2\theta} - 1\right) \quad \text{if } \sin^2\theta < \sin^2\Theta, \qquad (21)$$

where Θ is a filter value and $\mathcal{C}$ a user-defined torsion stiffness factor. It allows to take into account only the most critical angles. The maximum of the torsion spring for a given segment is then converted to a segment spring using the following relation:

$$\gamma = \frac{1}{\delta}\mathbf{C} = \frac{\mathcal{C}}{\delta}\left(\frac{1}{\sin^2\theta} - 1\right), \qquad (22)$$

where δ is the distance between the segment and the opposite vertex in 2D and the opposite edge in 3D.

The nonisotropic behaviour of the *stops* causes the problem to be nonlinear. A time advancing strategy must be implemented, with the *stops* relaxing during the evolution of the procedure. The force applied on the node i is then given by

$$\mathbf{F}_i = \sum_{j \in k(i)} \alpha_j \left(\mathbf{x}_j - \mathbf{x}_i\right) + \sum_{j \in k(i)} \gamma_{ij}(\tilde{\mathbf{x}}_i - \mathbf{x}_i) = \mathbf{0}, \qquad (23)$$

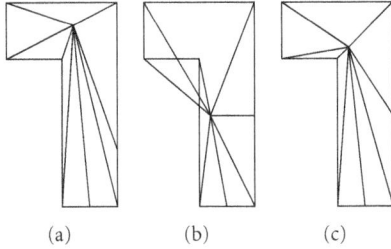

FIGURE 14: Torsion spring effect: (a) initial grid, (b) cell inversion, and (c) torsion spring.

which leads to the following formulation:

$$\mathbf{x}_i^{n+1} = \frac{\sum_{j \in k(i)} \alpha_j \mathbf{x}_j^n + \sum_{j \in k(i)} \gamma_{ij} \mathbf{x}_i^n}{\sum_{j \in k(i)} \alpha_j + \sum_{j \in k(i)} \gamma_{ij}}. \qquad (24)$$

Finally the effect of the torsion spring is shown in Figure 14.

We note that smoothing and stretching algorithms are global, and hence the parallelisation of the these algorithms requires a global renumbering of all the grid nodes. This means that each node in the overlapping region will be assigned to the update set of a unique processor before the smoothing and stretching can be performed. Apart from the renumbering phase, the communication required to solve (16) by the Jacobi method is essentially the same of the parallel matrix-vector product outlined in [3].

2.4. Parallelisation of Grid Adaptation Techniques. In order to perform the grid adaption procedures on a parallel computer, two approaches can be followed. The first one is the *master-slave* approach, in which a processor is responsible for the management of the grid data (and in general of I/O routines). In [20, 21], the authors have discussed the limitations and demonstrate the relative performances of unstructured calculations using the master-slave approach. Once a preliminary solution has been obtained, it is gathered from the slave processors to the master processor, where sequential grid adaptation is started. Then, the grid is partitioned using a graph partitioning algorithm and redistributed among the processors. This approach works quite well for "small" grid, and in general steady-state problems. For evolutionary problems and/or intensive calculations, a *no-master* approach is required, in which the completely parallel dynamic grid adaptation algorithm takes place entirely on the network of processors. This is precisely the approach we have followed. More detailed description of the no-master approach may be found in [22].

The paradigm we have adopted is based upon the concept of nonhierarchical grid adaptation; that is, the successive grids do not remember their original affiliation; see [23, 24]. This allows high flexibility and quality for the different stages of adaptation as the grid at a certain time does not rely on the background macrogrid. Hence, radical changes and optimisation are possible. Also, efficient automatic dynamic adaptation, which is particularly interesting for following

evolving or transitional phenomena, is facilitated. These concepts can also be developed for general element types when based on a concept of generalised elements consisting of the group of nearest neighbours called "shells," [16, 25]. This non-hierarchical technique, associated with an optimisation, produces similar grids to those obtained by the regriding. The drawbacks reside in the complexity of programming and the coherent reprojection on the geometry definitions defining the boundary surfaces.

Note that the incorporation of parallel grid adaptation within the solution process requires load balancing partitioning techniques to obtain well-balanced subdomains. This introduces other algorithmic concepts such as parallel sorting and renumbering techniques.

The grid adaptation techniques are applied globally throughout a prepartitioned mesh and require careful renumbering and reordering internally per processor (local) and globally of the addresses of the entities, cells, shells, faces, edges, nodes, and so forth, in order that the adaptation renders a global mesh that is in turn re-partitioned again. All this is dynamic and needs to have the partitioning procedure as an integral part of the adaptation procedure.

The parallelisation of the refinement leads to the tracking of nodes created on an updated segment which are considered as a new border (interface) node. For coarsening, the principle that when attempting to delete a border node, a border segment must be chosen was applied.

The parallelisation of the structural changes is one of the hardest points, especially for the choice of overlapping partitions. For these reasons the swapping and collapsing work most efficiently on the internal segments. However, diagonal swapping or face swapping is still straightforward across partition interfaces. Collapsing is often harder to control.

The smoothing procedures do not modify significantly the internal mesh topology; the parallelisation is hence straightforward as long as a coherent numbering of the nodes, segments, faces, and cells is employed.

2.5. Repartitioning. From the point of view of parallel computing, the grid adaptation procedure may result in an unbalanced distribution of the workload among the processors. Hence, the workload for each subdomain may be different, and this can produce an inefficient parallel performance. Effectively, the worker with the largest workload can delay the process. In fact, the starting domain decomposition was obtained to balance the workload for the initial grid (i.e., the same number of nodes on each subdomain and the minimum number of cut elements), whereas the adaptation algorithm could have generated many nodes on some subdomains (leading to more computing resources on the corresponding processors) and may have derefined in subdomains given to other processors (thus requiring less computational resources). Therefore, the computational domain is repartitioned dynamically within the parallel adaptation procedure using a parallel graph partitioning algorithm. For these purposes, the library ParMETIS [26] can be used dynamically within the source code, as well as home-made partitioners as in [22].

2.6. Reorder and Renumber. To complete the parallel adaptation procedure, fast efficient multiple renumbering techniques are necessary for the grid entities: elements, segments, faces, and nodes. For this MPI library routines are called explicitly and a fast dynamic binomial search tree to sort during renumbering procedures is implemented based on a balanced binomial search tree algorithm AVL (Adelson-Velskii and Landis) [27]. Consequently all the vectors and matrices used in the code may have to be reallocated in memory. In particular, the data structure for the parallel matrix-vector product must be recomputed.

An AVL tree is a dynamically balanced binary search tree that is height H balanced. Height balanced means that for every node in the tree, the heights of the left and right subtrees differ by at most one. The height of a tree is the number of nodes in the longest path from the root to any leaf. The implementation is as a recursive structure of interlinked nodes. The difference in height H between different branches is kept minimal by imposing that pairs of such subtrees of every node differ in height by at most 1. As it is a binary tree for n nodes, $H \leq n \leq 2^H - 1$ where the extrema correspond to a balanced tree.

When a new node is inserted into the tree, it appears at the root, then moves along the branches until it finds an attachment to the tree. Once the node is inserted, the tree balance is checked. If no imbalance is found, another node is inserted and the process continues. If an imbalance is found, the heights of some nodes are fixed and the process repeated. When a node is deleted, the root becomes unbalanced. The lookup is performed to balance again.

Lookup, insertion, and deletion operations are of $O(\log n)$, where n is the number of nodes in the tree, when the tree is balanced. Search steps $S(n)$ needed to find an item are bounded by $\log(n) \leq S(n) \leq n$.

3. Numerical Results

In order to assess the various functionalities of the techniques in place, a few test cases have been carried out in both two and three dimensions. For the two-dimensional case, the transonic flow over a NACA 0012 airfoil is used. For the three dimensional case we consider the supersonic flow over a forward wedge and different flow regimes for a concept aircraft. For all test cases, a parallel, unstructured grid, Euler solver THOR [3] was used.

3.1. NACA0012 Airfoil at $M_\infty = 0.80$ and $\alpha = 1°$. For this standard 2D test, the starting, nonadapted grid is composed of 2355 nodes. A first-order solution is computed on this grid; then, 4 steps of adaptation are performed, in order to improve the solution quality. The adaptation criteria are based on the density. Figure 15 shows the evolution of the grid, whose final size is of 3831 nodes and 7471 elements. The shock positions on the windward and leeward side are stabilised by the adaptation procedure. Note that the final grid is identical to the original one, except within the shock regions, where nodes have been added. The MURD scheme employed was a blended second-order scheme based on

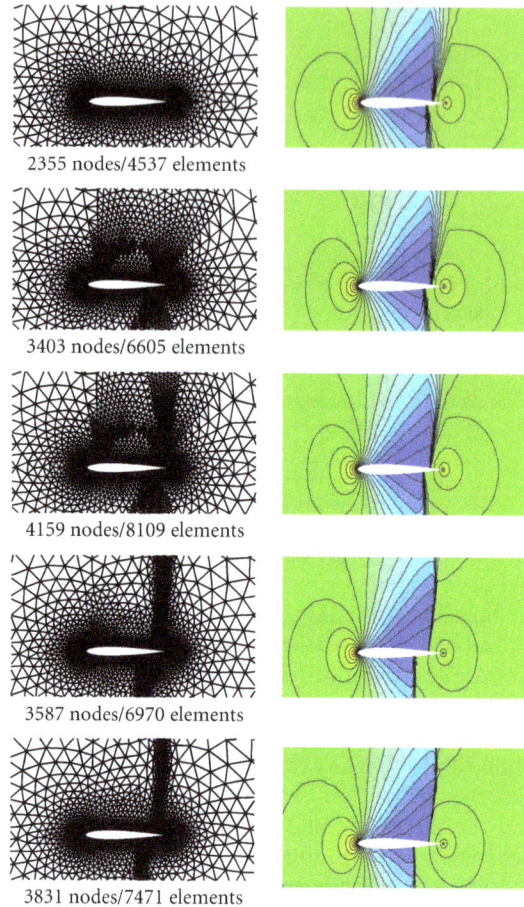

2355 nodes/4537 elements

3403 nodes/6605 elements

4159 nodes/8109 elements

3587 nodes/6970 elements

3831 nodes/7471 elements

FIGURE 15: Evolution of the NACA0012 grid and the corresponding solution field (density) for $M_\infty = 0.80$ and $\alpha = 1°$.

a strategic switch between the Lax-Wendroff and the PSI schemes (see [28] for further details).

3.2. Three Dimensional Forward Wedge at $M_\infty = 2.0$. In the second test case we present is a 3D forward wedge. We start from a rather coarse, hand-made grid, while the final adapted grid is composed by 80629 nodes and 480442 elements. The adaptive module is then used to refine the grid according to the physics of the solution field. This test case is interesting since despite its simple geometry, it presents shock reflections of different strength, which are to be captured by the adaptative procedure.

The successive grids and their corresponding solutions are presented in Figure 16. After each adaptation step, based on the gradient of the density, the number of nodes is roughly multiplied by a factor of 4.

The reasonably good quality of the last grid requires a large number of smoothing iterations. It is indeed essential to proceed carefully to avoid any element inversion. The solution scheme chosen is the standard N-scheme, which is a first-order scheme. Note that even if the starting grid is too coarse to permit an acceptable solution, adaptivity allows to obtain a solution which clearly captures the complex physics of the this problem.

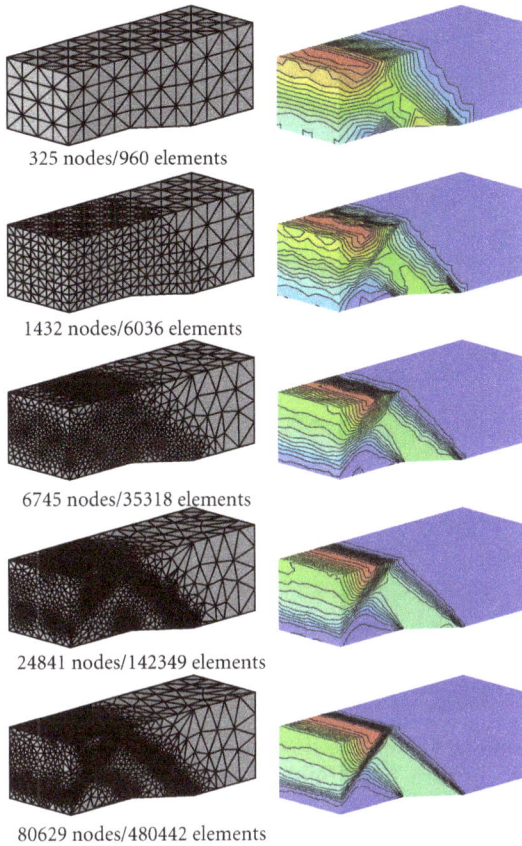

FIGURE 16: Evolution of the 3-dimensional wedge and the corresponding solution field (density) for $M_\infty = 2.0$.

3.3. Concept Aircraft.

The second, three-dimensional test case is represented by a concept aircraft, Smartfish [29]. The interest of the geometry in this work is the extremely changing and complex form of the airplane, which poses a challenge for the grid generation and adaptation.

Here we present the results of some adaptations with different initial grid sizes and adapted with different physical criteria. The tests are carried out at transonic Mach numbers and with nonzero angles of attack. In particular we first test a very coarse grid for this type of problem, with 274 899 elements and 48 481 nodes. The first adaptation is done with respect to the change in gradient of the Mach number, with two adaptation cycles. The mesh is heavily refined (Figure 17) but only along the leading edge and not much over the wing.

Moving onto a denser initial grid (742 294 elements and 129 865 nodes), the difference in the Mach number along a segment is considered. Here the adaptation gives a better result (Figure 18), mainly because of the better solution to which it was adapted, due to the finer starting mesh.

Finally a relatively fine grid was used to start the process (1 772 861 elements and 314 913 nodes). The initial conditions are of Mach number 0.9 and angle of attack of 4°. The grid is refined well in the area of the shock, above and below, as shown in Figure 19.

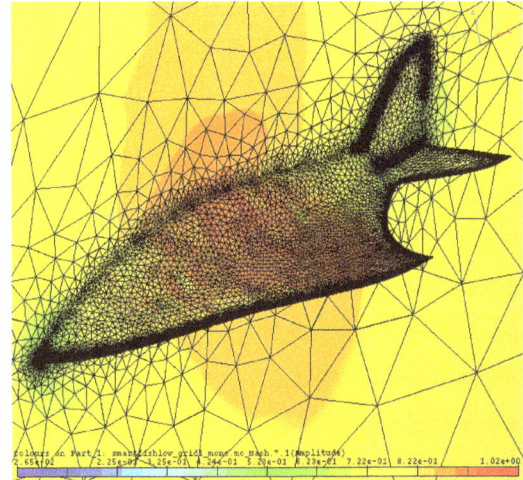

FIGURE 17: Adapted coarse grid with respect to the Mach gradient. 2 adaptation cycles only with refinement at the Mach number 0.8 and angle of attack 2°, 6 569 277 elements and 1 098 081 nodes.

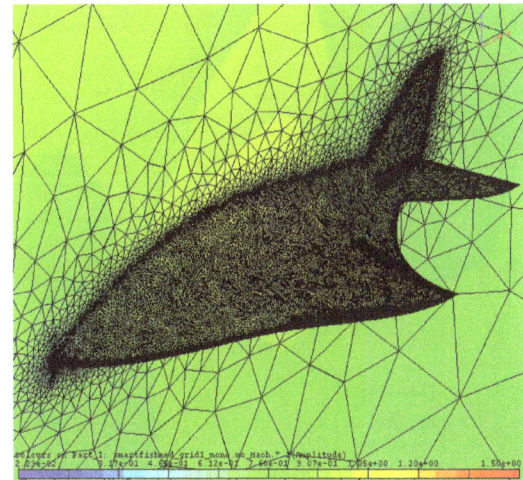

FIGURE 18: Adapted coarse grid with respect to the Mach difference. 1 adaptation cycle with refinement and derefinement at the Mach number 0.9 and angle of attack 4°, 1 795 794 elements and 302 723 nodes.

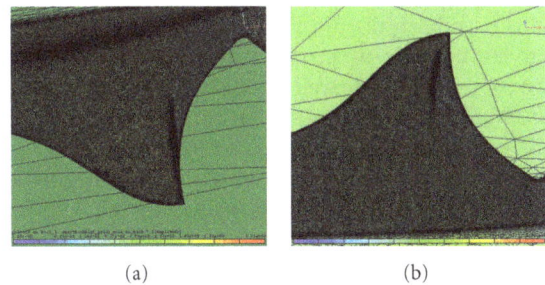

(a)   (b)

FIGURE 19: Adapted finer grid with respect to the Mach difference. 4 adaptation cycles at the Mach number 0.9 and angle of attack 4°, 3 303 715 elements and 555 347 nodes. Upper side of the wing and lower side.

TABLE 1: Final grid sizes for different number of processors after 4 adaptation steps. 2D NACA 0012 airfoil.

Processors	N_nodes	N_elements	N_surf_el
1	35 130	69 615	645
2	35 082	69 527	637
4	34 335	68 031	639
8	34 839	69 035	643

TABLE 2: Final grid sizes for different number of processors after 3 adaptation steps and after 1 adaptation step for 64 and 128 restarting from 32 final solution and grid. 3D wedge.

Processors	N_nodes	N_elements	N_surf_el
8	1 118 350	6 721 070	50 714
16	1 123 326	6 753 716	50 963
32	1 130 577	6 801 332	51 073
64	3 843 209	23 302 721	85 652
128	3 842 653	23 290 638	85 641

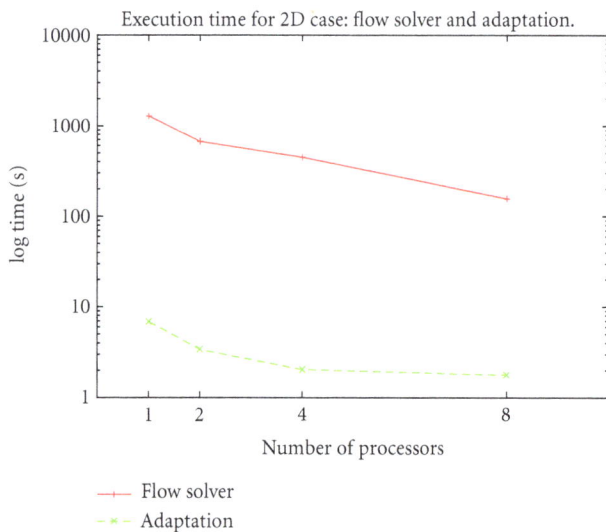

FIGURE 20: Execution time for flow solver and adaptation process on multiple processors. 2D NACA 0012, four adaptation steps.

4. Performance Aspects

In order to verify the performance of the adaptive procedures, some of the test cases presented in the previous sections have been re-run. First a 2D NACA 0012 airfoil is used to test the code on the Linux cluster (*Pleiades* [29]) used for the computations, as well as measuring the CFD code and the adaptation parts elapse time. In particular the nodes used for the computations reported here are biprocessor, bicore. Then various 3D test cases are used to measure the total time of the adaptation process with respect to the CFD time and the breakdown of the adaptation process stages.

4.1. 2D Results. In order to test the charge of the adaptation process, with respect to the total time of the CFD computation, a same NACA 0012 airfoil case was executed with a different number of processors. In particular the adaptation

process was run with refinement and coarsening procedures, adaptation with respect to the Mach gradient, 20 optimisation cycles (swapping and collapsing), and 20 smoothing cycles. Four adaptation steps were carried out; hence from an initial grid of 2 355 nodes, 4 537 triangular elements, and 173 boundary faces, the flow solver is run and the solution adapted in turn four times, and a final solution obtained from the final adapted grid. The final grid characteristics for the computations with different number of processors are reported in Table 1. In Figure 20 instead we report the total computational execution time of the flow solver and that of the adaptation process. As we can notice, the total time of adaptation can be considered negligible compared to the solution time, being at most less than 1.2% of the total CFD computation time.

4.2. 3D Results. In a similar way to that of the 2D case above, we first compare the execution time of the flow solver and that of the adaptation with a different number of processors. The test is carried out with the 3D wedge example, with an initial mesh of 306 415 tetrahedral elements, 54 370 nodes, and 12 906 boundary faces. The adaptation process was run with refinement only, with respect to the difference in density on the segments, 30 optimisation passes, and 10 smoothing cycles. Three adaptation steps were performed, each one after obtaining a partial solution with the flow solver, and a final solution obtained from the final adapted grid. This was carried out for 8, 16, and 32 processors as shown in Table 2 with the final adapted grids characteristics. The final mesh and solution obtained with 32 processors was then used for a single adaptation step, using 64 and 128 processors. In this last case only one solution step is required, as for the adaptation, since the grid is immediately adapted to the solution obtained with the previous computation.

Here the computations start from 8 processors, rather than 1, due to memory requirements for the grid obtained after the third adaptation step. Figures 21 and 22 show the computational times plotted for the three adaptation steps and for the restart, on a different number of processors. Once

TABLE 3: Step-by-step grid sizes for different number of processors after 1, 2, and 3 adaptation steps, and after 1 adaptation step for 64 and 128 restarting from 32 final solution and grid. 3D wedge.

Steps	Processors	N_nodes	$N\_elements$	N_surf_el
Step 1	4	140 180	811 918	20 902
	8	140 129	811 258	20 868
	16	140 133	810 941	20 915
	32	140 108	810 176	20 941
Step 2	4	362 718	2 145 897	31 844
	8	364 891	2 158 249	31 728
	16	365 409	2 161 629	31 911
	32	367 912	2 176 392	31 923
Step 3	8	1 118 350	6 721 070	50 714
	16	1 123 896	6 756 786	50 931
	32	1 131 649	6 806 732	50 949
Restart	64	3 843 363	23 310 788	85 648
	128	3 842 828	23 298 937	85 647

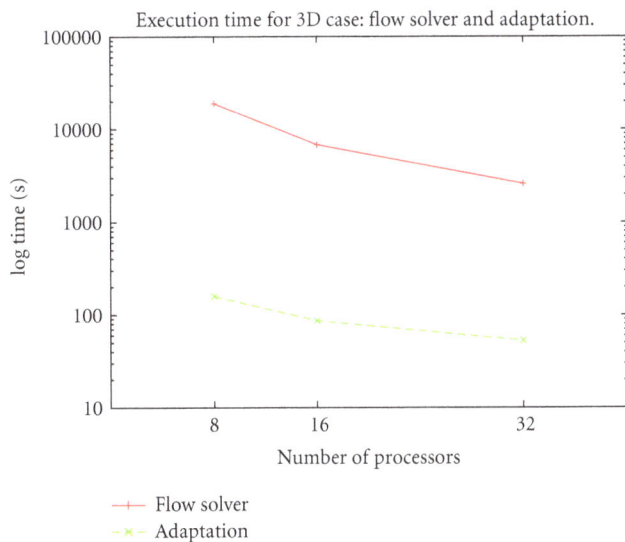

FIGURE 21: Execution time for flow solver and adaptation process on multiple processors. 3D wedge, three adaptation steps.

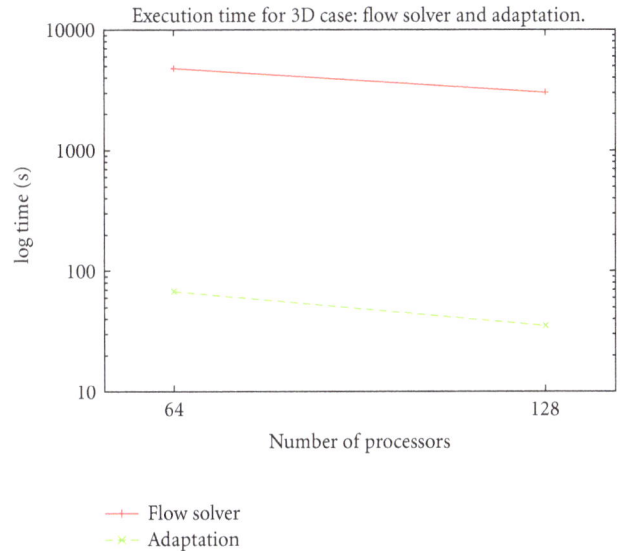

FIGURE 22: Execution time for flow solver and adaptation process on multiple processors. 3D wedge, one adaptation step after restart.

again the total adaptation time is negligible compared to that of the flow solver, reaching at most 2% of the total solution time.

4.3. Adaptation Execution Breakdown. Although the adaptation execution times are far less than the total computation, where the flow solver is accounted for, it is interesting to examine the various stages of the adaptation cycle and see the impact these have on the use of computational resources used. Therefore what follows is a breakdown of the adaptive cycle in three main blocks, with the refinement/derefinement and renumbering as a first block, swapping/collapsing and renumbering a second block, and smoothing being the third and last block.

The previous 3D wedge initial mesh was used to start a computation with two adaptation steps on 4 processors

and a third adaptation step for 8, 16, and 32 processors. The reason for this choice is that it is not possible to run three adaptation steps on 4 processors, due to memory constraints. The final solution and mesh of this last computation were once again used as a starting point for an adaptation step carried out with 64 and 128 processors. Adaptation conditions were maintained the same as for the previous case. Grids for all steps and number of processors are given in Table 3. Figures 23 and 24 show the breakdown of the adaptation process time for the first two steps with multiple processors and Figure 25 shows that of the third step. Figure 26 instead shows the breakdown for the restarted case. As we can see from the above examples, the two optimisation procedures are far more time consuming than the refinement/derefinement block for the case where the difference physical criteria is chosen.

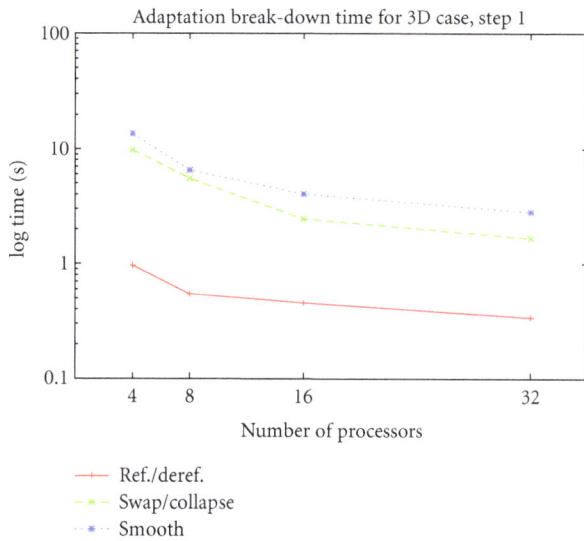

FIGURE 23: Execution time for adaptation blocks on multiple processors. 3D wedge, first adaptation step.

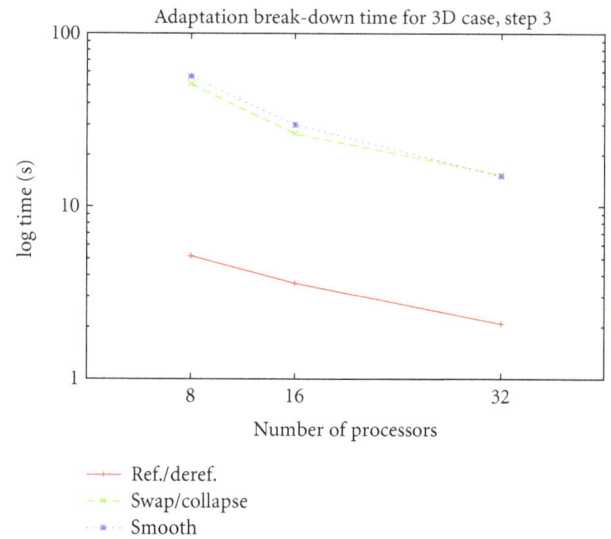

FIGURE 25: Execution time for adaptation blocks on multiple processors. 3D wedge, third adaptation step.

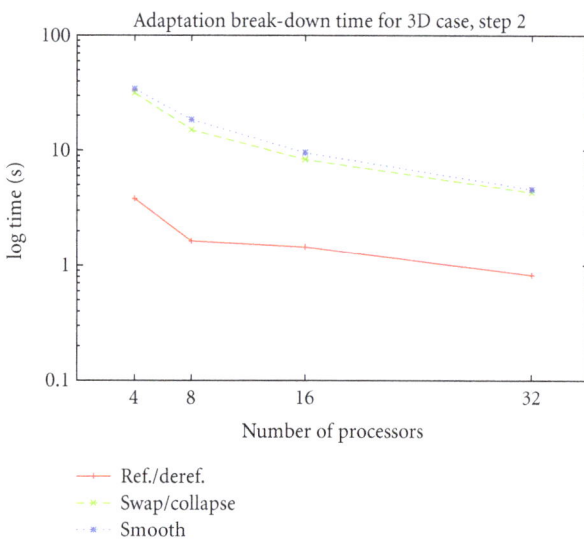

FIGURE 24: Execution time for adaptation blocks on multiple processors. 3D wedge, second adaptation step.

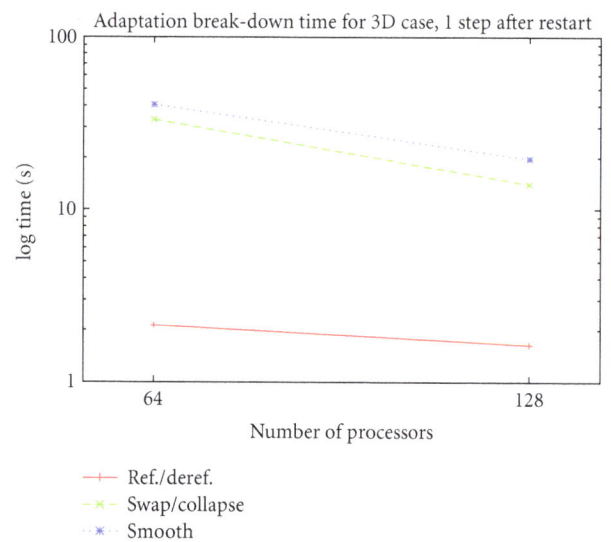

FIGURE 26: Execution time for adaptation blocks on multiple processors. 3D wedge, one adaptation step after restart.

5. Conclusions

In this paper mesh adaptation techniques based on physical phenomena are developed in a parallel environment. The various steps (refinement, coarsening, optimisation, smoothing, reordering, and renumbering) and their algorithms are described. The techniques are validated on extensive complex flow simulations. The parallel adaptation performances show the efficiency of the implementation of these methods.

Acknowledgments

The Swiss National Science Foundation (SNSF) and the Swiss Federal Office for Education and Science (OFES) are acknowledged for financial support.

References

[1] S. Z. Pirzadeh, "An adaptive unctructured grid method by grid subdivision, local remeshing, and grid movement," AIAA Paper 99-3255, 1999.

[2] R. Richter, *Schémas de Capture de Discontinuités en Maillage Non-Structuré avec Adaptation Dynamique: Applications aux écoulements de l'aérodynamique [Ph.D. thesis]*, Ecole Polytechnique Fédérale de Lausanne, Lausanne, Switzerland, 1993.

[3] M. Sala, P. Leyland, and A. Casagrande, "A parallel adaptive Newton-Krylov-Schwarz method for the 3D compressible flow simulations," *Modelling and Simulation in Engineering*. In press.

[4] K. Eriksson, C. Johnson, and J. Lennblad, "Error estimates and automatic time and space step control for linear parabolic problems," *SIAM Journal on Scientific Computing*, 1990.

[5] P. Leyland, F. Benkhaldoun, N. Maman, and B. Larrouturou, "Dynamical mesh adaptation criteria for accurate capturing of stiff phenomena in combustion," *International Journal of Numerical Methods in Heat and Mass Transfer*, 1993, (INRIA Report 1876).

[6] P. G. Ciarlet, "The finite element method for elliptic problems," *Classics in Applied Mathematics*, vol. 40, pp. 1–511, 2002.

[7] M. Sala, *Domain decomposition preconditioners: theoretical properties, application to the compressible euler equations, parallel aspects [Ph.D. thesis]*, Ecole Polytechnique Fédérale de Lausanne, Lausanne, Switzerland, 2003.

[8] B. Rivière, M. F. Wheeler, and V. Girault, "A priori error estimates for finite element methods based on discontinuous approximation spaces for elliptic problems," *SIAM Journal on Numerical Analysis*, vol. 39, no. 3, pp. 902–931, 2002.

[9] R. Hartmann and P. Houston, "Adaptive discontinuous Galerkin finite element methods for nonlinear hyperbolic conservation laws," *SIAM Journal on Scientific Computing*, vol. 24, no. 3, pp. 979–1004, 2003.

[10] W. G. Habashi, M. Fortin, J. Dompierre, M. G. Vallet, and Y. Bourgault, "Anisotropic mesh adaptation: a step towards a mesh-independent and user-independent cfd," in *Barriers and Challenges in Computational Fluid Dynamics*, pp. 99–117, Kluwer Academic, 1998.

[11] L. Formaggia and S. Perotto, "New anisotropic a priori error estimates," *Numerische Mathematik*, vol. 89, no. 4, pp. 641–667, 2001.

[12] T. Apel, *Anisotropic Finite Elements: Local Estimates and Applications. Advances in Numerical Mathematics*, Habilitationsschrift, Teubner, Germany, 1999.

[13] D. J. Mavriplis, "Unstructured mesh generation and adaptivity," Tech. Rep. TR-95-26, ICASE-NASA, 1995.

[14] G. Warren, W. K. Anderson, J. L. Thomas, and S. L. Krist, "Grid convergence for adaptive methods," AIAA Paper 91-1592, 1991.

[15] Y. Savoy and P. Leyland, "Adaptive module," Tech. Rep. TR5.1, IDeMAS, 2000.

[16] G. F. Carey, *Computational Grids: Generation, Adaptation and Solution Strategies*, Taylor & Francis, 1997.

[17] J. T. Batina, "Unsteady Euler airfoil solutions using unstructured dynamic meshes," *AIAA Journal*, vol. 28, no. 8, pp. 1381–1388, 1990.

[18] C. Degand and C. Farhat, "A three-dimensional torsional spring analogy method for unstructured dynamic meshes," *Computers and Structures*, vol. 80, no. 3-4, pp. 305–316, 2002.

[19] F. J. Blom and P. Leyland, "Analysis of fluid-structure interaction by means of dynamic unstructured meshes," *Journal of Fluids Engineering*, vol. 120, no. 4, pp. 792–798, 1998.

[20] R. Richter and P. Leyland, "Distributed CFD using auto-adaptive finite element," in *ICASE/LaRC Workshop on Adaptive Grid Methods*, 1994.

[21] D. J. Mavriplis, "Three-dimension high-lift analysis using a parallel unstructured multigrid solver," Tech. Rep. TR-98-20, ICASE, 1998.

[22] P. Leyland and R. Richter, "Completely parallel compressible flow simulations using adaptive unstructured meshes," *Computer Methods in Applied Mechanics and Engineering*, vol. 184, no. 2–4, pp. 467–483, 2000.

[23] R. Richter, *Schémas de Capture de Discontinuités en Maillage Non-Structuré avec Adaptation Dynamique: Applications aux écoulements de l'aérodynamique*, Ecole Polytechnique Fédérale de Lausanne, Lausanne, Switzerland, 1993.

[24] R. Richter and P. Leyland, "Entropy correcting schemes and non-hierarchical auto-adaptive dynamic finite element-type meshes: applications to unsteady aerodynamics," *International Journal for Numerical Methods in Fluids*, vol. 20, no. 8-9, pp. 853–868, 1995.

[25] Y. Savoy and P. Leyland, "Parallel mesh adaptation for unstructured grids within the IDeMas project," Tech. Rep., IMHEF-DGM EPFL, 2000.

[26] G. Karypis and V. Kumar, "ParMETIS: parallel graph partitioning and sparse matrix ordering library," Tech. Rep. 97-060, Department of Computer Science, University of Minnesota, 1997.

[27] M. A. Weiss, *Data Structure and Algorithm Analysis*, The Benjamin Cummings Publishing Company, 1992.

[28] J. Bastin and G. Rogé, "A multidimensional fluctuation splitting scheme for the three dimensional Euler equations," *Mathematical Modelling and Numerical Analysis*, vol. 33, no. 6, pp. 1241–1259, 1999.

[29] R. Gruber and V. Keller, *HPC@ Green IT: Green High Performance Computing Methods*, Springer, 2010.

15

Corporate-Feed Multilayer Bow-Tie Antenna Array Design Using a Simple Transmission Line Model

S. Didouh, M. Abri, and F. T. Bendimerad

Telecommunications Laboratory, Faculty of Technology, Abou-Bekr Belkaid University, 13000 Tlemcen, Algeria

Correspondence should be addressed to M. Abri, abrim2002@yahoo.fr

Academic Editor: S. Taib

A transmission line model is used to design corporate-fed multilayered bow-tie antennas arrays; the simulated antennas arrays are designed to resonate at the frequencies 2.4 GHz, 5 GHz, and 8 GHz corresponding to RFID, WIFI, and radars applications. The contribution of this paper consists of modeling multilayer bow-tie antenna array fed through an aperture using transmission line model. The transmission line model is simple and precise and allows taking into account the whole geometrical, electrical, and technological characteristics of the antennas arrays. The proposed transmission line model showed its interest in the design of different multilayered bow-tie antennas and predicted the correct resonance frequency for different applications in telecommunications. To validate the proposed transmission line model, the simulation results obtained are compared with those obtained by the method of moments. The results of simulations are presented and discussed. Using this transmission line approach, the resonant frequency, input impedance, and return loss can be determined simultaneously. The paper reports several simulation results that confirm the validity of the developed model. The obtained results are then presented and discussed.

1. Introduction

Microstrip antenna arrays are exploited in a vast number of engineering applications due to their ease of manufacturing, low cost, low profile, and light weight [1, 2].

Antenna arrays are used to scan the beam of an antenna system, increase the directivity, and perform various other functions which would be difficult with any single element. In the microstrip array, elements can be fed by a single line or by multiple lines in a feed network arrangement. Based on their feeding method [3–5] the array is classified in series-feed network or corporate-feed network.

Corporate-feed network is general and versatile because it offers the designer more freedom in controlling the feed of each element (amplitude and phase). Although it leads to performance degradation due to radiation, its constructional simplicity and low cost are still considered. This method has more control of the feed of each element and is ideal for scanning phased arrays, multiband arrays. Thus it provides better directivity as well as radiation efficiency and reduces the beam fluctuations over a band of frequencies compared to the series-feed array. The corporate-feed network is used to provide power splits of $2n$ (i.e., $n = 2, 4, 8, 16$, etc.). This is accomplished by using either tapered lines or using quarter wavelength impedance transformers.

In this paper, a transmission line model is used to design corporate-feed multilayer antennas arrays to resonate at the frequencies of 2.4 GHz, 5 GHz, and 8 GHz corresponding to RFID, WIFI, and radars applications, and the patches chosen as radiating elements for these arrays are in the bow-tie shape. The obtained simulation results are compared with those obtained by the moment's method (MoM).

2. Transmission Line Model Analysis

The preferred models for the analysis of microstrip patch antennas are the transmission line model, cavity model, and full wave model (which include primarily integral equations/Moment Method). The transmission line model is the simplest of all and it gives good physical insight, but it is less accurate.

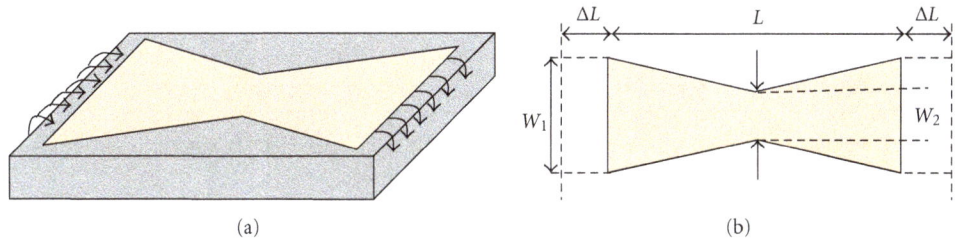

FIGURE 1: Bow-tie antenna and its effective length.

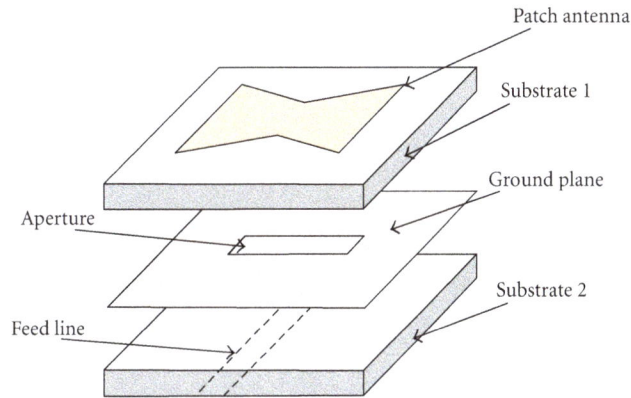

FIGURE 2: Configuration of bow-tie antenna fed by aperture coupled.

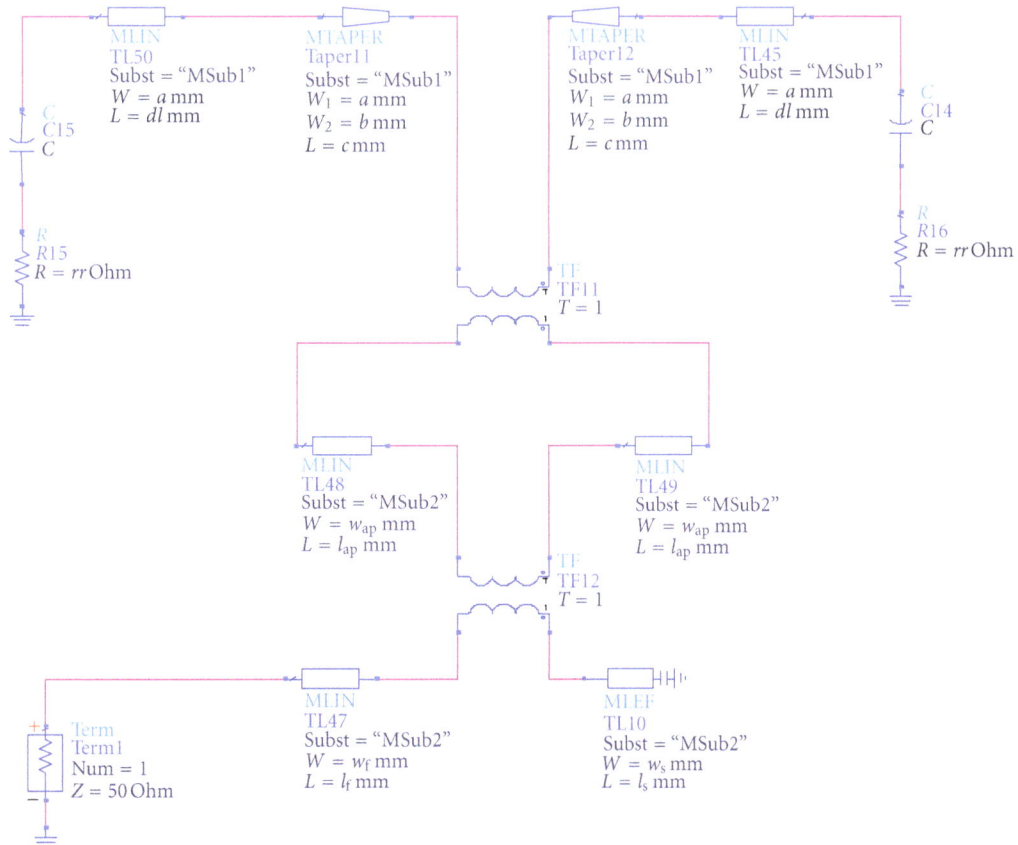

FIGURE 3: Equivalent circuit of the proposed antenna.

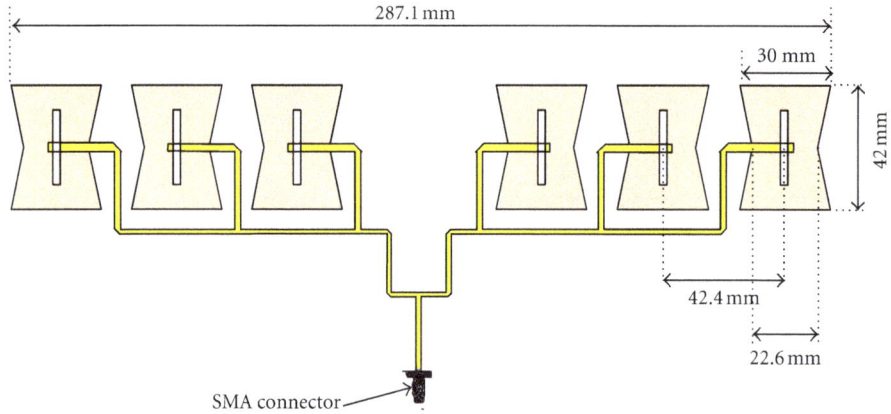

FIGURE 4: Mask of the multilayered bow-tie antenna array operating at the frequency 2.4 GHz.

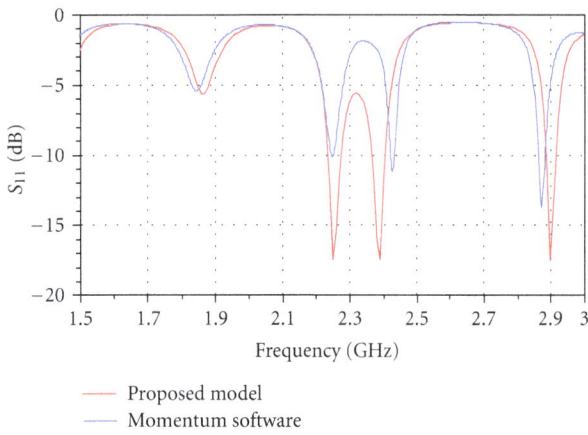

FIGURE 5: Simulated input antenna array return loss.

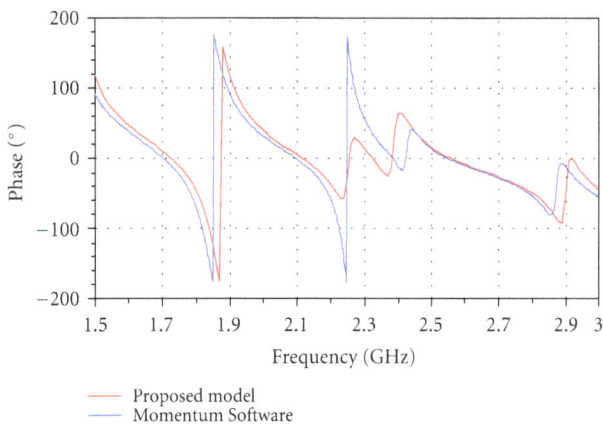

FIGURE 6: Reflected phase at the antenna array input.

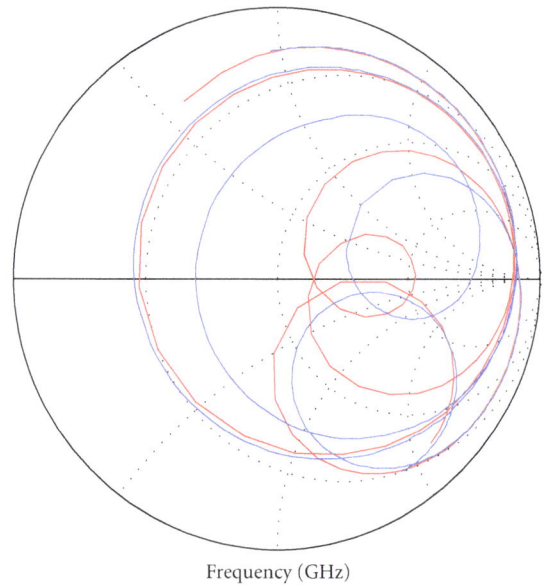

FIGURE 7: Smith's chart of the input impedance return losses.

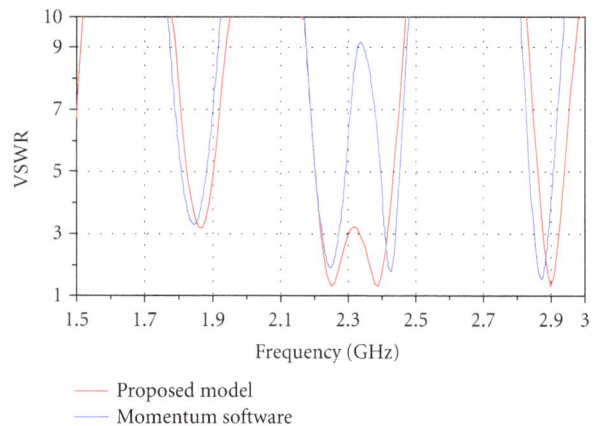

FIGURE 8: Bow-tie antennas array VSWR.

In this study, six bow-tie microstrip radiating elements are used to design the corporate-fed array antenna. A bow-tie microstrip radiating patch which is shown in Figure 1 can be considered as an open-ended transmission line of length L and width W.

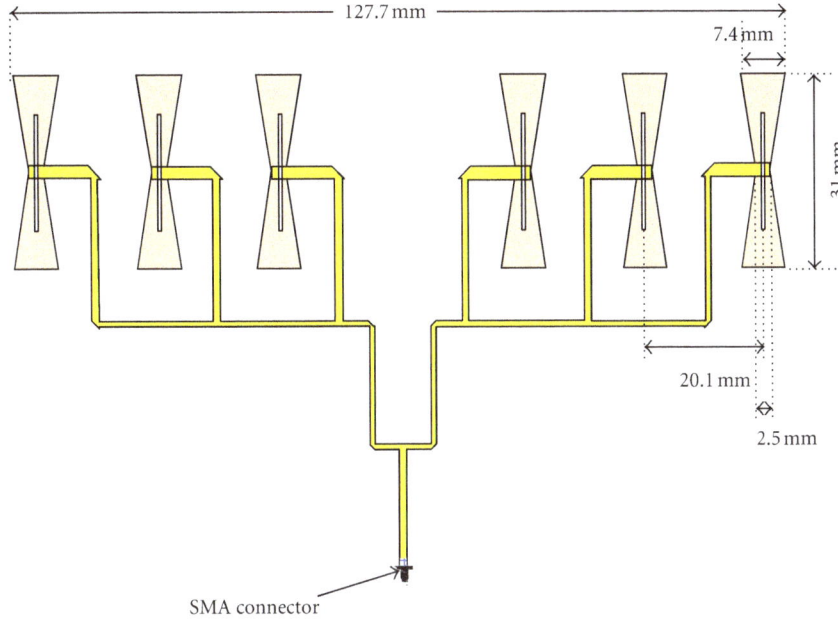

FIGURE 9: Mask of the multilayered bow-tie antenna array operating at the frequency 5 GHz.

3. Microstrip Corporated-Feed Array Antenna

The printed array to be considered is one using aperture-coupled bow-tie microstrip patches. The aperture-coupled patch element [4, 6] consists of two substrates, with a ground plane in between. As shown in the geometry for a single aperture-coupled patch in Figure 2, a microstrip feed line is printed on the bottom (feed) substrate, while the patch element is printed on the top (antenna) substrate. Coupling between the feed line and the radiating element is through a small slot in the ground plane below the patch.

The proposed transmission line equivalent circuit for an aperture coupled bow-tie antennas fed via microstrip line is shown in Figure 3.

In this equivalent circuit, two ideal transformers are assumed between the slot ground plane and both sides of the line. The energy is transferred and stored in these two transformers in terms of load susceptance. In fact, all the energy passes the slot aperture and delivers to the patch for radiating. The ratios of these two transformers can be determined using [4]:

$$n_1 = \frac{L_{ap}}{L}, \quad (1)$$

where L_{ap} is the length of the slot.

While the second transformation ratio n_2 can be approximated by the expression:

$$n_2 = \frac{L_{ap}}{\sqrt{W \cdot h}}, \quad (2)$$

where h is the thickness of the substrate, the capacitance C is calculated using the following equation as in [7]:

$$C(\varepsilon) = \frac{\varepsilon_0 \varepsilon_r A}{h \gamma_n \gamma_m} + \frac{1}{2\gamma_n} \left(\frac{\varepsilon_{reff}(\varepsilon_r, h, W)}{c_0 Z(\varepsilon_r = 1, h, W)} \right) - \frac{\varepsilon_0 \varepsilon_r A}{h}, \quad (3)$$

$$\gamma_j = \begin{cases} 1, & j = 0 \\ 2, & j \neq 0. \end{cases} \quad (4)$$

4. Results and Discussions

The validity of the suggested model is highlighted by comparing the results of the return loss, the input phase, input antenna VSWR, and input impedance locus to those obtained by the moment's method of the Momentum Software. The simulated antennas arrays are designed to resonate, respectively, at the frequencies 2.4 GHz and 8 GHz.

4.1. Bow-Tie Antenna Array Operating at the Resonant Frequency 2.4 GHz. The configuration of the proposed array is shown in Figure 4, which consists of 6 identical bow-tie patch elements in parallel or corporate feed to cover 2.4 GHz operating frequency. The corporate feed has a single input port and multiple feed lines in parallel constituting the output ports. Each of these feed lines terminates at an individual radiating element and therefore transfers all its energy into the element.

The antenna array is to be designed on substrate which has a relative permittivity ε_r of 2.54, a dielectric thickness h of 1.6 mm, and a loss tangent of about 0,019 and 0.05 mm conductor thickness. A rectangular slot with $L_{ap} = 26$ mm and width $W_{ap} = 2.6$ mm is used for coupling the patch to a microstrip line of length $L_f = 20$ mm, etched on substrate which has a relative permittivity ε_r of 2.54, a dielectric

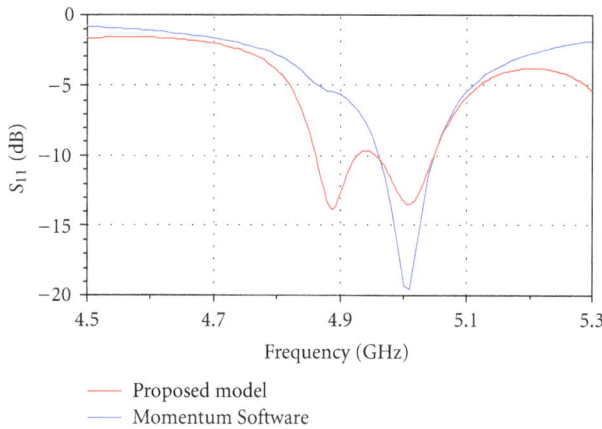

FIGURE 10: Simulated input antenna array return loss.

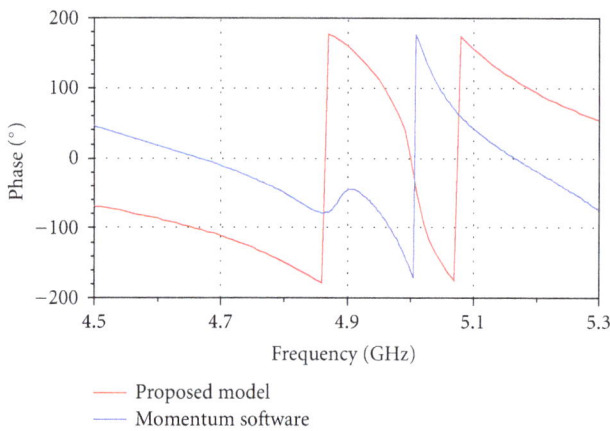

FIGURE 11: The antenna array input-reflected phase.

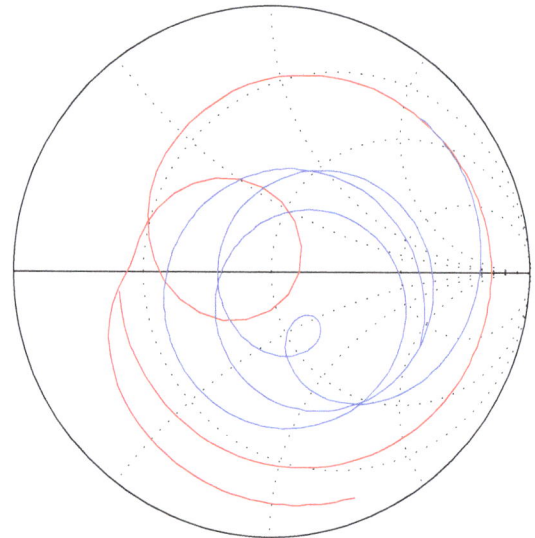

FIGURE 12: Smith's chart of the input impedance locus.

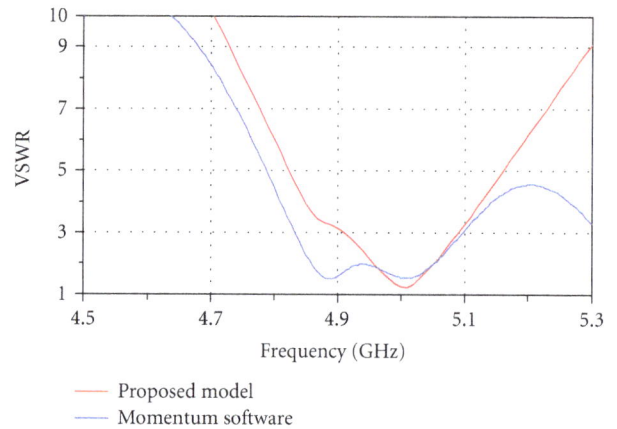

FIGURE 13: Bow-tie antenna array VSWR.

thickness h of 1.6 mm, and a loss tangent of about 0.019 and 0.05 mm conductor thickness.

The mask of the multilayer bow-tie antenna array with dimensions is shown in Figure 4.

The simulated input return loss of multilayer bow-tie antenna array is displayed at the frequency 2.4 GHz in Figure 5.

The representation of the reflection coefficient as a function of the resonance frequency is shown by the appearance of several resonance frequencies, which is a characteristic of the multiband antenna array.

The results show the appearance of a resonant mode at the frequency 2.4 GHz and a good agreement by the proposed model and the Momentum software. It appears that a peak of -17.52 dB using transmission line model with a light shift by the moment method provides a return loss of -11.24 dB at the frequency 2.42 GHz.

The moments results and those obtained from transmission line model of the input phase of return loss for this antenna array are shown in Figure 6.

From Figure 6, both models have the same shape and we note very well that the phase is null by the two models at the resonant frequencies, which means a perfect adaptation.

The impedance locus of the antennas array from 1.5 to 3 GHz is illustrated on Smith's chart in Figure 7.

It can be seen from Figure 7 that both models represent the locations of input impedances in a manner almost identical; this justifies the good agreement between the two models.

From Figure 8, there is a good agreement between the two models; the level of VSWR is close to unity, implying a good adaptation of the antenna array and precision of the model line transmission.

4.2. Bow-Tie Antenna Array Operating at the Resonant Frequency 5 GHz.
The selected configuration is shown in Figure 9 and consists of six bow-tie identical patches multilayer operating at the resonant frequency 5 GHz.

The antenna array is to be designed on substrate which has a relative permittivity ε_r of 2.54, a dielectric thickness h

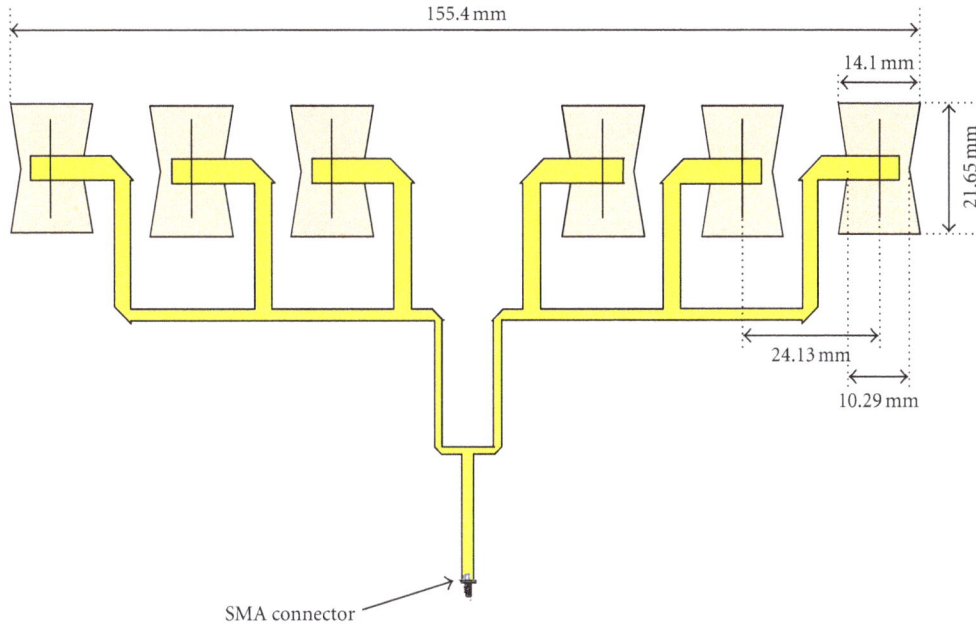

FIGURE 14: Mask of the multilayered bow-tie antenna array operating at the frequency 8 GHz.

of 1.6 mm, and a loss tangent of about 0,019 and 0.05 mm conductor thickness. A rectangular slot with L_{ap} = 18 mm and width W_{ap} = 0.6 mm is used for coupling the patch to a microstrip line of length L_f = 10 mm, etched on substrate which has a relative permittivity ε_r of 2.54, a dielectric thickness h of 1.6 mm, and a loss tangent of about 0.019 and 0.05 mm conductor thickness.

The multilayer bow-tie antenna array designed with dimensions in millimeter is represented in Figure 9.

The mask of the multilayer bow-tie antenna array with dimensions is shown in Figure 10.

From Figure 10, the resonance of antenna array is correctly predicted by both models to be 5 GHz, and as a result we note a peak of about −13.54 dB obtained by transmission line model and of about −19.6 dB by the moment's method.

The moments results and those obtained from transmission line model of the input phase of return loss for this antenna array are shown in Figure 11.

The reflected phase is null by the two models in spite of the shift observed by transmission line model.

The input impedance locus of the multilayer bow-tie antenna array is illustrated on Smith's chart in Figure 12.

It is observed that the curves of the two models pass by the axis of 50 Ω. The simulated VSWR is represented on Figure 13.

According to Figure 13, there is good agreement between the two models (transmission line model and the moment method). Around the resonant frequency the VSWR is close to unity implying a good adaptation of the network.

4.3. Bow-Tie Antenna Array Operating at the Resonant Frequency 8 GHz.

In this section, other geometry is analyzed by using the method proposed in this paper. The antenna array

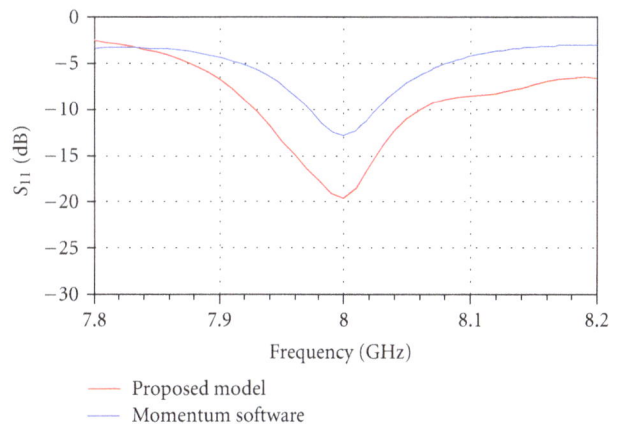

FIGURE 15: Simulated input antenna array return loss.

consists of six bow-tie identical multilayer patches, as shown in Figure 14 and is designed to operate at 8 GHz frequency.

The antenna array is to be designed on substrate which has a relative permittivity ε_r of 2.54, a dielectric thickness h of 1.6 mm, and a loss tangent of about 0,019 and 0.05 mm conductor thickness. A rectangular slot with L_{ap} = 16 mm and width W_{ap} = 2.6 mm is used for coupling the microstrip line of 10 mm length to the patch, etched on a substrate which has a relative permittivity ε_r of 2.54, a dielectric thickness h of 1.6 mm, and a loss tangent of about 0,019 and 0.05 mm conductor thickness.

Figure 14 presents the mask layout for multilayer bow-tie antenna array at the resonant frequency 8 GHz.

The simulated input return loss of multilayer bow-tie antenna array is displayed at the frequency 8 GHz in Figure 15.

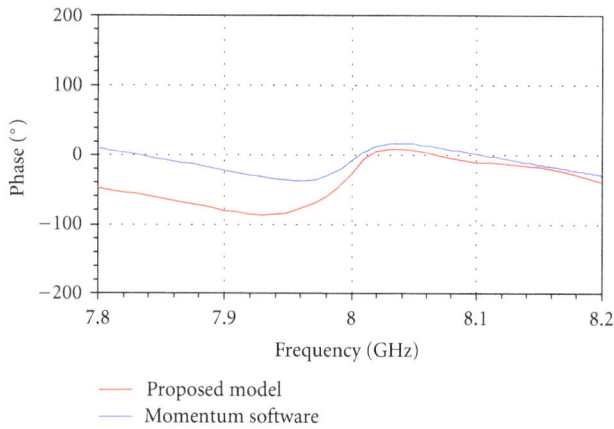

FIGURE 16: The antenna array input-reflected phase.

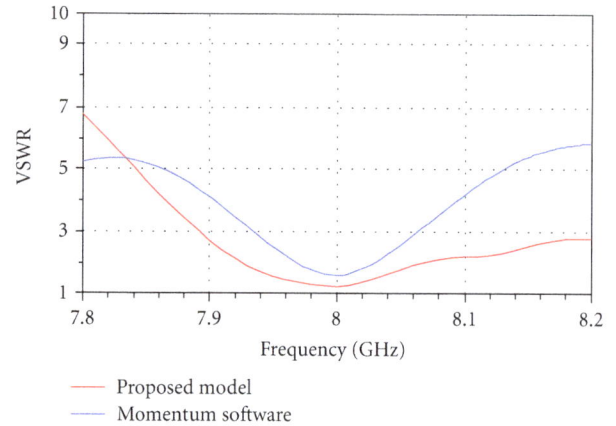

FIGURE 18: Bow-tie antenna array VSWR.

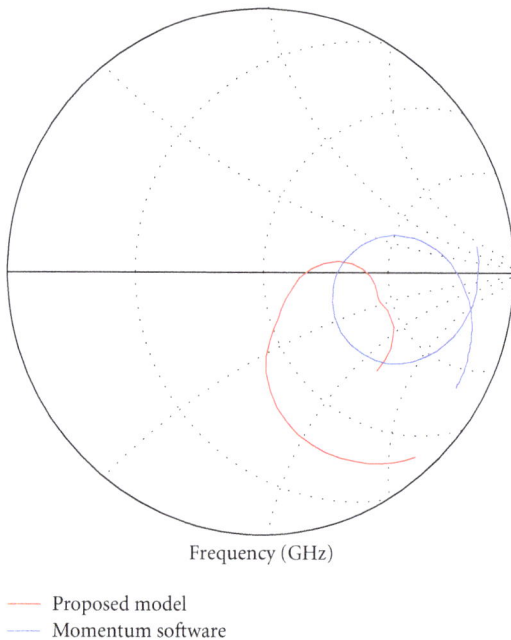

FIGURE 17: Smith's chart of the input impedance return losses.

TABLE 1: Comparison between transmission line model and method of moments.

Antennas arrays	Model	Return loss (dB)	Resonant frequency (GHz)	Frequency shift (%)
2.4 GHz (RFID)	MLT	−17.52	2.4	0.8%
	MoM	−11.24	2.42	
5 GHz (WIFI)	MLT	−13.54	5	0.2%
	MoM	−19.60	5.01	
8 GHz (RADAR)	MLT	−19.60	8	0%
	MoM	−12.84	8	

From Figure 15, it is observed that the resonance of the antenna array is correctly predicted to 8 GHz by the two models. It shows a peak of −19.60 dB using transmission line model and a peak of 12.84 dB by the moment method.

The moments results and those obtained from transmission line model of the input phase of return loss for this antenna array are shown in Figure 16.

As shown in Figure 16, a good agreement between the transmission line model and moment's method, the simulation results also show that the phase is null by the two models.

The impedance locus of the multilayer bow-tie antenna array is illustrated on Smith's chart in Figure 17.

The input impedance or the antenna has been calculated over a frequency range of 7.8–8.2 GHz. It can be seen from

Figure 17 that the curves of the two models pass by the axis of 50 Ω.

From Figure 18, in the vicinity of the resonant frequency the VSWR is close to the unit which corresponds to an ideal matching.

To better illustrate the results obtained in terms of adaptation, the comparison of the results in terms of return loss and resonant frequency between the transmission line model (MLT) and the method of moments (MoM) is summarized in Table 1.

Table 1 shows that the largest amount of frequency shift is produced by antenna 2.4 GHz which produced a resonance frequency of 2.4 GHz by MLT and 2.42 GHz by MoM, a shift of about 0.8% from 2.4 GHz. The lowest frequency shift is shown by antenna 8 GHz, a shift of about 0% from 8 GHz.

The return losses generated by all antennas arrays, which are all in the magnitudes less than −11 dB, show that a good impedance matching has been achieved in both models.

5. Conclusion

In this paper three multilayered bow-tie antennas arrays have been designed, which consist of 6 identical bowtie patch elements in parallel or corporate feed to resonate at the frequencies 2.4 GHz, 5 GHz, and 8 GHz corresponding to RFID, WIFI, and radars applications using an equivalent circuit. The transmission line model can be successfully used to design the corporate-fed multilayer bow-tie antennas arrays, and even though the model is conceptually simple, it still produces accurate results in a relatively short period of computing time. The proposed transmission line model showed its interest in the design of different multilayered bow-tie antennas arrays feed in parallel, predicting the correct resonance frequency for different applications in telecommunications. The results obtained highlighted a good agreement between the transmission line model and the moment's method. A comparison of the results produced by the final model with the moment's method showed the validity of the proposed model.

References

[1] M. Abri, N. Boukli-hacene, and F. T. Bendimerad, "Application du recuit simulé à la synthèse d'antennes en réseau constituées d'éléments annulaires imprimés," *Annales Des Télécommunications*, vol. 60, no. 11-12, pp. 1424–1440, 2005.

[2] G. Dubost, "Broadband circularly polarized flat antenna," in *Proceedings of the International Symposium on Antennas and Propagat*, pp. 89–92, Sendai, Japan, 1978.

[3] M. M. Alam, "Design and performance analysis of microstrip array antenna," in *Progress in Electromagnetic Research Symposium Proceedings*, Moscow, Russia, August 2009.

[4] M. Abri, N. Boukli-Hacene, and F. T. Bendimerad, "Weighted array design of an aperture coupled printed antennas," in *Proceedings of the Mosharaka Multi-Conference on Communications, Signals and Control (MM-CSC '07)*, Amman, Jordan, 2007.

[5] M. F. Bendahmane, M. Abri, F. T. Bendimerad, and N. Boukli-Hacene, "A simple modified transmission line model for inset fed antenna design," *International Journal of Computer Science Issues*, vol. 7, no. 5, pp. 331–335, 2010.

[6] M. Himdi, J. P. Daniel, and C. Terret, "Transmission line analysis of aperture-coupled microstrip antenna," *Electronics Letters*, vol. 25, no. 18, pp. 1229–1230, 1989.

[7] C. A. Balanis, *Antenna Engineering*, Willey, 2nd edition, 1982.

16

A Study of the Location of the Entrance of a Fishway in a Regulated River with CFD and ADCP

Anders G. Andersson,[1] Dan-Erik Lindberg,[2] Elianne M. Lindmark,[1] Kjell Leonardsson,[2] Patrik Andreasson,[1] Hans Lundqvist,[2] and T. Staffan Lundström[1]

[1] *Division of Fluid Mechanics, Luleå University of Technology, 971 87 Luleå, Sweden*
[2] *Department of Wildlife, Fish and Environmental Studies, Swedish University of Agricultural Sciences, 901 83 Umeå, Sweden*

Correspondence should be addressed to Anders G. Andersson, aneane@ltu.se

Academic Editor: Guan Heng Yeoh

Simulation-driven design with computational fluid dynamics has been used to evaluate the flow downstream of a hydropower plant with regards to upstream migrating fish. Field measurements with an Acoustic Doppler Current Profiler were performed, and the measurements were used to validate the simulations. The measurements indicate a more unstable flow than the simulations, and the tailrace jet from the turbines is stronger in the simulations. A fishway entrance was included in the simulations, and the subsequent attraction water was evaluated for two positions and two angles of the entrance at different turbine discharges. Results show that both positions are viable and that a position where the flow from the fishway does not have to compete with the flow from the power plant will generate superior attraction water. Simulations were also performed for further downstream where the flow from the turbines meets the old river bed which is the current fish passage for upstream migrating fish. A modification of the old river bed was made in the model as one scenario to generate better attraction water. This considerably increases the attraction water although it cannot compete with the flow from the tailrace tunnel.

1. Introduction

Computational fluid dynamics (CFD) is used to simulate the flow within a tailrace channel of a hydropower plant with the purpose to scrutinize alternative positions of an entrance to a fishway. The simulations are carried out on full scale implying a length of the virtual model of 320 m, a typical width of 75 m, a typical depth of 10 m, and a maximum inlet flow rate of 1000 m³/s. A numerical challenge with the large scale is to fulfill conditions of a sufficiently resolved flow structure at locations with high gradients (e.g., at boundaries) and a good mesh overall with a decent usage of computational resources (Marjavaara and Lundström [1]). Another challenge is the validation of the simulations which is here done by measurement with an Acoustic Doppler Current Profiler (ADCP).

Studies of tagged Atlantic salmon and sea trout in the unregulated river Vindelälven in northern Sweden during 1995–2005 have shown that only a third of the upstream migrating fish find their way to their natural spawning grounds (Lundqvist et al. [2]). The main reason for this is the Stornorrfors power plant located downstream the confluence between the rivers Vindelälven and Umeälven, the latter being a regulated river. A major issue at the power plant is that the fish are attracted into the tailrace channel from the turbines rather than migrating up through the old river bed that offers a fishway around the turbines (Rivinoja et al. [3]). The flow rate from the turbines is typically 20 times larger than the flow rate from the old river bed and its entrance into the confluence is very wide. Hence, fluid flow conditions for the old river bed to attract fish are limited. The fact that migrating fish are attracted to the tailrace of the turbines instead of the weaker current from the fishway is a common problem (Arnekleiv and Kraabøl [4], Webb [5]). The difficulties of upstream migrating fish coming across in regulated rivers in northern Sweden have been documented by, for example, Rivinoja [6], Lindmark [7], and Lindmark and Gustavsson [8].

There are two major measures that are being considered for improving the upstream migration of fish at the Stornorrfors power plant. One is to construct a new fishway in the form of a fish ladder from the tailrace channel since a majority of the fish reside there for a long period of time during the migration season. The other alternative is to create better attraction water from the old river bed into the confluence area. The alternatives are here modeled with CFD, and the attraction water created using given configurations is examined. The interest in numerical simulations of flows in rivers is increasing, and, due to the rapid development of user-friendly efficient codes and computer power in recent years, more advanced models than before can be applied in areas such as fish migration, habitat modeling, sedimentation transport, erosion, and dam safety. Olsen and Stokseth [9] created a model of the Sokna River in Norway where they applied a k-ε turbulence model and a porosity-based model for large roughness elements in the river bed showing good resemblance with observed data. The SSIIM model suggested by Olsen has been validated against LDA measurements in a meandering channel on a lab scale (Wilson et al. [10]) where the model showed the ability to capture secondary currents. CFD has also been applied to the River Cole, Birmingham, UK (Clifford et al. [11]), and the River Thame, Birmingham, UK (Booker [12]), where the potential for use in habitat modelling was discussed. A numerical model of a 4 km stretch of the Columbia River downstream of the Wanapum Dam has been performed and calibrated against measured data highlighting the importance of bed roughness for accurate flow predictions (Sinha et al. [13]). Dargahi [14] used the commercial code Fluent to model fluid flow and sediment transport in the River Klarälven, Sweden, and validated the results with ADCP measurements. The design of a submerged flow guiding device to increase the survivability of downstream migrating fish has been performed in the commercial code CFX-10 (Lundström et al. [15]). Simulations of flow in an ice-covered channel with the k-ω turbulence model with different roughness values for river bed and ice-cover resulted in a 16% increase in mean flow depth of the channel (Yoon et al. [16]). The effects of submerged weirs in natural channels to improve the navigation conditions for barges have been investigated numerically (Jia et al. [17, 18]).

Rakowski et al. [19] used field-measured data with ADCP to validate their CFD simulations of a 2.7 km reach starting downstream of the Bonneville powerhouse and spillway with total river flows between 3275 m³/s and 11328 m³/s. The velocities were measured and averaged over a 10 minutes period to get adequate representation of the mean velocity. When comparing the CFD simulations (steady state, k-ε turbulence model) to ADCP data, the modeled velocity was slightly lower than the measured, but within the standard deviation of the field velocity. Viscardi et al. [20] also used ADCP measurements to validate CFD simulations in a 3 km stretch of the Paraná de las Palmas River with flow rates ranging from 2200 to 5000 m³/s (steady state, k-ε turbulence model, rigid lid, bed roughness Manning n = 0.025). In their case, the velocities were averaged over 2 seconds in each vertical sample in order to minimize the effect of the tidal change and the velocities correspond reasonable accurate.

To summarize, two-equation turbulence models are in most cases used to simulate the flow in rivers and no one is considering how attraction water from a fishway competes with the flow in the river.

2. Geometry

The actual geometry in the present study consists of four parts, the tunnel from the turbines, the tailrace channel, the old river bed, and the confluence area, see Figure 1 where the tailrace channel and the old river bed are defined. The confluence area is located where the water from the old river bed and the tailrace channel meet, while the tunnel from the turbines is located upstream the tailrace channel. CFD calculations are performed on all parts except the old river bed, while velocity measurements are only reported for a couple of transects within the tailrace channel.

3. Experimental

To measure topology and water velocity downstream Stornorrfors power plant, an ADCP was used. The ADCP has four transducers directed into the water. The transducers send out sound waves that reflect on small particles traveling with the water, and the transducers detect the Doppler frequency of the reflected sound waves. These frequencies are proportional to the velocity of the water (the particles). ADCP is a relatively fast way of measuring velocities in field and to calculate river discharge. The ADCP used in this case is a RiverBoat RioGrande, and the data processing was performed with the software Winriver II, both from RD Instruments.

The bathymetry in the area was measured using two setups. The ADCP was dragged besides a motorboat with a pole and rope, which enabled measurements close to the shoreline. By combining the bottom-tracking feature of the ADCP with GPS data, a point cloud consisting of ADCP provided depths at specific satellite coordinates was obtained. The ADCP however fails to find the bottom of the deepest area in the tailrace channel; hence, a SIMRAD EY60, GPT 200 kHz, split beam echo sounder with the transducer mounted vertically on the boat was used near the tailrace tunnel outlet. There is a small shallow part of the tailrace that is located above and behind the tunnel outlet. The GPS reception was very low this far into the spillway due to the surrounding terrain which caused large uncertainties in the acquired coordinates. The length of this region is approximately 50 m, and observations suggest that it consists of a slow circulating flow and the assumption was made that this part of the spillway does not have any significant effect on the flow in the remaining channel. It was thus omitted from the numerical model, and the entrance to this innermost part was excluded from the geometry. The points of measurements are shown in Figure 1(a).

For the velocity measurements, a steel wire was stretched across the tailrace channel and the ADCP was tethered to it. A manual winch enabled the ADCP to travel across the channel and capture the velocities in the entire cross-section. The transect T1 in Figure 1(a) was measured on several

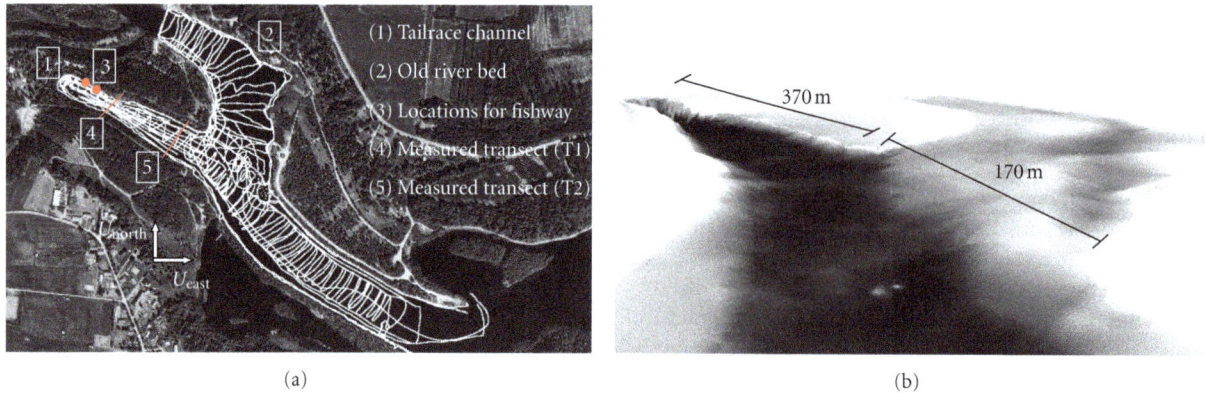

(a)   (b)

FIGURE 1: (a) Aerial photograph of tailrace channel and confluence area downstream Stornorrfors power plant. White lines (points) represent data points used in geometry creation. Please notice the cardinal direction. (b) Visualization of the confluence area, tailrace channel, and old river bed looking upstream. Darker grey represents a larger channel depth.

occasions at different turbine discharges and a minimum of four times at each flow. Three vertical profiles in the T1 transect and three in the T2 transect were measured during a minimum of 600 s. The profiles were collected when the flow rate through the power plant was ~500 m³/s (according to the discharge calculation in WinRiver and data from the hydropower company). Profiles were measured with a time difference of 0.95 s between ensembles. During measurements, the distance to the shore was measured with a laser distance meter. The total width of the T1 section was measured to 40 m, and the profiles were located at 16, 23, and 32 m from the south shore. Transect T2 had a measured width of 85 m, and the verticals were located 30, 44, and 59 m from the north shore.

The accuracy of the ADCP depends on many factors, such as side-lobe interference, ringing, and ADCP-flow inter- action that exclude the ADCP from doing any measurements near the water surface or close to the bottom of the river (Simpson and Oltmann [21]). Nystrom et al. [22] compared ADCP accuracy with an Acoustic Doppler Velocimeter (ADV) in a lab flume within which the turbulence intensity of the flow was 0.1. The ADCP measurements were carried out during 15 min, and the error was less than 3% in the areas away from the boundaries not being affected by ringing, side lobe interference, and flow disturbance.

4. Numerical Setup

The point cloud collected with ADCP and SIMRAD seen in Figure 1 was converted to a bottom surface in the software Imageware 13. The surface was imported to Ansys Icem Cfd 11 where a solid model was created. The formed geometry was divided in two parts, the tailrace channel and the confluence area between the channel and the old river bed. The simulation volumes were discretized as tetrahedral elements in the CFD model. Local refinements of the grid were carried out in areas of simulated attraction water to increase the resolution in the most interesting parts of the flow. All simulations were carried out with the commercial software CFX11 from Ansys Inc. A mesh sensitivity study of

the tailrace channel was performed with different numerical grids, ranging from 239 k to 7389 k nodes. The velocity in the east direction was evaluated at T1 and T2 for the different grids and the conclusion was that the coarsest mesh did not capture the flow field with sufficient accuracy. A mesh with 526 k nodes however produced a very similar flow field to that of the 7389 k mesh with significantly lower computational cost. Hence, the final grids for the tailrace channel consisted of ~500 k nodes and the confluence area of ~600 k nodes.

In reality, the water from the power plant goes through an approximately 4 km long tunnel before entering the tailrace channel. To create a realistic inlet boundary condition for the simulations, this tunnel was modeled separately and the velocity profile at the end of the tunnel was used at the inlet of the tailrace channel simulations. The tunnel was given a sufficient length to give a fully developed velocity profile, and the tunnel walls were given a wall roughness of a typical excavated rock.

The high-resolution advection scheme was used for solving the equations of fluid motion and turbulence closure. The high-resolution scheme uses a close to second-order solution in areas with low variable gradients, and, in areas where the gradients change sharply, it will be close to a first-order solution (ANSYS [23]). The incompressible Reynolds-Averaged Navier Stokes equation and the continuity equation are expressed as

$$\frac{\partial U_i}{\partial t} + U_j \frac{\partial U_i}{\partial x_j} = -\frac{1}{\rho} \frac{\partial P}{\partial x_i} + \nu \nabla^2 U_i - \frac{\partial}{\partial x_j}\left(\overline{u_j u_i}\right),$$

$$\frac{\partial U_i}{\partial x_i} = 0, \tag{1}$$

where U is the mean part of the velocity component, P is the pressure, ν is the viscosity of the fluid, ρ is the fluid density, and $\overline{u_j u_i}$ are the Reynolds stresses. All simulations were run with the k-ε turbulence model with scalable wall functions.

In the k-ε turbulence model, Reynolds stresses are linearly related to the strain:

$$-\overline{u_j u_i} = 2\nu_T S_{ij} - \frac{2}{3}k\delta_{ij}, \qquad (2)$$

where k is the turbulent kinetic energy, ν_T is the eddy viscosity, and S_{ij} is the mean strain tensor defined as

$$S_{ij} = \frac{1}{2}\left(\frac{\partial U_j}{\partial x_i} - \frac{\partial U_i}{\partial x_j}\right). \qquad (3)$$

The eddy viscosity is modeled as

$$\nu_T = C_\mu \frac{k^2}{\varepsilon}, \qquad (4)$$

where $C\mu$ is a model constant and ε is the turbulent dissipation rate. For more information, see Launder and Spalding [24]. The RMS residual target for all simulations was set to 10^{-6}. This convergence target could not be achieved with a steady-state solver due to initial fluctuations of the flow. Simulations were instead run on a transient solver until it approached a steady solution and the final values from the simulation were used. A physical time step of 0.5–2 s was selected depending on grid size, and the solution was considered steady when the velocity in 18 monitored points throughout the domain had been virtually constant for at least 1000 time steps.

The water surface was modeled as a rigid lid with zero friction. This approximation is viable when the surface level variation is smaller than 10% of the total channel depth (Rodriguez et al. [25]). The outlets are given a pressure type boundary condition. The bottom surface of the numerical model was defined as a rough wall. A scalable wall function that uses an extension of the method suggested in [24] was selected for near wall modeling. The dimensionless velocity u^+ in the logarithmic layer close to the rough wall is typically written as

$$u^+ = \frac{U_t}{u_\tau} = \frac{1}{\kappa}\ln y^+ + B - \Delta B, \qquad (5)$$

where

$$y^+ = \frac{u_\tau \Delta y}{\nu}, \quad u_\tau = \sqrt{\frac{\tau_\omega}{\rho}}, \qquad (6)$$

and where u_τ is the friction velocity, U_t is the velocity tangent to the wall at a distance Δy from the wall, y^+ is the nondimensional wall unit, τ_ω is the wall shear stress, κ is the von Karman constant, B is a constant, and ΔB is the so-called roughness characterization function. Since the roughness of the channel is not well documented, a global representation of roughness was selected. The wall roughness can be described as an equivalent sand-grain roughness, k_s (ANSYS [23]). With this formulation, the roughness characterization function can be described as (White [26])

$$\Delta B = \frac{1}{\kappa}\ln(1 + 0.3k_s^+), \qquad (7)$$

where the dimensionless roughness height k_s^+ is defined as

$$k_s^+ = \frac{u_\tau k_s}{\nu}. \qquad (8)$$

To obtain a realistic value for the equivalent sand-grain roughness of the channel, an empirical Gauckler-Manning coefficient n that describes the channel is selected (Arcement and Schneider [27]). The advantage of using the Manning's n instead of other coefficients is that n is nearly constant regardless of flow depth, Reynolds number (Re = $4UR_h/\nu$), or k_s/R_h for fully developed turbulent flow over a rough surface (Yen [28]). The selected n is used to calculate a Darcy friction factor f given by

$$f = \frac{8gn^2}{R_h^{1/3}}. \qquad (9)$$

The friction factor obtained is then used to find k_s from the Colebrook-White formula (Colebrook [29])

$$\frac{1}{\sqrt{f}} = -2\log_{10}\left(\frac{k_s}{3.71 \cdot 4R_h} + \frac{2.51}{\text{Re}\sqrt{f}}\right), \qquad (10)$$

and the derived k_s is finally used for input into the simulations. In the present case, the second term in (10) can be neglected due to the high Reynolds number of the flow.

The wall function approach is common in river simulations since the scales of roughness are very costly to model physically in problems of such large scales. The limitations of this method are discussed by Patel [30]. Since the roughness is only an approximation and is difficult to measure in reality, a parameter study was here performed in the numerical model with a flow rate of 350 m³/s from the turbines.

With no surface roughness, the jet leaving the tunnel barely leaves the bottom of the channel which does not seem likely with regards to the characteristics of free surface channel flow, see Figure 2. With a k_s value of 0.3 m (Manning $n \approx$ 0.033) which can be considered typical for a rock excavated channel such as the tailrace channel, the flow characteristics change considerably. The jet emerging from the tunnel now moves towards the free surface of the channel. Increasing k_s to 0.5 m (Manning $n \approx$ 0.037) did not affect the solution in any major way, and all following simulations on the tailrace channel, were run with $k_s = 0.3$ m.

Two ways of improving the upstream fish migration around the power plant were studied: a new fishway in the tailrace channel and higher attraction to the old river bed. In the tailrace channel, two positions and two angles of a new fishway entrance were evaluated. The positions were selected from previous observations of fish during the migration season. The dimensions of the entrance were 2×2.7 m², and the flow rate used was 10 m³/s. The two inlet angles of the fishway entrance were perpendicular and 45° to the main flow.

The modification of the confluence area to improve the attraction to the old river bed was realized by adding a wall at a distance from the river bank directing nearly all the flow in the old river bed to a narrow open channel between the wall and the shoreline. In such opening the water may

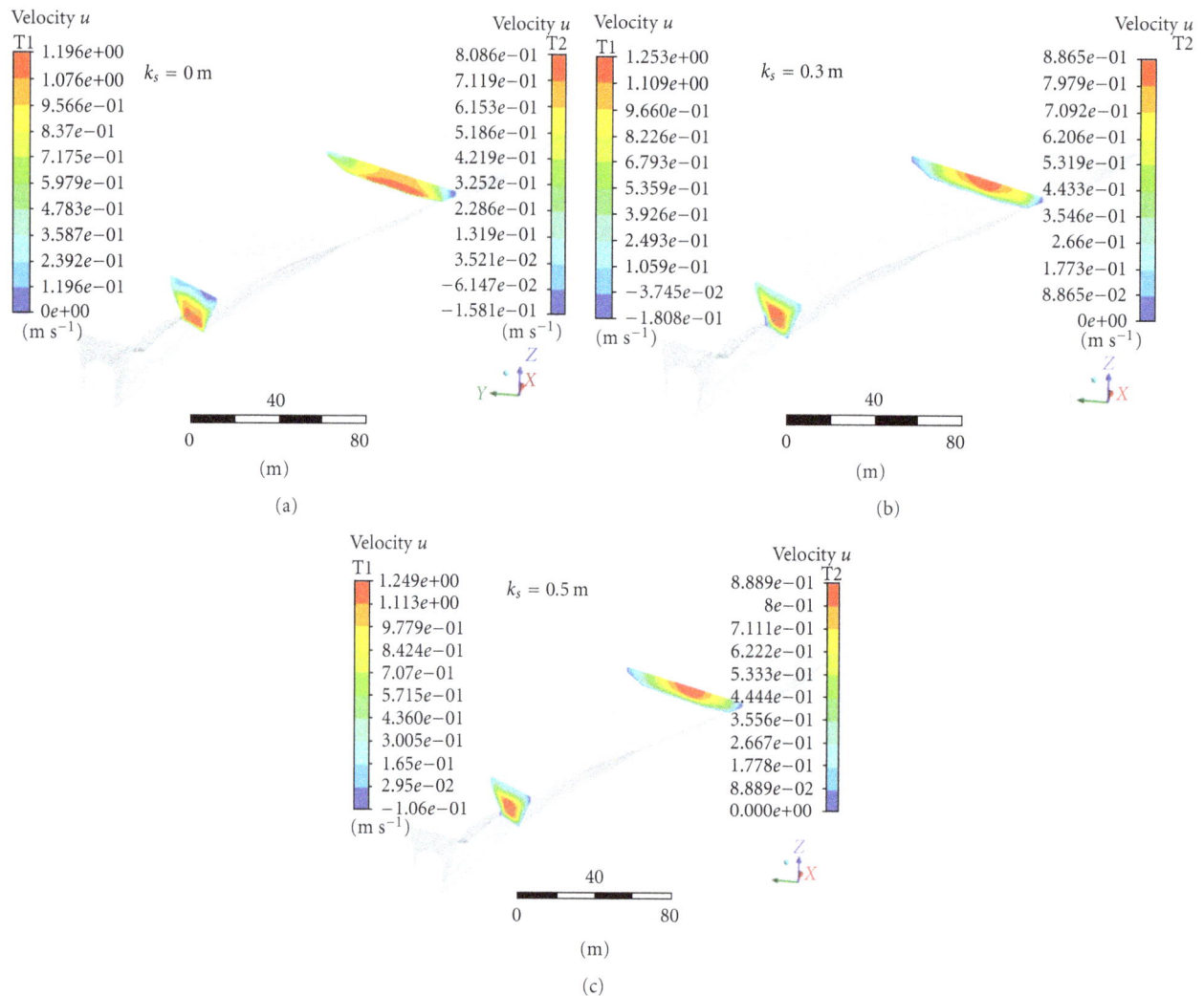

FIGURE 2: Parameter study of the wall roughness in the tailrace channel showing the development of the velocity profile at two different cross-sections for three different roughness values. The characteristics of the flow change completely when the equivalent sand-grain roughness increases from 0.0 to 0.3 m increasing it further to 0.5 m giving no noticeable additional effect.

be accelerated with a ramp, for instance, as suggested in Lindmark and Gustavsson [8] and Green et al. [31]. The flow in the old river bed was set to 20 m³/s and that from the tailrace tunnel to 500, 750, and 1000 m³/s, representing a low flow, a normal flow, and a flow close to the maximum flow, respectively.

5. Results and Discussion

The characteristics of the flow will be described followed by a comparison to experimental data, and finally the results from simulation of attraction water will be presented.

5.1. Characterization of the Flow in the Simulations. The flow exiting the tunnel takes the form of a jet that gradually develops into an open channel flow profile. Approximately after 2/3 of the channel length, the jet maximum velocity is at the surface of the water as can be seen in Figure 3.

The jet however influences the surface orientation flow much earlier in the channel as revealed by plots of vorticity and turbulence intensity in a plane at 1 m depth, see Figures 4(a) and 5(a). High-vorticity areas are found at the bottom of the channel near the edges of the channel and at one large area of recirculation after the expansion, near the north shore, but there is also a noticeable rotation of the flow close to the surface near the inlet of the tailrace channel, see Figures 4(a) and 4(b). This is also reflected by relatively high turbulence intensity in this area, see Figures 5(a) and 5(b).

5.2. Comparison to Experiments. The results from ADCP measurements in the tailrace channel yield an unstable behavior of the flow, see Figure 6 showing a $12 \times 12\,\text{m}^2$ section in the middle of the T1 transect at five different times where the raw data from the ADCP has been averaged to $1 \times 1\,\text{m}^2$ cells. The measurements were taken in succession, and the velocities have been normalized with transect average

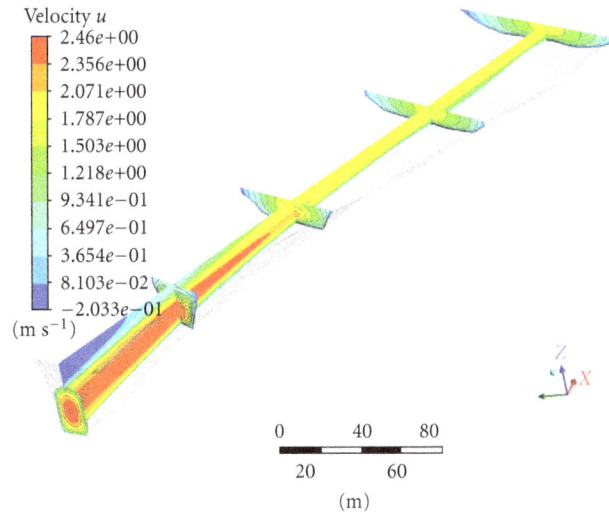

FIGURE 3: Development of the turbine jet as seen in a section along the channel, cross-sections T1, T2, one intermediate cross-section, inlet, and outlet.

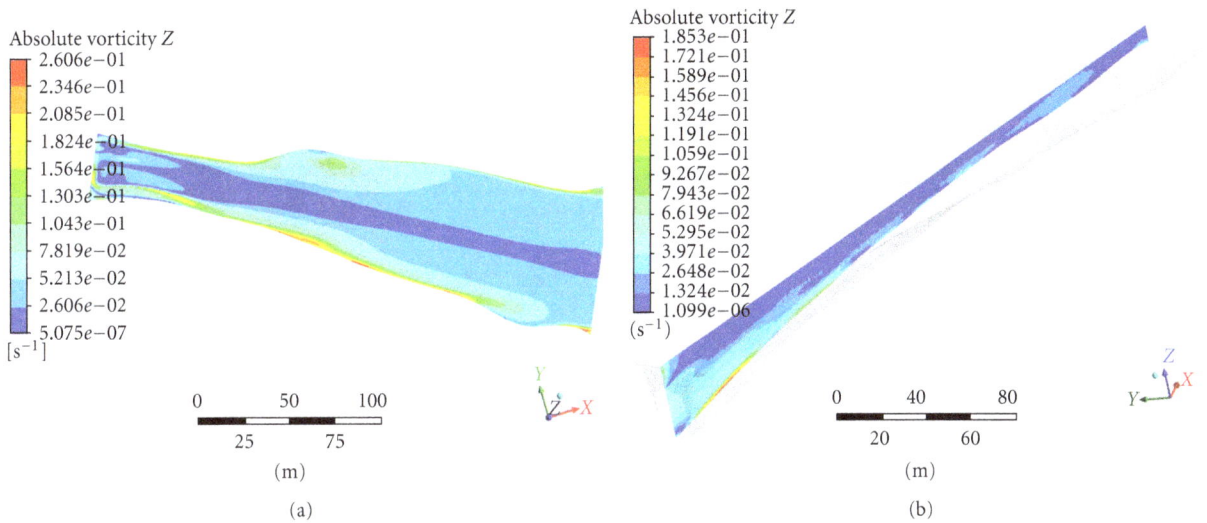

(a)

(b)

FIGURE 4: (a) Magnitude of the vertical vorticity in a plane at 1 m depth. (b) Magnitude of the vertical vorticity in a section along the channel.

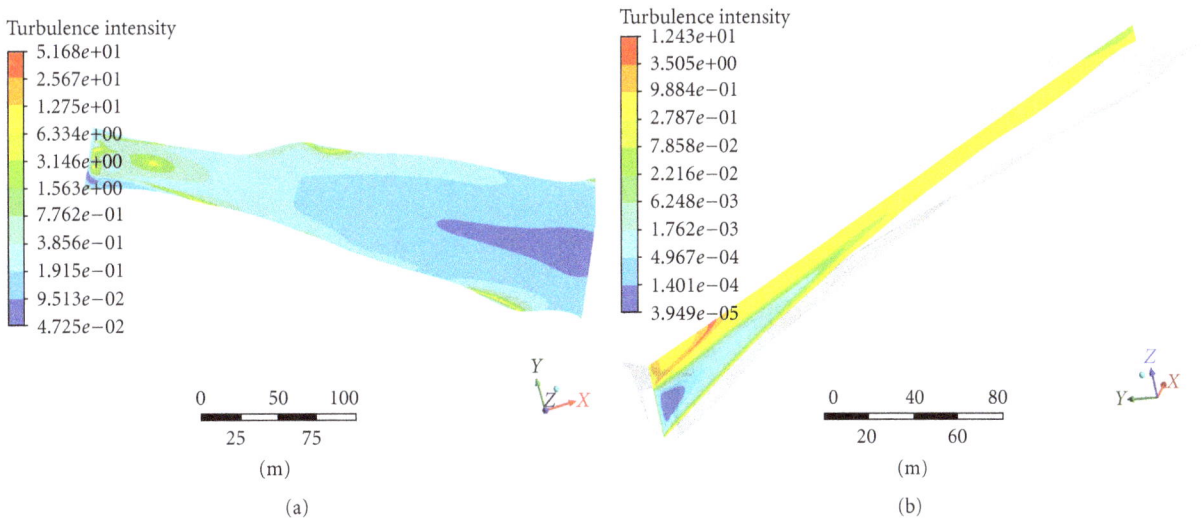

(a)

(b)

FIGURE 5: (a) Turbulence intensity (logarithmic scale) in a plane at 1 m depth. (b) Turbulence intensity (logarithmic scale) in a section along the channel.

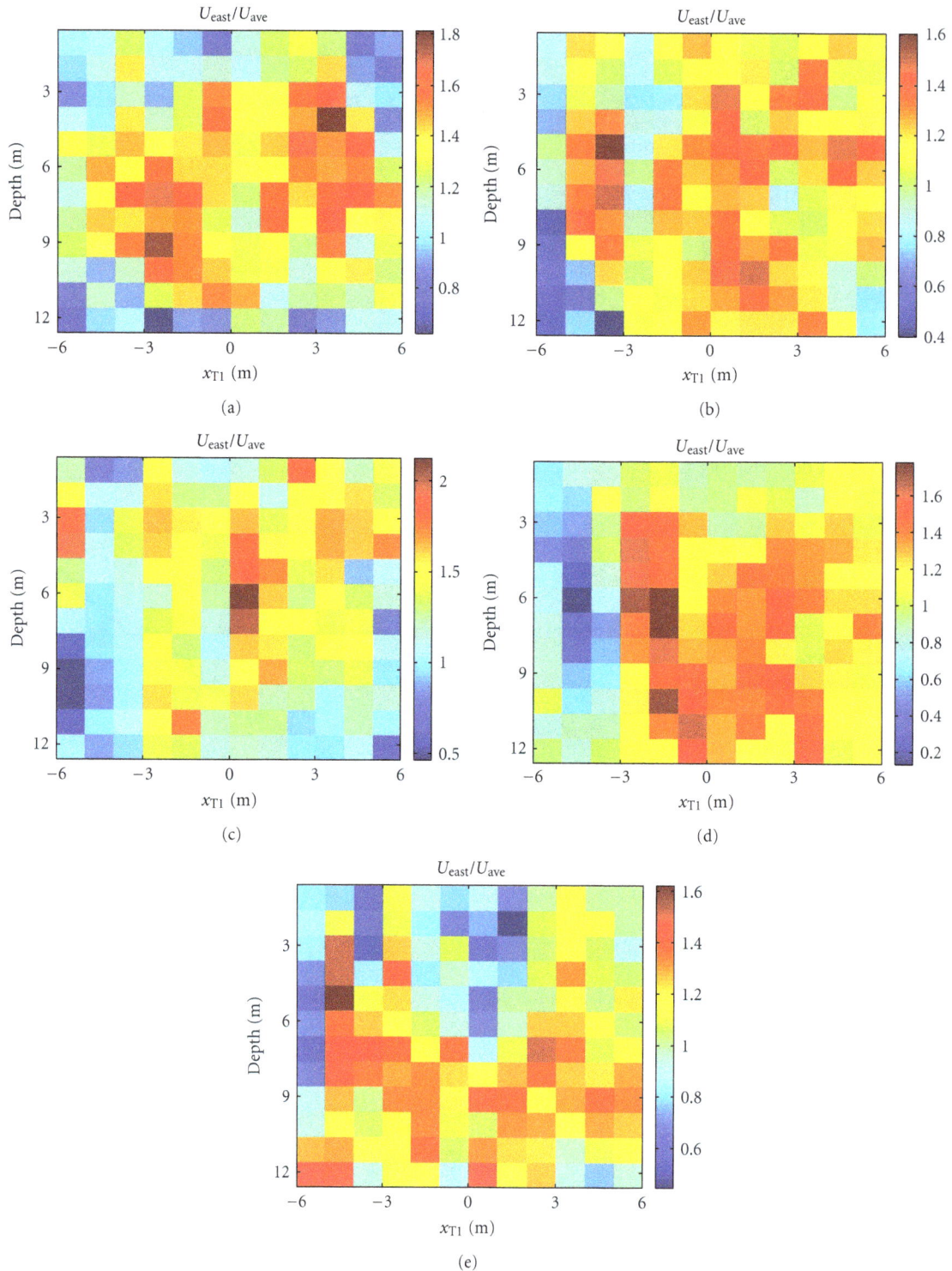

FIGURE 6: Five individual measurements for a section of 12×12 m^2 from transect T1. Red indicates high velocity and blue low velocity. The area with high velocity changes position as a function of time.

velocity to account for minor differences in total flow from the tunnel. The jet stemming from the tunnel is apparent in all transects but not as well defined as in the simulations, compare Figures 3 and 6. To examine the time dependence of the flow, the ADCP was kept in the same position and the velocity was measured during a longer period of time. Three vertical profiles at 15, 22, 31 m from the south shore were measured. The standard deviation from the mean distance was 0.01-0.02 m. The results from the measurements show a highly fluctuating flow. Initial frequency analysis does not

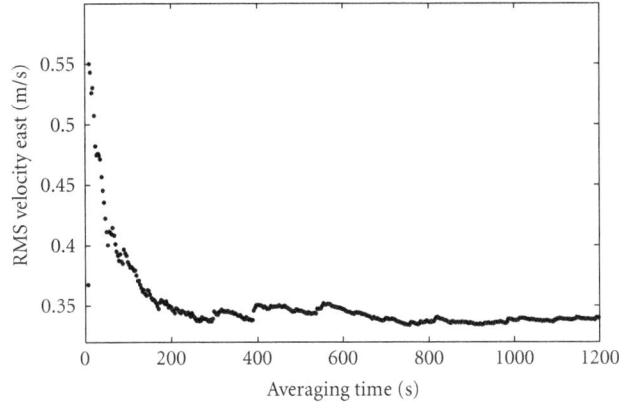

FIGURE 7: RMS of the east velocity component as a function of the averaging time. The velocities are from the profile 22 m from the south shore at 5 m depth in transect T1.

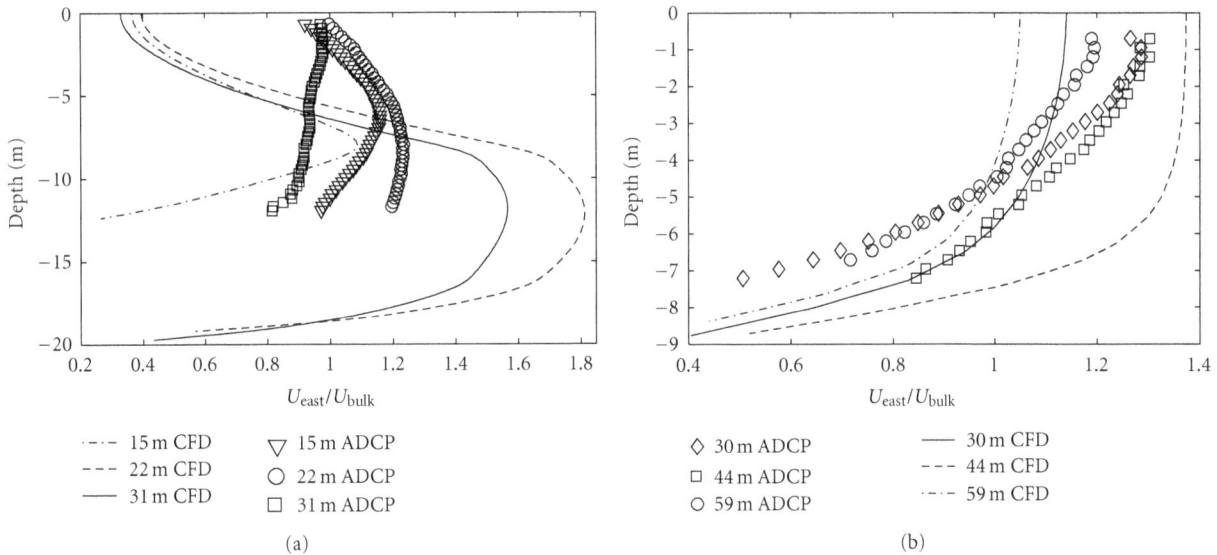

FIGURE 8: (a) Comparison between vertical velocity profiles in experiments and simulation in verticals at 15 m, 22 m, and 31 m from the south shore, respectively, for transect T1. U_{east} is the velocity in the east direction. (b) Comparison between simulations and measurements in verticals at 30 m, 44 m, and 59 m from the north shore, respectively, for transect T2. U_{east} is the velocity in the east direction.

indicate any periodicity; however, it cannot be excluded that fluctuations are influenced by large-scale structures of the flow, originating from upstream instabilities. How the RMS velocity (east) stabilizes with time is shown for the profile at 22 m in Figure 7. From the results, it is concluded that, to measure representative mean velocities, each profile must be measured during at least 600 s. The measurements over a *complete* transect presented in Figure 6 took about 120 s which means that these measurements by no means represent the mean velocity in that transect which explains the different velocity patterns.

To validate the simulations, time-averaged velocities of fixed-point measurements at both T1 and T2 are derived. The agreement between simulation and experiment at T1 is rather poor, see Figure 8(a) where normalized velocity profiles are compared. The velocity is normalized with the bulk velocity $U_{bulk} = Q/A_{T1}$, where Q is the flow rate and A_{T1}

is the area of the T1 transect being 516 m^2 as derived from the virtual model. The jet that exits the tunnel appears closer to the water surface in the measurements than in the simulations, and it is much more diffuse in the measurements. This is most apparent for the measurements at 31 m at T1 where measurements indicate a plug flow while the simulations yield a sinus-shaped profile. Hence, there is a discrepancy at the surface and at the bottom and the jet penetrates the surface much earlier in reality as compared to the simulations. For T2, the agreement between simulations and experiments is better especially close to the free surface, see Figure 8(b). The maximum velocity of the flow in the middle of the channel is lower in the experiments than in reality, while it is actually higher towards the shores indicating a more diffusive flow also in this transect. One reason for the differences, especially apparent in T1, might be the inlet boundary condition in the simulations, which is described

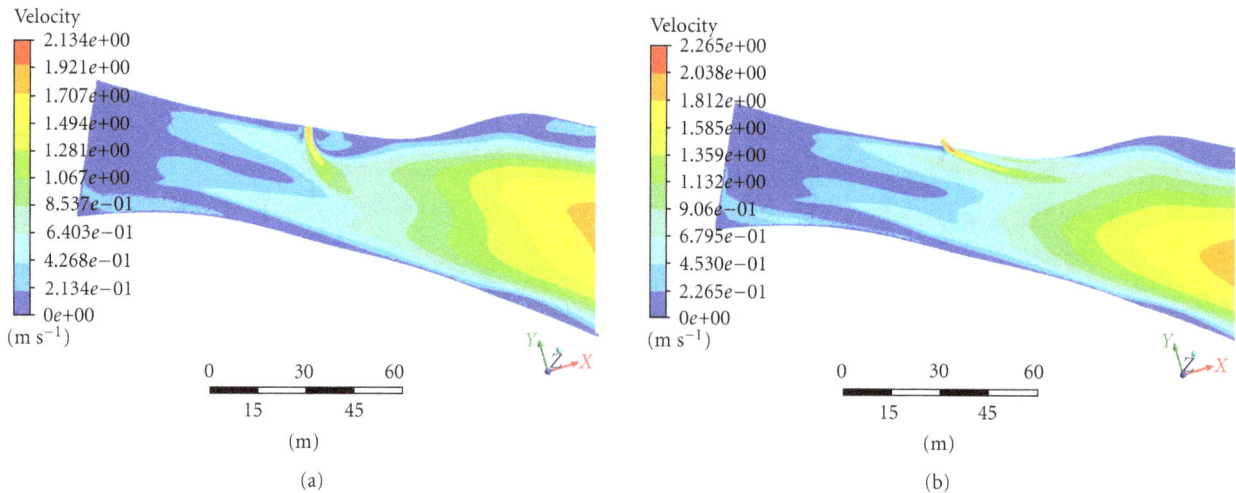

FIGURE 9: Fishway inlet at position 1 with 0° and 45° angle. The flow rate through the power plant is 750 m³/s, and the velocities are shown at 1 m depth.

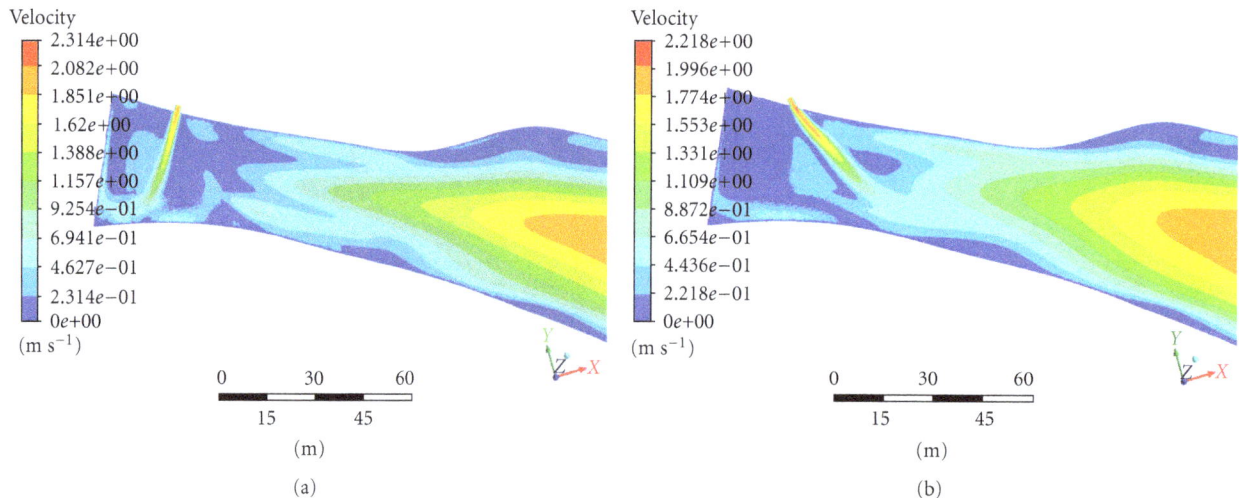

FIGURE 10: Fishway inlet at position 2 with 0° and 45° angle. The flow rate through the power plant is 750 m³/s, and the velocities are shown at 1 m depth.

as a stationary velocity profile where in reality effects of the turbines, larger discrete wall roughness elements or sudden changes in discharge may come into play. Other contributing factors may be difference between model geometry and real geometry as to surface roughness, for instance, and oversimplified modeling of turbulence or that the rigid lid assumption creates unphysical behavior when the jet from the tailrace tunnel approaches the water surface of the tailrace. It is also likely that the flow field is smeared out by the method to measure the velocity field. The discrepancy between simulations and measurements is a subject for future research as to turbulence intensity, for instance. When later on discussing the results from the simulations with the fishway entrances, it should be remembered that the jet is more diffuse and surface orientated in reality as compared to the simulations.

5.3. *Simulation of Attraction Water.* For position 1 in the tail race channel, the perpendicular entrance gives a noticeable jet that stretches to the center of the tail race channel while the angled inlet gives a jet that aligns with the flow from the tailrace tunnel and reaches further downstream, see Figure 9. Even better attraction water is created at the second position as shown in Figure 10. Since the small jet from the fishway does not collide with the large jet from the tailrace tunnel, the generated attraction water stretches further out in the channel, see Figure 11. Noticeable attraction water was created even at the highest flow (1000 m/s) from the turbines, see Figure 12. The relatively high-vorticity levels and turbulence intensities are thus too weak to influence the attraction water to any larger extent. Hence, position 2 is, as to generation of attraction water, a better choice than position 1. This conclusion is strengthening by the fact that, in

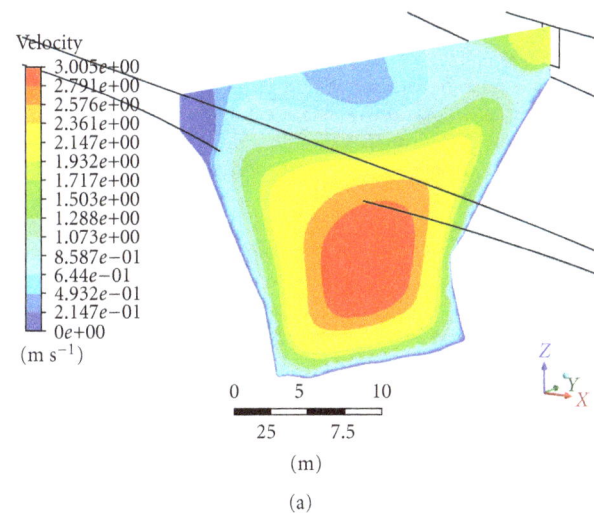

(a)

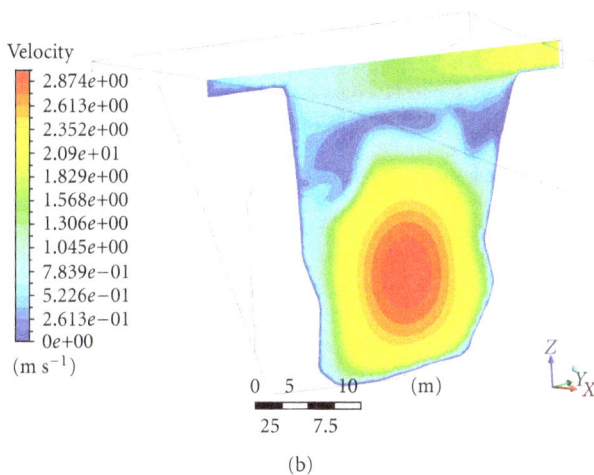

(b)

FIGURE 11: Fishway outlet (upper right corner) at positions 1 and 2 with 0° angle. The flow rate through the power plant is 750 m³/s.

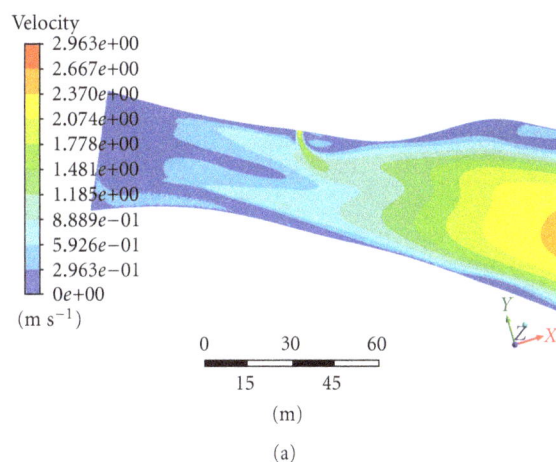

(a)

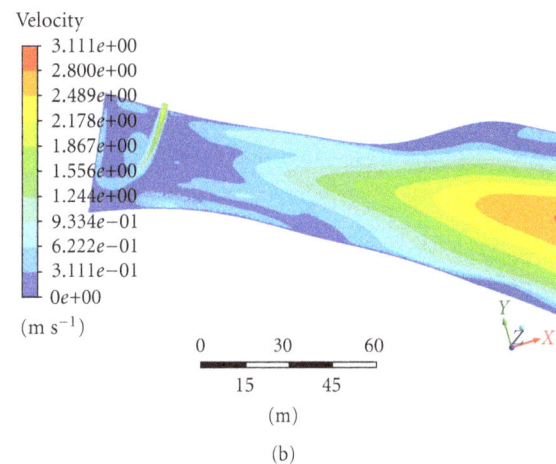

(b)

FIGURE 12: Fishway outlet at positions 1 and 2 with 0° angle. The flow rate through the power plant is 1000 m³/s.

reality, the jet from the turbines is more surface oriented which probably will make the attraction water created at position 1 less prominent than obtained in the simulations and from the turbines stressing the fact that the fishway should be placed as long into the tail race channel as possible for optimum generation of attraction water.

When scrutinizing possible improvement of the attraction water from the old river bed, the simulations yield a rather different result. The attraction water cannot compete with the flow from the tail race channel except in an area quite close to the shore, see Figure 13 where the simulated attraction water competes with two flow rates from the turbines (500 and 750 m³/s). As compared with the current situation, this modification of the confluence would still provide considerable improved attraction water along the north side (the right-hand side in the simulated results in Figure 13 and see Figure 1 for cardinal directions). This should improve the probability that fish migrating upstream on the north side of the river or fish exiting the tailrace tunnel on the north side find the fish passage in the old river bed.

6. Conclusion

The measurements show that the flow is considerably more unstable in reality as compared to the simulations. The flow fields in the simulations are therefore less diffuse as to time-averaged quantities, and the tailrace jet from the tunnel outlet is stronger but less surface oriented in the simulations as compared to reality. Keeping this in mind, a number of additional conclusions can be made from the work here presented. A fishway in the tailrace channel can generate noticeable attraction water for all relevant flows from the turbines. Of the cases studied, the simulations show that position two gives considerably stronger attraction water as compared to position one. It is likely that this difference is enlarged by the diffusivity of the tailrace jet existing in reality. By a concentration of the flow from the old river bed, noticeable attraction water can be created at the confluence area. In this case, the attraction water only stretches a short distance into the tailrace flow since this flow is completely surface oriented in the confluence area. However, if the fish migrate along the north shore, they will sense the attraction water.

(a)

(b)

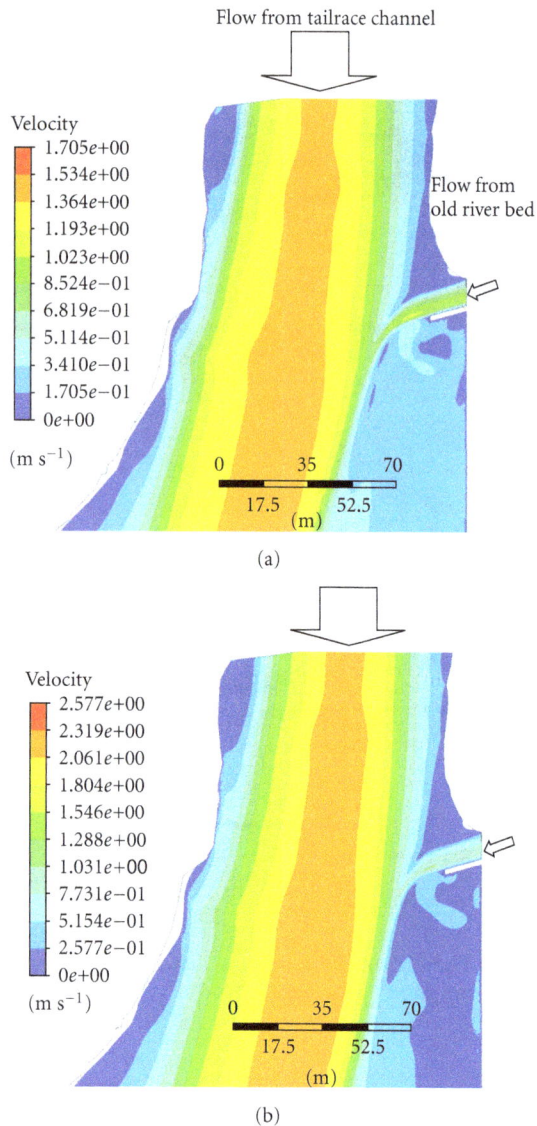

Figure 13: Confluence area with flow rate from the turbines of $500 \, \text{m}^3/\text{s}$ and $750 \, \text{m}^3/\text{s}$ and flow rate in the old river bed is $20 \, \text{m}^3/\text{s}$. A wall is inserted 10 m from the north shore.

Acknowledgments

This work was financed by Vattenfall AB and Umeå Kommun. The authors also acknowledge Vindeln Utveckling that always kept their administration in phase with the real EU time and the National Board of Fisheries for their suggestions on the design of this study.

References

[1] D. Marjavaara and S. Lundström, "Response surface-based shape optimization of a Francis draft tube," *International Journal of Numerical Methods for Heat and Fluid Flow*, vol. 17, no. 1, pp. 34–45, 2007.

[2] H. Lundqvist, P. Rivinoja, K. Leonardsson, and S. McKinnell, "Upstream passage problems for wild Atlantic salmon (*Salmo salar* L.) in a regulated river and its effect on the population," *Hydrobiologia*, vol. 602, no. 1, pp. 111–127, 2008.

[3] P. Rivinoja, S. Mckinnell, and H. Lundqvist, "Hindrances to upstream migration of atlantic salmon (*Salmo salar*) in A Northern Swedish River caused by a hydroelectric power-station," *River Research and Applications*, vol. 17, no. 2, pp. 101–115, 2001.

[4] J. V. Arnekleiv and M. Kraabøl, "Migratory behaviour of adult fast-growing brown trout (*Salmo trutta*, L.) in relation to water flow in a regulated Norwegian river," *Regulated Rivers: Research and Management*, vol. 12, no. 1, pp. 39–49, 1996.

[5] J. Webb, "The behaviour of adult Atlantic salmon ascending the rivers Tay and Tummel to Pitlochry dam," Scottish Fisheries Research Report 48, 1990.

[6] P. Rivinoja, *Migration problems of Atlantic Salmon (Salmo salar L.) in Flow Regulated Rivers*, Ph.D. thesis, Swedish University of Agricultural Sciences, Department of Aquaculture, Umeå, Sweden, 2005.

[7] E. M. Lindmark, *Flow design for migrating fish*, Ph.D. thesis, Luleå University of Technology, Division of Fluid Mechanics, Luleå, Sweden, 2008.

[8] E. Lindmark and H. Gustavsson, "Field study of an attraction channel as entrance to fishways," *River Research and Applications*, vol. 24, no. 5, pp. 564–570, 2008.

[9] N. R. B. Olsen and S. Stokseth, "Three-dimensional numerical modelling of water flow in a river with large bed roughness," *Journal of Hydraulic Research*, vol. 33, pp. 571–581, 1995.

[10] C. A. M. E. Wilson, J. B. Boxall, I. Guymer, and N. R. B. Olsen, "Validation of a three-dimensional numerical code in the simulation of pseudo-natural meandering flows," *Journal of Hydraulic Engineering*, vol. 129, no. 10, pp. 758–768, 2003.

[11] N. J. Clifford, N. G. Wright, G. Harvey, A. M. Gurnell, O. P. Harmar, and P. J. Soar, "Numerical modeling of river flow for ecohydraulic applications: Some experiences with velocity characterization in field and simulated data," *Journal of Hydraulic Engineering*, vol. 136, no. 12, pp. 1033–1041, 2010.

[12] D. J. Booker, "Hydraulic modelling of fish habitat in urban rivers during high flows," *Hydrological Processes*, vol. 17, no. 3, pp. 577–599, 2003.

[13] S. K. Sinha, F. Sotiropoulos, and A. J. Odgaard, "Three-dimensional numerical model for flow through natural rivers," *Journal of Hydraulic Engineering*, vol. 124, no. 1, pp. 13–23, 1998.

[14] B. Dargahi, "Three-dimensional flow modelling and sediment, transport in the River Klarälven," *Earth Surface Processes and Landforms*, vol. 29, no. 7, pp. 821–852, 2004.

[15] T. S. Lundström, J. G. I. Hellström, and E. M. Lindmark, "Flow design of guiding device for downstream fish migration," *River Research and Applications*, vol. 26, no. 2, pp. 166–182, 2010.

[16] J. Y. Yoon, V. C. Patel, and R. Ettema, "Numerical model of flow in ice-covered channel," *Journal of Hydraulic Engineering*, vol. 122, no. 1, pp. 19–26, 1996.

[17] Y. Jia, S. H. Scott, Y. Xu, S. Huang, and S. S. Y. Wang, "Three-dimensional numerical simulation and analysis of flows around a submerged weir in a channel bendway," *Journal of Hydraulic Engineering*, vol. 131, no. 8, pp. 682–693, 2005.

[18] Y. Jia, S. Scott, Y. Xu, and S. S. Y. Wang, "Numerical study of flow affected by bendway weirs in victoria bendway, the mississippi river," *Journal of Hydraulic Engineering*, vol. 135, no. 11, pp. 902–916, 2009.

[19] C. L. Rakowski, L. L. Ebner, and M. C. Richmond, "Fast-track design efforts using CFD: Bonneville second powerhouse," in *World Water and Environmental Resources Congress: Critical Transitions in Water and Environmental Resources Management*, pp. 1790–1798, July 2004.

[20] J. M. Viscardi, A. Pujol, V. Weitbrecht, G. H. Jirka, and N. R. B. Olsen, "Numerical simulations on the Paraná de las Plamas River," in *Proceedings of the 3rd International Conference on Fluvial Hydraulics, River Flow*, Lisbon, Portugal, 2006.

[21] M. R. Simpson and R. N. Oltmann, "Discharge-measurement system using an acoustic Doppler current profiler with applications to large rivers and estuaries," *US Geological Survey Water-Supply Paper*, vol. 2395, 1993.

[22] E. A. Nystrom, C. R. Rehmann, and K. A. Oberg, "Evaluation of mean velocity and turbulence measurements with ADCPs," *Journal of Hydraulic Engineering*, vol. 133, no. 12, pp. 1310–1318, 2007.

[23] ANSYS, *Ansys CFX User manual Ver. 11*, Ansys, Inc, 2007.

[24] B. E. Launder and D. B. Spalding, "The numerical computation of turbulent flows," *Computer Methods in Applied Mechanics and Engineering*, vol. 3, no. 2, pp. 269–289, 1974.

[25] J. F. Rodriguez, F. A. Bombardelli, M. H. García, K. M. Frothingham, B. L. Rhoads, and J. D. Abad, "High-resolution numerical simulation of flow through a highly sinuous river reach," *Water Resources Management*, vol. 18, no. 3, pp. 177–199, 2004.

[26] F. M. White, *Viscous Fluid Flow*, McGraw Hill, New York, NY, USA, 1991.

[27] G. J. Arcement and V. R. Schneider, "Guide for selecting Manning's roughness coefficients for natural channels and flood plains," *US Geological Survey Water-Supply Paper*, vol. 2339, 1989.

[28] B. C. Yen, "Open channel flow resistance," *Journal of Hydraulic Engineering*, vol. 128, no. 1, pp. 20–39, 2002.

[29] C. F. Colebrook, "Turbulent flow in pipes with particular reference to the transition region between the smooth- and rough-pipe laws," *Journal of the Institution of Civil Engineers*, vol. 11, pp. 133–156, 1939.

[30] V. C. Patel, "Perspective: flow at high reynolds number and over rough surfaces—Achilles Heel of CFD," *Journal of Fluids Engineering*, vol. 120, no. 3, pp. 434–444, 1998.

[31] T. M. Green, E. M. Lindmark, T. S. Lundström, and L. H. Gustavsson, "Flow characterization of an attraction channel as entrance to fishways," *River Research and Applications*, vol. 27, pp. 1290–1297, 2011.

17

Radar Cross-Section Formulation of a Shell-Shaped Projectile Using Modified PO Analysis

Mohammad Asif Zaman and Md. Abdul Matin

Department of Electrical and Electronic Engineering, Bangladesh University of Engineering and Technology, Dhaka 1000, Bangladesh

Correspondence should be addressed to Mohammad Asif Zaman, asifzaman13@gmail.com

Academic Editor: Azah Mohamed

A physical optics based method is presented for calculation of monostatic Radar Cross-Section (RCS) of a shell-shaped projectile. The projectile is modeled using differential geometry. The paper presents a detailed analysis procedure for RCS formulation using physical optics (PO) method. The shortcomings of the PO method in predicting accurate surface current density near the shadow boundaries are highlighted. A Fourier transform-based filtering method is proposed to remove the discontinuities in the approximated surface current density. The modified current density is used to formulate the scattered field and RCS. Numerical results are presented comparing the proposed method with conventional PO method. The results are also compared with published results of similar objects and found to be in good agreement.

1. Introduction

Prediction and measurement of Radar Cross-Section (RCS) have been a significant area of research for scientists and engineers for many years. The widespread uses of radar technology since the Second World War have demanded accurate and efficient prediction of fields scattered by radar targets. The knowledge of echo characteristics of radar targets is of great importance for the design of high-performance radars, as well as low visibility stealth targets [1, 2].

In radar technology, an antenna radiates electromagnetic (EM) energy. When an object is illuminated by the radar EM field, it reflects back some EM energy, which is received by an antenna. In monostatic radar system, the transmitting and reception of the EM energy are done by the same antenna or by multiple antennas located very close to each other [1, 3]. In bistatic radar system, separate antennas are used for transmitting and receiving, and the antennas are usually far away from each other [1]. The radar field reflective nature of an object is specified in terms of RCS. The RCS is the area that a target would have to occupy to produce the amount of reflected power that is detected back at the radar [2, 4]. The RCS of an object depends on the viewing angles, the size, geometry, and composition of the object, frequency and polarization of the radar signal, and so forth [3, 4]. Stealth technology concentrates on reducing the RCS of airplanes and missiles to make them invisible to radars. Conversely, radar engineers are developing more sensitive radars that can detect low RCS targets. In both cases, accurate numerical simulation methods are essential for design purposes.

Numerical simulation of RCS of an object requires calculation of the scattered field from the object for a given incident field. Several numerical methods exist for scattered EM field calculations. Pure numerical methods such as Method of Moments (MoM), Finite-Difference Time-Domain (FDTD) method, Fast Multipole Method (FMM), and Transmission-Line Matrix (TLM) have been successfully used in predicting RCS of radar targets [4, 5]. Conformal FDTD-based methods have also appeared in literature [6]. Recently, some variants of MoM have been developed for monostatic RCS formulation [7, 8] and other scattering related problems [9]. These methods do not depend on the geometry and can be used for any arbitrary shaped objects. However, these methods are computationally demanding, and the high simulation time for electrically large objects is not always acceptable for design and optimization problems.

High-frequency asymptotic methods, such as Geometrical Optics (GO), Physical Optics (PO), Uniform Geometrical Theory of Diffraction (UTD), and Physical Theory of Diffraction (PTD), have also been used for RCS formulation [4]. These methods are based on local interaction of EM fields. Therefore, they are computationally less demanding than the pure numerical methods, and they require far less simulation time [10]. However, these methods are geometry dependent. For complex shaped objects, the scattered field formulation can be tedious, especially when using UTD method [11]. GO and PO methods do not suffer to the same extent as UTD for complex geometry cases. The GO method is the fastest among the high frequency techniques, but it is relatively less accurate [11]. The PO method gives much more accurate results compared to GO. It is a well-accepted method for formulating scattered field from electrically large objects [12, 13].

Because of its relatively high accuracy, the PO method has been widely used for RCS formulation [14, 15]. The PO method is also computationally more efficient than pure numeric methods, making it relatively faster which makes it an essential tool for aerospace designers [16]. Several modified versions of PO have been developed to further increase the speed and accuracy of the PO method [17, 18]. In this paper, a new modified PO method is used to formulate the monostatic RCS of a shell-shaped projectile.

This paper presents a detailed description and procedure of RCS calculation method of an object using PO method and modified PO method. The procedure includes the geometrical modeling of the shell-shaped radar object, approximation of the induced surface current on the object, and formulation of the PO radiation integral in parametric space. Although the paper concentrates on RCS formulation of a specific shell-shaped radar target, the analysis procedure is general and can be applied to objects of any geometry. In spite of the presence of many research and review articles [4], a complete description of PO method-based procedure for RCS calculation is rare in literature. In addition to the complete calculation procedure, this paper also presents a new modified PO method. The PO method approximates the induced surface current on the radar target and uses this current to calculate the scattered field [11, 19]. This approximation leads to discontinuous surface current across the target surface near the shadow boundaries [19]. In the proposed method, a better approximate of the induced surface currents is used to remove the unnatural discontinuities and improve the accuracy in calculation of the scattered field. The proposed method incorporates filtering inspired Fourier transform-based methods to remove the current discontinuity.

The paper is arranged as follows: in Section 2, the geometrical modeling of the shell-shaped projectile is described. Differential geometry-based definition of the surfaces and normals is derived here. Section 3 contains approximation of the induced surface current on object and formulation of the scattered field and RCS. Numerical results are presented in Section 4. Finally, concluding remarks are given in Section 5.

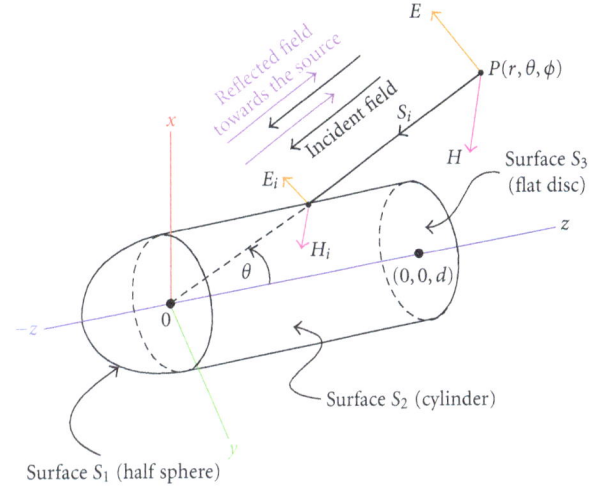

FIGURE 1: Three dimensional geometry of the shell-shaped projectile, and the coordinate system along with incident field vectors.

2. Geometrical Modeling

The first step in simulating the RCS of an object is to accurately model the surface of the object. This paper concentrates on the RCS of a shell-shaped object. A shell-shaped object is selected as most projectiles represent this basic shape. A schematic diagram of the object and the three-dimensional coordinate system is shown in Figure 1. The object can be modeled by three different canonical surfaces: a half sphere (surface S_1), a cylinder (surface S_2), and a flat disc (surface S_3). Most objects can be similarly modeled by a few common canonical shapes. For this reason, scattering from common shapes such as flat plates, cones, and cylinders has received attention in literature since the 1960s [20, 21].

The origin of the coordinate system is taken at the center of the half sphere, and z axis is taken as the axis of the cylinder. The three surfaces can be expressed by the following equations:

$$S_1 : x^2 + y^2 + z^2 = b^2, \quad -b \leq z \leq 0,$$
$$S_2 : x^2 + y^2 = b^2, \quad 0 \leq z \leq d, \quad (1)$$
$$S_3 : x^2 + y^2 = \rho^2, \quad 0 \leq \rho \leq b, z = d.$$

Here, b= radius of the half-sphere = radius of the cylinder, and d = length of the cylinder. For scattering problems and RCS formulation, it is often more convenient to express the surfaces in differential geometry format rather than coordinate geometry format [11]. To express the surfaces in differential geometry format, the following parameters are used:

$$S_1 : x = \rho_s \cos \phi_s, \quad y = \rho_s \sin \phi_s, \quad z = -\sqrt{b^2 - \rho_s^2},$$
$$0 \leq \rho_s \leq b, 0 \leq \phi_s \leq 2\pi,$$
$$S_2 : x = b \cos \phi_s, \quad y = b \sin \phi_s, \quad z = \rho_s,$$
$$0 \leq \rho_s \leq d, 0 \leq \phi_s \leq 2\pi,$$

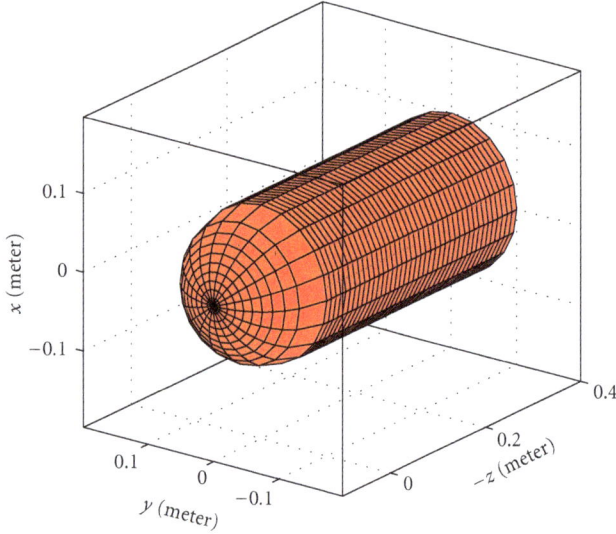

FIGURE 2: Computer-generated three-dimensional model of the shell-shaped projectile.

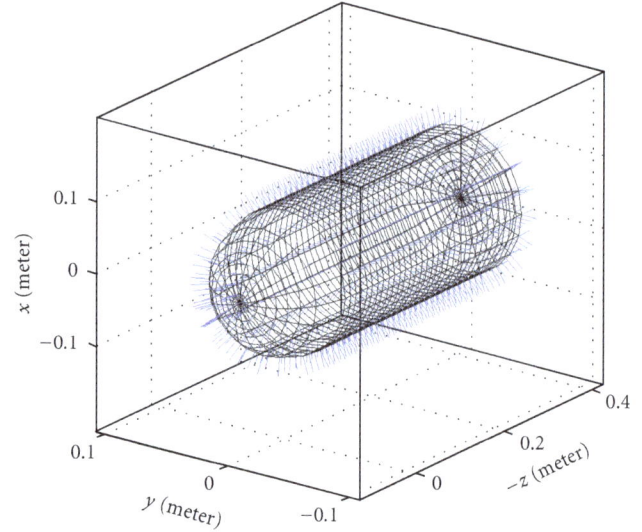

FIGURE 3: Normal vectors (blue lines) on the object surfaces.

$$S_3 : x = \rho_s \cos\phi_s, \quad y = \rho_s \sin\phi_s, \quad z = d,$$

$$0 \le \rho_s \le b, \ 0 \le \phi_s \le 2\pi. \tag{2}$$

Here, ρ_s and ϕ_s are parameters. Note that the ρ_s and ϕ_s for S_1 are not the same as the ρ_s and ϕ_s for S_2 or S_3. For each surface the parameters are different and unrelated. Same parameter names are used for simplicity only. Using these parameters, the differential geometric expressions of the surfaces are [22]

$$S_1 : \mathbf{r} = \rho_s \cos\phi_s\ \hat{\mathbf{x}} + \rho_s \sin\phi_s\ \hat{\mathbf{y}} - \sqrt{b^2 - \rho_s^2}\ \hat{\mathbf{z}},$$

$$S_2 : \mathbf{r} = b \cos\phi_s\ \hat{\mathbf{x}} + b \sin\phi_s\ \hat{\mathbf{y}} + \rho_s\ \hat{\mathbf{z}}, \tag{3}$$

$$S_3 : \mathbf{r} = \rho_s \cos\phi_s\ \hat{\mathbf{x}} + \rho_s \sin\phi_s\ \hat{\mathbf{y}} + d\ \hat{\mathbf{z}}.$$

The limits of the parameters are given in (2). Using these equations, it is possible to construct a three-dimensional model of the shell-shaped projectile using computer coding. The computer generated model is shown in Figure 2. For computation, $b = 10$ cm and $d = 40$ cm are used.

Once the surfaces of the object are defined, it is necessary to define normal vectors on each point of the surface. These normal vectors are necessary for GO-, UTD-, or PO-based scattering formulations [19]. The normal vectors, $\hat{\mathbf{n}}$, can be calculated from the differential geometric expression of the surfaces using the following equation [11, 22]:

$$\hat{\mathbf{n}} = \pm \frac{(\partial\mathbf{r}/\partial\rho_s) \times (\partial\mathbf{r}/\partial\phi_s)}{|(\partial\mathbf{r}/\partial\rho_s) \times (\partial\mathbf{r}/\partial\phi_s)|}. \tag{4}$$

The sign of the normal vectors is selected so that they always point away from the surface. Using (4), the normal vectors on the three surfaces of the object are calculated to be

$$S_1 : \hat{\mathbf{n}} = \frac{\rho_s}{b}\cos\phi_s\ \hat{\mathbf{x}} + \frac{\rho_s}{b}\sin\phi_s\ \hat{\mathbf{y}} - \frac{\sqrt{b^2 - \rho_s^2}}{b}\ \hat{\mathbf{z}},$$

$$S_2 : \hat{\mathbf{n}} = \cos\phi_s\ \hat{\mathbf{x}} + \sin\phi_s\ \hat{\mathbf{y}}, \tag{5}$$

$$S_3 : \hat{\mathbf{n}} = \hat{\mathbf{z}}.$$

Using (5), the normal vectors are plotted over the wire frame of the object using computer coding. The results are shown in Figure 3. From visual inspection it is verified that the normals are perpendicular to the surface and points away from it. This verifies the geometrical modeling performed in this section.

3. RCS Formulation Using Modified PO Method

To formulate the RCS of the object, the incident field must be defined first. As the radar and the target are usually very far away from each other, the incident field can be modeled as a plane wave, implying that the direction of the wave, the direction of the electric field, and the direction of the magnetic field are perpendicular to each other. With respect to a coordinate system defined at the source point of the wave, if the wave travels at z direction, and the electric field is assumed to be polarized along x direction, then the magnetic field will be polarized along y direction. However, in this case, the coordinate system is defined with respect to the object. So, coordinate transformation must be performed to find the expression of the incident field with respect to the defined coordinate system [23, 24]. For a wave that is incident on the angle (θ, ϕ), an x direction polarized electric field converts to a θ polarized wave [24]. So, the incident electric field can be expressed as [11, 24]

$$\mathbf{E_i} = |\mathbf{E_i}|\hat{\theta} = \cos\theta\cos\phi\ \hat{\mathbf{x}} + \cos\theta\sin\phi\ \hat{\mathbf{y}} + \sin\theta\ \hat{\mathbf{z}}. \tag{6}$$

Here, the amplitude of the incident electric field is assumed to be 1. The direction of the incident ray is given by

$$\hat{\mathbf{s_i}} = \sin\theta\cos\phi\ \hat{\mathbf{x}} + \sin\theta\sin\phi\ \hat{\mathbf{y}} - \cos\theta\ \hat{\mathbf{z}}. \tag{7}$$

The incident ray direction, along with electric and magnetic field polarizations, is shown in Figure 1.

For monostatic RCS calculation, the reflected ray is

$$\hat{\mathbf{s}}_{\mathbf{r}} = -\hat{\mathbf{s}}_{\mathbf{i}}. \tag{8}$$

PO method uses approximate expression of the surface current density induced on the surface of the object due to the incident field to find the scattered field. The surface current density depends on the incident magnetic field. The incident magnetic field is given by [11, 19]

$$\mathbf{H}_{\mathbf{i}} = \frac{1}{\eta}(\hat{\mathbf{s}}_{\mathbf{i}} \times \mathbf{E}_{\mathbf{i}}). \tag{9}$$

Here, $\eta = 120\pi$ = intrinsic impedance of air. Equation (9) is valid only for plane wave incidence. As the object can be considered to be very far away from the source for most radar applications, this assumption is justified. The PO approximates the surface current density, $\mathbf{J}_{\mathbf{PO}}$, as [18]

$$\mathbf{J}_{\mathbf{PO}} = \begin{cases} 2\hat{\mathbf{n}} \times \mathbf{H}_{\mathbf{i}}, & \text{for illuminated surface,} \\ 0, & \text{otherwise.} \end{cases} \tag{10}$$

Equation (10) holds for a perfectly conduction object. For calculating RCS of a metallic object, this approximation is justified. Using (7), (9) along with (5) in (10), it is possible to calculate $\mathbf{J}_{\mathbf{PO}}$. However, it is necessary to identify which part of the surface is illuminated by the incident field and which part is not for calculation of (10). This can be accomplished by noticing the angle between $\hat{\mathbf{s}}_{\mathbf{i}}$ and $\hat{\mathbf{n}}$. When the vectors are perpendicular to each other, the incident field is tangent on the surface [22, 23]. These surface points indicate the shadow boundary, beyond which the surface points will not be illuminated. So, surface points on which the angle between the normal vector and incident ray vector is greater than 90° are in the shadowed region. This statement can be mathematically expressed as

$$\hat{\mathbf{n}} \cdot \hat{\mathbf{s}}_{\mathbf{i}} \leq 0 \ \text{ for illuminated region,}$$
$$\hat{\mathbf{n}} \cdot \hat{\mathbf{s}}_{\mathbf{i}} > 0 \ \text{ for shadowed region.} \tag{11}$$

Thus using (10) and (11), the PO approximate of the surface current density can be formulated on S_1, S_2, and S_3 surfaces. It should be mentioned that (11) is true only for objects that only have convex surfaces. If one region of the object creates shadow for another region, then (11) cannot be used to identify the illuminated region and shadowed region. For the relatively simple geometry presented in this paper, (11) is sufficient for calculating the shadow boundaries.

From $\mathbf{J}_{\mathbf{PO}}$, the magnetic vector potential, $\mathbf{A}$ can be calculated using the radiation integral [12, 19]:

$$\mathbf{A} = \int \mathbf{J}_{\mathbf{PO}}\frac{e^{-jkR}}{R}dS = \int \mathbf{J}_{\mathbf{PO}}\frac{e^{-jkR}}{R}dS_1 + \int \mathbf{J}_{\mathbf{PO}}\frac{e^{-jkR}}{R}dS_2$$
$$+ \int \mathbf{J}_{\mathbf{PO}}\frac{e^{-jkR}}{R}dS_3. \tag{12}$$

Here, R is the distance from the object to the receiver. As monostatic radar cross-section is considered, the source and

the receiver are located at the same position. The position of the source point is expressed in polar coordinates as $P(r, \theta, \phi)$ with $r = 1000$ meter. R can easily be calculated from the coordinates of the source and the object. The surface integrals can be performed over parametric space. The differential surface element dS can be expressed in terms of the parameters as [22, 23]

$$dS = \left| \frac{\partial \mathbf{r}}{\partial \rho_s} \times \frac{\partial \mathbf{r}}{\partial \phi_s} \right| d\rho_s \, d\phi_s. \tag{13}$$

Evaluating (13) for the three surfaces and using it in (12):

$$\mathbf{A} = \int_{\phi_s=0}^{2\pi} \int_{\rho_s=0}^{b} \mathbf{J}_{\mathbf{PO}}(\rho_s, \phi_s) \, \frac{e^{-jkR}}{R} \frac{\rho_s b}{\sqrt{b^2 - \rho_s^2}} \, d\rho_s \, d\phi_s$$
$$+ \int_{\phi_s=0}^{2\pi} \int_{\rho_s=0}^{d} \mathbf{J}_{\mathbf{PO}}(\rho_s, \phi_s) \, \frac{e^{-jkR}}{R} b \, d\rho_s \, d\phi_s \tag{14}$$
$$+ \int_{\phi_s=0}^{2\pi} \int_{\rho_s=0}^{b} \mathbf{J}_{\mathbf{PO}}(\rho_s, \phi_s) \, \frac{e^{-jkR}}{R} \rho_s \, d\rho_s \, d\phi_s.$$

The scattered field, $\mathbf{E}_{\mathbf{s}}$, can be formulated from $\mathbf{A}$ as [12]

$$\mathbf{E}_{\mathbf{s}} = -j\frac{60\pi}{\lambda}\mathbf{A}. \tag{15}$$

Here, λ = wavelength of the incident field. The RCS, σ of the object is calculated from the scattered field using (6) and (15) [1, 4]:

$$\sigma = \lim_{R \to \infty} 4\pi R^2 \frac{|\mathbf{E}_{\mathbf{s}}|^2}{|\mathbf{E}_{\mathbf{i}}|^2}. \tag{16}$$

The method described so far is the conventional PO method. Although sufficiently accurate, one of the weak points of this method is the approximation of the surface current density, $\mathbf{J}_{\mathbf{PO}}$. From (10), it can be seen that $\mathbf{J}_{\mathbf{PO}}$ abruptly changes to zero in the boundary between illuminated and shadow region. Practical surface current densities do not have this discontinuous nature. This discontinuity arises because PO method does not take into account the *creeping waves* which exist in the boundary between shadowed region and illuminated region. These waves gradually decrease with distance from the shadow boundaries, and the resulting surface current density decreases gradually.

So, to increase the accuracy of the PO method, the discontinuity $\mathbf{J}_{\mathbf{PO}}$ in must be removed. In this paper, a filtering-based approach is proposed to remove the discontinuity in current distribution. The spatial variation in the current distribution over the surface of the object can be compared with temporal variation of an analog signal in time domain. In Fourier expansion of time signals, the time signal is imagined as superimposition of many sinusoidal signals, termed Fourier spectral components. A sharply varying time signal has high Fourier spectral components. A discontinuous signal has large high Fourier spectral components. Using this analogy, the spatial variation of the surface current in scattering problems can be imagined as superimposition of many surface current components with sinusoidal variation.

These components can be isolated using Fourier transformation. It can be imagined that the rapidly varying high components of the surface current are responsible for the sharp discontinuous spatial distribution. The high spectral components add in opposite phase with the other terms in the shadow region to create destructive interference. In absence of these high spectral components, the other components will not completely cancel each other, and therefore there will be an oscillating distribution of surface current in the shadow region. These oscillating distributions can be compared to the creeping waves. Thus filtering out high spectral components should create a smooth distribution of surface current densities which may accurately model the actual current distribution with a higher degree of accuracy.

To perform filtering operations to make J_{PO} continuous, Fourier transform is performed in the parametric space. This produces Fourier domain representation of the PO surface current, J_{FPO}. This can be obtained using the following relation [24]:

$$
\begin{aligned}
J_{FPO}(\rho_s, \zeta) &= F_{\phi_s}[J_{PO}(\rho_s, \phi_s)] \\
&= \frac{1}{\sqrt{2\pi}} \int_{-\infty}^{\infty} J_{PO}(\rho_s, \phi_s) e^{-j\zeta\phi_s} d\phi_s.
\end{aligned}
\tag{17}
$$

Here, Fourier transform is performed with respect to ϕ_s, and ζ is another parameter corresponding to the frequency term in conventional Fourier transform of time series data. The complete information of J_{PO} is implicit within J_{FPO}. The discontinuity in J_{PO} with respect to ϕ_s will result in high J_{FPO} values for large ζ. This is analogous to high frequency terms in discontinuous time series data [25]. As these high ζ components contribute to the discontinuity, removing them should result in a smoother surface current density. How many high ζ value components need to be removed to produce an accurate surface current density cannot be analytically calculated due to the complex nature of the mathematics. Observing the discontinuity in the PO current in many canonical problems found in literature [11] and testing the filter for different parameters, it is found through trial and error that removing 70% of high ζ value components results in a relatively accurate surface current distribution. A modified current density, J_{SPO}, is constructed in ζ space by removing 70% of high ζ value components:

$$
J_{SPO}(\rho_s, \zeta) = \begin{cases} 0, & \zeta \geq 0.7\zeta_{max} \\ J_{FPO}(\rho_s, \zeta), & \text{otherwise.} \end{cases}
\tag{18}
$$

Now, the modified smoother current density, J_{MPO}, in $\rho_s - \phi_s$ space can be obtained by using inverse Fourier transform [24]:

$$
\begin{aligned}
J_{MPO}(\rho_s, \phi_s) &= F_{\zeta}^{-1}[J_{SPO}(\rho_s, \zeta)] \\
&= \frac{1}{\sqrt{2\pi}} \int_{-\infty}^{\infty} J_{PO}(\rho_s, \zeta) e^{j\zeta\phi_s} d\zeta.
\end{aligned}
\tag{19}
$$

Equation (19) is applicable for canonical geometries and surface of revolution. Now, the discontinuity in J_{PO} with respect to ϕ_s should no longer be present in J_{MPO}. However,

J_{MPO} may still be discontinuous with respect to ρ_s. A similar analysis can be performed to remove this discontinuity by taking Fourier transform of J_{MPO} with respect to ρ_s, eliminating high components and then taking inverse Fourier transform. Then a smoothed surface current density, J_M, can be obtained which is continuous across the surface of the object.

It is noted that that as some components of the current densities are filtered out, the energy of the conventional PO current may not be equal to the energy of the modified PO current. This may not be consistent with conservation of energy. To rectify this problem, the filtered current densities must be multiplied by a constant scalar so that the resulting current densities have the same energy as the PO current. The value of the constant can easily be calculated by calculating the energy of the conventional PO current and the unscaled modified PO current. This scaling ensures that no changes have occurred in total energy of the surface currents.

Due to harmonic oscillating nature of Fourier transform, J_M has oscillating characteristics near the shadow boundary which accurately represents diffraction patterns. Using J_M instead of J_{PO} in (12)–(16) will result in a more accurate estimate of RCS.

4. Numerical Results

For numerical analysis, the frequency of the incident field is taken to be 10 GHz. The radius of the sphere and cylinder, $b = 10$ cm, and the length of the cylinder, $d = 40$ cm, are assumed. To perform numerical analysis, the surface of the object must be divided into discrete points. Discrete values of the parameters ρ_s and ϕ_s are taken to create discrete points on the surface. Large number of points increase accuracy but also take considerable simulation time. Here, 22 values ρ_s per wavelength are considered, and 204 values of ϕ_s per ρ_s are considered. These values fall in the range of typical selected values for numerical analysis using PO type method [19]. All numerical results are obtained using computer coding.

Figures 4 and 5 show the surface current density for two different incident field orientations. These current densities are obtained using conventional PO methods using (10). The discontinuity in the current densities can easily be spotted in both figures. All the current densities are plotted after normalization process. The normalization is performed by dividing the current densities by the maximum value of the current density. Due to this division, the normalized current density is unitless. The resulting normalized current densities are expressed in decibels by taking logarithm.

Using the proposed modified PO method, the surface current densities are smoothed. Fourier transform is used with respect to both parameters. The resulting smoothed current densities are shown in Figures 6 and 7. The difference in current densities can easily be observed. Comparing Figure 4 with Figure 6 and Figure 5 with Figure 7, it can be seen that the current densities decrease gradually when the proposed method is used whereas the conventional PO method creates sharp discontinuities. The oscillating pattern of the currents near the shadow boundary is created when the proposed method is used. These accurately describe realistic diffracted fields [11].

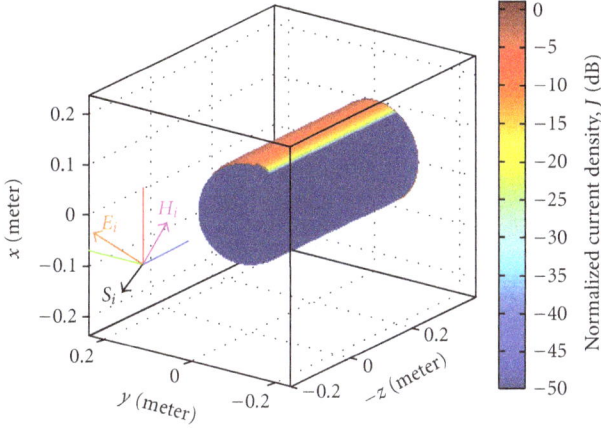

FIGURE 4: Surface current density using PO method for an incident field $\theta = 20°$, $\phi = 60°$.

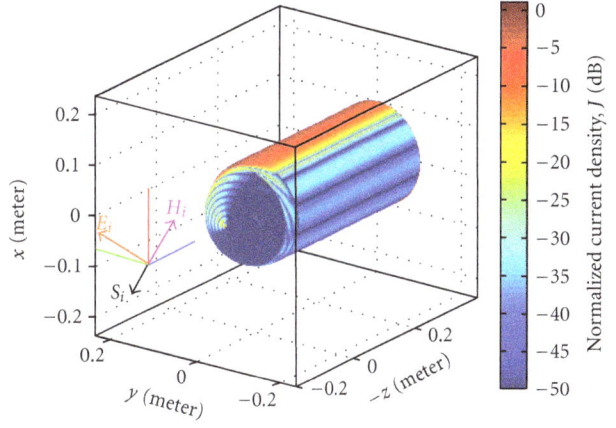

FIGURE 6: Surface current density using modified PO method for an incident field $\theta = 20°$, $\phi = 60°$.

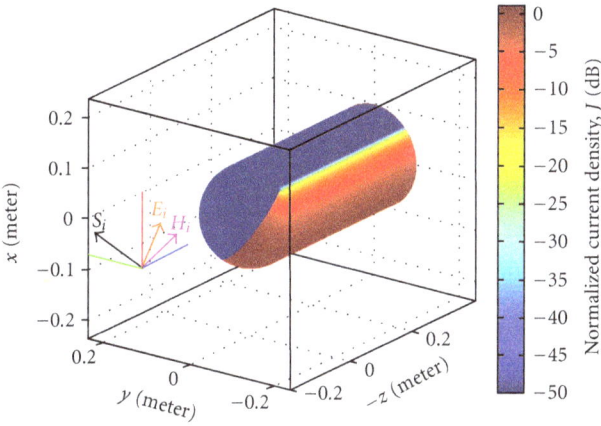

FIGURE 5: Surface current density using PO method for an incident field $\theta = 60°$, $\phi = 45°$.

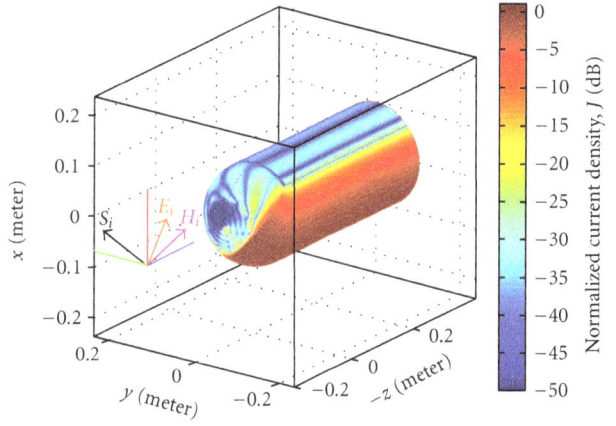

FIGURE 7: Surface current density using modified PO method for an incident field $\theta = 60°$, $\phi = 45°$.

To verify accuracy of the surface current densities approximated in the proposed filtering method, a simplified canonical problem of scattering from an infinite cylinder is considered. Considering an infinitely long cylinder with axis parallel to the z axis is illuminated by a plane wave. The orientation of the magnetic field of the incident wave is considered along the z axis, and the orientation of the electric field is assumed to be along the x axis. The wave propagates along the y axis. The PO current given by (10) is constant throughout the illuminated region of the cylinder. There will be no variation of the surface current along the z axis. The plane wave will illuminate only half the surface of the cylinder. If an angle, χ, is defined along the xy plane as the angle from the y axis, then for a plane wave propagating along the y axis, the illuminated region spans from $\chi = -90°$ to $90°$. The PO current abruptly falls to zero outside this angle. The proposed filtering method adjusts this discontinuity and makes the current smooth. The actual angular distribution of the current density can be formulated using a Method of Moments (MoM), which is a well-known benchmark method [19]. The current distribution obtained

from the PO method, the proposed modified PO method, and MoM are shown in Figure 8. It can be clearly observed that the proposed method produces a current distribution which resembles the results obtained from MoM much closely compared to the PO method. This higher degree of accuracy in estimating the surface current ensures that the proposed method will generate more accurate RCS results compared to the conventional PO method.

The RCS of the shell-shaped object is formulated using both conventional PO and the modified PO method. The RCS at $\phi = 0°$ plane as a function of zenith angle, θ, is shown in Figure 9.

RCS is expressed in dB with respect to $1\,\text{m}^2$ area. This unit is represented as $\text{dB}\,\text{m}^2$ or $\text{dB}\,\text{sm}$ [1, 4]. As the object is circularly symmetric around the z axis, the scattered field and RCS are independent of the azimuth angle, ϕ [11, 26]. It is seen that both methods give similar results in most angular regions. The variation in result comes for angular regions where the spherical surface is illuminated. This is expected as the spherical surfaces are affected by the diffracted rays and creeping waves more than the other shaped surfaces

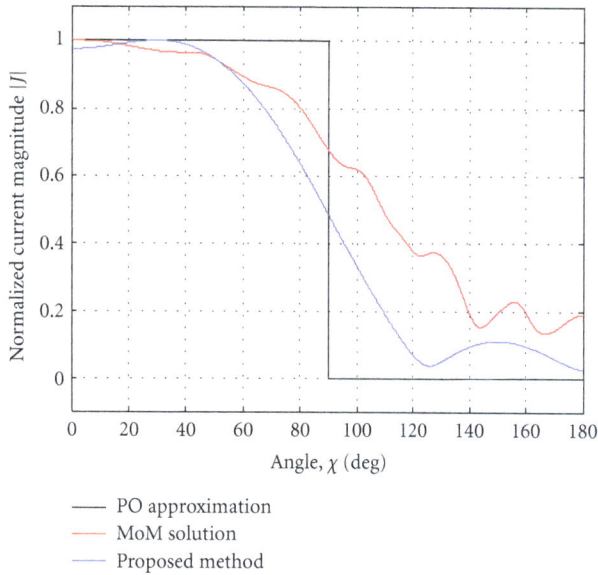

FIGURE 8: Surface current densities on an infinite cylinder surface at $f = 3\,\text{GHz}$.

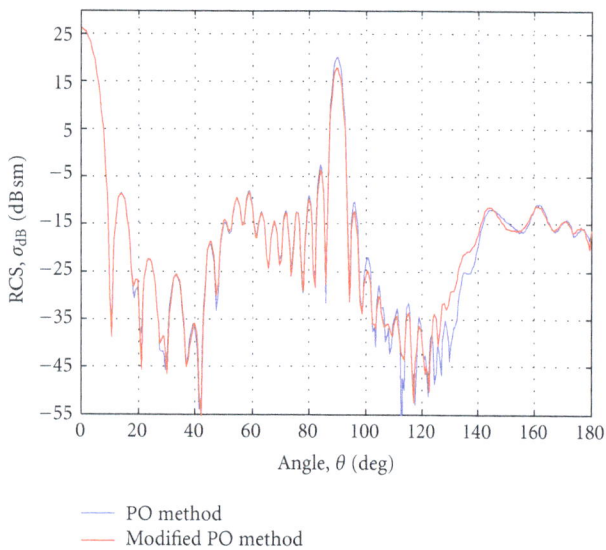

FIGURE 10: Monostatic RCS of the shell-shaped projectile at 5.5 GHz as a function of θ.

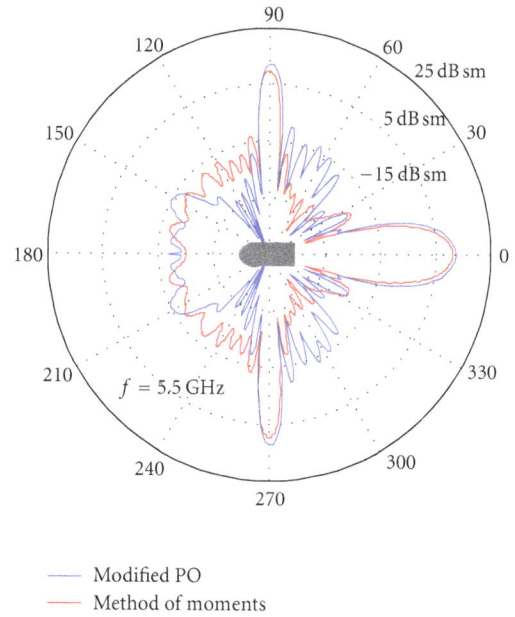

FIGURE 9: Monostatic RCS of the shell-shaped projectile at 10 GHz as a function of θ.

[11]. Due to the better approximate of the surface current, the results obtained from the proposed method are more accurate. It is noted that at $\theta = 180°$ only the half-sphere surface is illuminated. The RCS at this angle is around $-15\,\text{dB sm}$. The projected area of the half sphere $= \pi b^2 = 0.0314\,\text{m}^2 = -15\,\text{dB sm}$, which is equal to the RCS. This is expected for a spherical shaped object. This consistency verifies the numerical analysis.

The RCS of the projectile at 5.5 GHz as a function θ is shown in Figure 10. The results from the proposed method and MoM are shown in the same plot for comparison. It can be observed that the results obtained from the proposed

method matches with the results obtained from MoM closely for most values of θ.

For the angles where RCS is high, the results of both methods are almost identical. The deviation arises for angles where RCS value is low. The higher degree of accuracy obtained from MoM comes at a cost of complex computation and extensive simulation time. The errors produced by the proposed method are relatively small and acceptable for many applications. In most cases the accuracy from PO method is sufficient, and the proposed method is expected to be more accurate than the PO method.

The RCS of the shell-shaped projectile for frequencies 3 GHz, 6 GHz, 10 GHz, and 14 GHz as a function θ is shown in Figure 11. The obtained RCS pattern is similar to the RCS patterns of objects of similar shape described in [4, 16].

The computer simulation was performed on an Intel Core i5-2430M 2.4 GHz CPU with 2.94 GB usable RAM. The simulation time of conventional PO and modified PO is compared in Table 1.

From Table 1, it can be seen that for lower frequency, the simulation time of conventional PO and the proposed modified PO is comparable. For higher frequency, the modified PO method takes 80% to 90% longer time to simulation compared to the conventional PO method. This excess simulation time can be expressed by the fact that for high-frequency simulation, the number of discrete surface points selected is larger. As there are 22 values ρ_s selected per wavelength and the wavelength is smaller for high-frequency simulation, the overall number of point increases. The filtering of the surface currents requires this additional simulation time due to large number of points. However, the proposed method is much faster compared to Method of Moments. For each simulation, the Method of Moments requires over 500 seconds of simulation time.

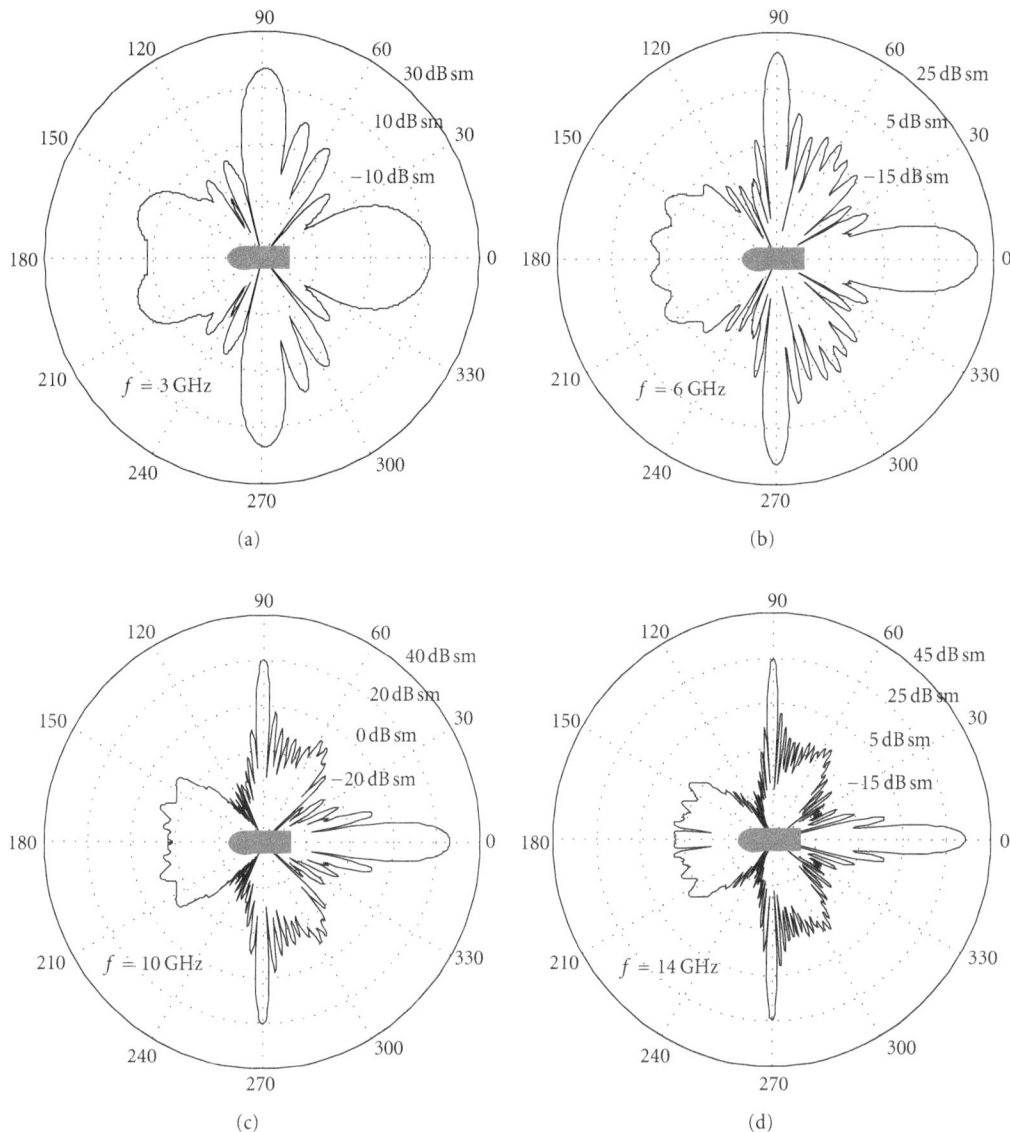

FIGURE 11: Monostatic RCS of the projectile at 3 GHz, 6 GHz, 10 GHz, and 14 GHz as a function of θ.

TABLE 1: Comparison of CPU simulation time.

Frequency (GHz)	Simulation time for PO method (sec)	Simulation time for modified PO method (sec)
3	16.83	26.09
6	32.09	50.79
10	53.19	87.01
14	74.15	144.23

5. Conclusion

RCS formulation procedure using PO method is described in detail in this paper. The paper covers geometrical modeling of shell-shaped projectile using differential geometry, formulation of surface current density, evaluation of the scattered field integral in parametric space, and monostatic RCS formulation. A modified PO method is presented which approximates surface current density more accurately. The additional computational steps require Fourier transform and inverse Fourier transform only. These can be easily incorporated in computer code using well-known Fast Fourier Transform (FFT) algorithm. Thus the accuracy is increased without significant increase in computational complexity. The obtained results using the modified PO method are consistent with similar results found in literature.

References

[1] E. F. Knott, J. F. Shaeffer, and M. T. Tuley, *Rada Cross Section*, Scitech Publishing, 2nd edition, 2004.

[2] D. C. Jenn, *Radar and Laser Cross Section Engineering*, American Institute of Aeronautics and Astronautics (AIAA), 2nd edition, 2005.

[3] C. A. Balanis, *Antenna Theory Analysis and Design*, John Wiley & Sons, 3rd edition, 2005.

[4] C. Uluişik, G. Çakir, M. Çakir, and L. Sevgi, "Radar cross section (RCS) modeling and simulation, part 1: a tutorial review of definitions, strategies, and canonical examples," *IEEE Antennas and Propagation Magazine*, vol. 50, no. 1, pp. 115–126, 2008.

[5] G. Çakir, M. Çakir, and L. Sevgi, "Radar cross section (RCS) modeling and simulation, part 2: a novel FDTD-based RCS prediction virtual tool for the resonance regime," *IEEE Antennas and Propagation Magazine*, vol. 50, no. 2, pp. 81–94, 2008.

[6] C. M. Kuo and C. W. Kuo, "A new scheme for the conformal FDTD method to Calculate the radar cross section of perfect conducting curved objects," *IEEE Antennas and Wireless Propagation Letters*, vol. 9, pp. 16–19, 2010.

[7] J. Ling, S. X. Gong, X. Wang, B. Lu, and W. T. Wang, "A novel two-dimensional extrapolation technique for fast and accurate radar cross section computation," *IEEE Antennas and Wireless Propagation Letters*, vol. 9, pp. 244–247, 2010.

[8] J. Ling, S. X. Gong, W. T. Wang, X. Wang, and Y. J. Zhang, "Fast monostatic radar cross section computation using Maehly approximation," *IET Science, Measurement and Technology*, vol. 5, no. 1, pp. 1–4, 2011.

[9] J. Bao, D. Wang, and E. K. N. Yung, "Electromagnetic scattering from an arbitrarily shaped bi-isotropic body of revolution," *IEEE Transactions on Antennas and Propagation*, vol. 58, no. 5, pp. 1689–1698, 2010.

[10] J. M. Rius, M. Ferrando, and L. Jofre, "High-frequency RCS of complex radar targets in real-time," *IEEE Transactions on Antennas and Propagation*, vol. 41, no. 9, pp. 1308–1319, 1993.

[11] D. A. McNamara, C. W. I. Pistorius, and J. A. G. Malherbe, *Introduction to The Uniform Geometrical Theory of Diffraction*, Artech House, 1990.

[12] L. Angermann, *Numerical Simulations—Applications, Examples and Theory*, Intech, 2011.

[13] X. J. Chen and X. W. Shi, "An expression for the radar cross section computation of an electrically large perfect conducting cylinder located over a dielectric half-space," *Progress in Electromagnetics Research*, vol. 77, pp. 267–272, 2007.

[14] S. Blume and G. Kahl, "The physical optics radar cross section of an elliptic cone," *IEEE Transactions on Antennas and Propagation*, vol. 35, no. 4, pp. 457–460, 1987.

[15] C. Bourlier and P. Pouliguen, "Useful analytical formulae for near-field monostatic radar cross section under the physical optics: far-field criterion," *IEEE Transactions on Antennas and Propagation*, vol. 57, no. 1, pp. 205–214, 2009.

[16] D. M. Elking, J. M. Roedder, D. D. Car, and S. D. Alspach, "A review of high-frequency radar cross section analysis capabilities at McDonnell Douglas Aerospace," *IEEE Antennas and Propagation Magazine*, vol. 37, no. 5, pp. 33–43, 1995.

[17] J. Lee, M. Havrilla, M. Hyde, and E. J. Rothwell, "Scattering from a cylindrical resistive sheet using a modified physical optics current," *IET Microwaves, Antennas and Propagation*, vol. 2, no. 5, pp. 482–491, 2008.

[18] T. Shijo, L. Rodriguez, and M. Ando, "The modified surface-normal vectors in the physical optics," *IEEE Transactions on Antennas and Propagation*, vol. 56, no. 12, pp. 3714–3722, 2008.

[19] R. C. Johnson, *Chapter 20: Phased Arrays, Antenna Engineering Handbook*, McGraw-Hill, 4th edition, 2007, Edited by: J. L. Volakis.

[20] R. Ross, "Radar cross section of rectangular flat plates as a function of aspect angle," *IEEE Transactions on Antennas and Propagation*, vol. 14, no. 8, pp. 329–335, 1966.

[21] S. D. Weiner and S. L. Borison, "Radar scattering from blunted cone tips," *IEEE Transactions on Antennas and Propagation*, vol. 14, no. 6, pp. 774–781, 1966.

[22] D. V. Widder, *Advanced Calculus*, Prentice Hall, 2nd edition, 2004.

[23] T. M. Apostol, *Calculus Volume II*, John Wiley & Sons, 2nd edition, 1969.

[24] A. D. Polyanin and A. V. Manzhirov, *Handbook of Mathematics for Engineers and Scientists*, Chapman & Hall, 2007.

[25] S. S. Soliman and M. D. Srinath, *Continuous and Discrete Signals and Systems*, Prentice Hall, 2007.

[26] M. Andreasen, "Scattering from bodies of revolution," *IEEE Transactions on Antennas and Propagation*, vol. 13, no. 2, pp. 303–310, 1965.

18

Theoretic and Experimental Studies on the Casting of Large Die-Type Parts Made of Lamellar Graphite Grey Pig Irons by Using the Technology of Polystyrene Moulds Casting from Two Sprue Cups

Constantin Marta, Ioan Ruja, Cinca Ionel Lupinca, and Monica Rosu

Department of Engineering, University "Eftimie Murgu" of Resita, Piaţa Traian Vuia, nr.1-4, 320085 Resita, Jud. Caras-Severin, Romania

Correspondence should be addressed to Constantin Marta, c.marta@uem.ro

Academic Editor: Philippe Boisse

This paper presents a comparative analysis between the practical results of pig iron die-type part casting and the results reached by simulation. The insert was made of polystyrene, and the casting was downward vertical. As after the part casting and heat treatment cracks were observed in the part, it became necessary to locate and identify these fissures and to establish some measures for eliminating the casting defects and for locating them. The research method was the comparisons of defects identified through verifications, measurements, and metallographic analyses applied to the cast part with the results of some criteria specific to simulation after simulating the casting process. In order to verify the compatibility between reality and simulation, we then simulated the part casting respecting the real conditions in which it was cast. By visualising certain sections of the cast part during solidification, relevant details occur about the possible evolution of defects. The simulation software was AnyCasting, the measurements were done through nondestructive methods.

1. Introduction

The pearlitic grey pig irons constitute of family of ferrous materials with a wide range of mechanical properties, being using in the manufacture of bed frames, cylinder covers, dies, mechanisms bodies, pistons, cylinders [1]. The properties of strength, tenacity, and plasticity of the lamellar graphite pig irons are relatively low. These alloys exhibit unique proprieties of use due to the presence of graphite under the form of lamellas, such as: capacity of vibration damping, "lubricant" character; high processability; good resistance to abrasion; good corrosion resistance; preservation of properties in the temperature range from $-100°C$ to $+350°C$; very good resistance to heat shocks, very good castability. The parts are produced by casting in moulds, which makes them usable in the manufacture of mechanical components. Other advantages are the low cost and the wide scope of uses. The cooling rate of cast parts may affect the toughness and structure of the materials. In the present case we executed a die of lamellar graphite pig iron, toughness >150 <300 HB mechanic strength >500 <1000 $N \cdot mm^{-2}$. This group of pig irons represents the quality of high strength generally obtained through a pearlitic die, which is relatively tough and has low tenacity and good processability. The casting of large die-type parts, approximately 11,000 kg raises a problem of the design and elaboration technology are not respected, especially the procedures of pig irons modification. The technology of casting in polystyrene moulds with two runner basins exhibits some particularities with major consequences on the parts quality.

2. Analysis of the Cast Part

Table 1 shows the chemical composition in % weight [1, 2], as well as the execution of the part according to the own

TABLE 1: Chemical compositon of material, (wt.%).

	C	Si	Mn	P	S	Cu
EN GJL 200	3.15–3.45	2.00–1.40	0.60–0.80	max 0.200	0.05–0.100	—

casting technology, specific to the casting in polystyrene moulds. The charge was cast through two runner basins.

After casting the visual verification of the cast surface and the location of possible casting defects were carried out. The defect in the cast part occurs under the form of a transversal crack in the median area of the part, on the horizontal base plate, 2 to 2.5 mm width and 1500 mm long, defect which appears in Figure 1. Figure 1 shows the areas where one has extracted the metallographic samples, marks: A, B, and F.

The analysis contains study of the design of the part geometry by visual analysis and extraction of metallurgic samples from the area of the defect. From the part design viewpoint, we find that the joining angle between ribs (fins) and the body of the part is 90° without junctions, and the ribs have widths ranging between 25 mm and 90 mm without respecting the basic principles of the casting moulds [3, 4].

3. Macro- and Microstructural Analysis

According to Figure 1, three samples were extracted from three different zones: (1) F-position middle zone where the crack occurred, (2) final zone A, and (3) end zone B.

3.1. Macro- and Microstructural Analysis of Area F. Figure 2 shows the sample extracted from area F, that is, the middle area of the part where the crack occurs. This area is where the two metal fronts merged. The casting was done by using two sprue cups placed at the ends of the casting mould.

Figure 2 presents the middle area, that is, the zone where the crack occurs and at the same time the joining zone of the two jets (flowing currents) of the liquid pig iron. This zone exhibits nonmetallic inclusions such as oxides, sulphides, and small gaseous inclusions (air bubbles) which may constitute crack triggers in the conditions of the existence of very high internal strains generated by a braked contraction during solidification and cooling. One should remark the area of primary graphitisation in the insufficiently developed eutectic cells, due to the low casting temperature (1220°C), and to a accelerated rate cooling area, as well as to a weak modification of the pig iron. By exceeding the time between modification and casting by over 15 minutes, the effect of the modification is very much attenuated [4]. There occur also air bubbles and oxides, as the middle area is the joining zone of the jets of liquid alloy come from the two sprue cups. Due to the fact that the part casting was made with two sprue cups we encounter two liquid fronts with different temperatures (of which one has the temperature below the technological casting level), as well as the differences in chemical composition. All these factors are potentially probabilistically generators of the defects from the above macroscopic image. The area where the defect is not oxidised which shows that the crack occurred at cold.

In Figure 3 one observed a microstructure of hypereutectic grey pig with metallic mass of pearlitic base, according to

the chemical composition analysed, confirms the weak evolution of the separation of primary graphite, in the process of passing through the solidification range. Obviously the casting temperature was lower than the technological one, as well as the effect of the modification which is inefficient. The zone exhibits separations of lamellar graphite of semiarched shape and nest-shaped graphite separations. One witnesses the occurrence of only very small modified lamellar graphite (glm) separations. The second sample, Figure 4, confirms the cause of the material defect. The degenerate shape of primary graphite can be remarked. It may occur in the structure of the pig iron of the second fusion only if the loading of the elaboration furnace was made with furnace raw pig iron, but not of high purity; the primary, hereditary pig iron separates following the incomplete dissolution of the graphite from the first-fusion pig iron-furnace raw iron pig, at the melting of the metallic charge; the seeds or grains of primary hereditary graphite may lead to the increase of primary hereditary graphite separations, at the solidification of the second-fusion grey pig iron; we witness a directing of carbon atoms from the melt toward theses grains with a distribution towards the limit of the eutectic cell very close to the typical shape of the interdendritic graphite.

The microstructure in Figure 5 indicates a correct separation of secondary graphitisation, with around 50 μm-long graphite lamellas, but not with rounded points, as well as a rather inappropriate length, considering the part mass (11t), which at a slow cooling would have led to the enlargement of the graphite lamellas to around 100–150 μm. They are shapes of separations of modified lamellar graphite. In Figure 6 we remark that the metallic basic mass is pearlite.

3.2. Macro- and Microstructural Analysis of the Part in Zone A. In Figure 7 the macroaspect is appropriate, typical for pearlitic grey iron pig with a granulation specific to the studied make of pig iron.

The sample in Figure 8 is without reactive attack and allows the observation of the graphite distribution, which is lamellar, with appropriate length, but with smaller widths and without rounded points (it produces the effect of growth in the basic metallic mass). This leads to the conclusion that we need a better modification in the casting ladle, with FeSi75, and to the setting of an appropriate casting temperature. The following areas examined in Figure 8 (microstructure in zone A) and Figure 9 (microstructure in zone A) validate the conclusions. The analysis with reactives highlights a pearlitic base metallic mass, Figure 9.

3.3. Macro- and Microstructural Analysis of the Part in Zone B. In Figure 10 the macro aspect is appropriate, typical for the pearlitic grey pig iron with a granulation specific to the studied make of pig iron.

In Figure 11 we remark the size of the graphite lamellas, smaller than in the pig iron from the other runner basin,

Theoretic and Experimental Studies on the Casting of Large Die-Type Parts Made of Lamellar Graphite Grey Pig Irons by
Using the Technology of Polystyrene Moulds Casting from Two Sprue Cups

177

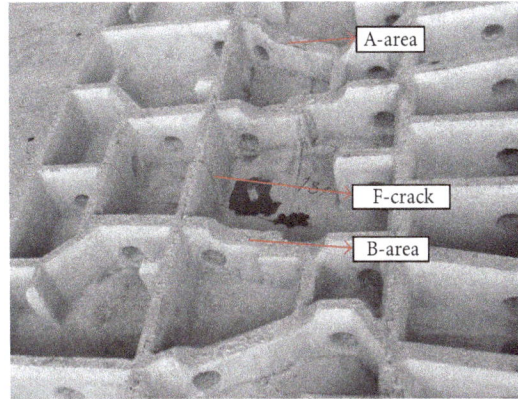

FIGURE 1: Areas where the metallographic samples were extracted (A, B) and crack in the part (F).

FIGURE 2: Zone F macroaspect of the sample in the cracked area.

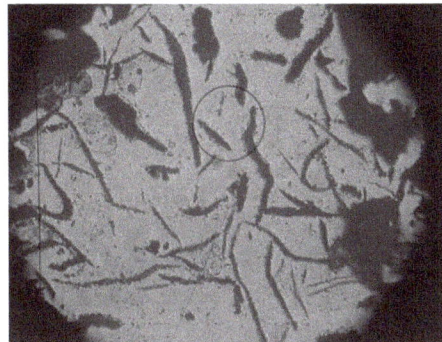

FIGURE 3: Microstructure in zone F 100x.

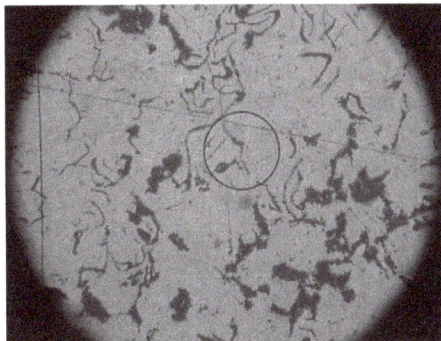

FIGURE 4: Microstructure in zone F 100x.

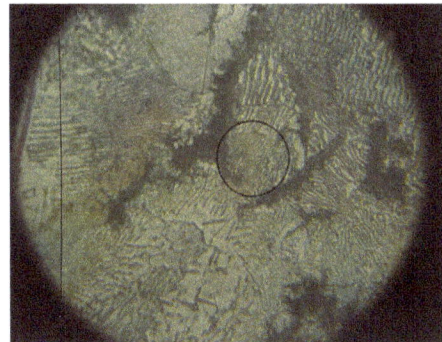

FIGURE 5: Microstructure in zone F 100x.

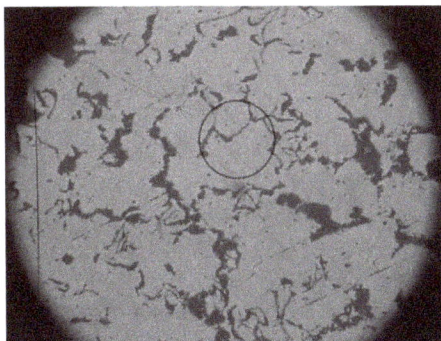

FIGURE 6: Microstructure in zone F 500x.

FIGURE 7: Macroaspect of the sample in zone A.

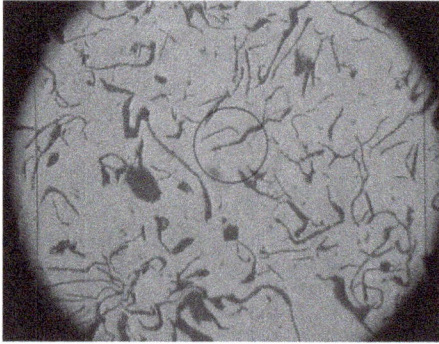

FIGURE 8: Microstructure in zone A 100x.

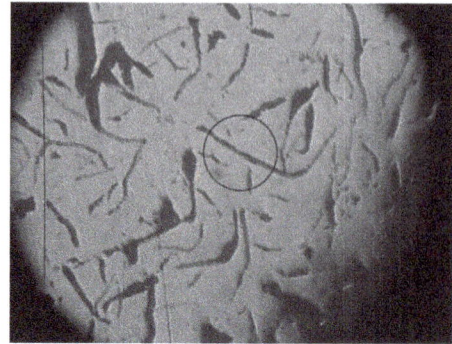

FIGURE 9: Microstructure in zone A 500x.

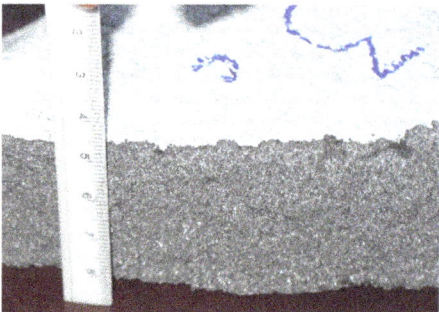

FIGURE 10: Macroaspect of the sample in zone B.

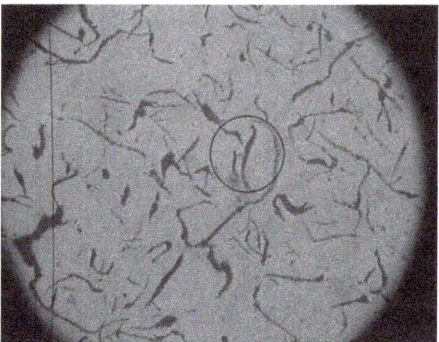

FIGURE 11: Microstructure in zone B 100x.

FIGURE 12: Microstructure in zone B 100x.

possibly more weakly modified and colder cast. The micro-analysis in the adjacent area, Figure 12, exhibits a better microstructure than the previous one, not by much, but enough to draw the conclusion of a weaker thermal-chemical homogenisation, at sprue cup no. 1. The microanalysis with reactives shows a pearlitic base metallic mass, with areas of binary phosphorus eutectic.

4. Conclusions

From the macroanalysis of the part we remarked the deficiencies of design, moulding and elaboration which trigger the occurrence of the defect:

(i) T-joinings between the walls of the part with great differences in widths (ribs and base plates);

(ii) the crack is caused by a braked contraction triggered by the ribs-walls and the use of certain cores—the mixture mould which does not enable the free contraction of the part (noncompressible moulding-coring mixtures);

(iii) a high amount of internal strains was accumulated on the part base plate as a result of the braked contraction at solidification and cooling;

(iv) it is possible that the crack started from some triggers - internal casting defects: nonmetallic inclusions or air bubbles and even the shape of the phosphorous eutectic, which may constitute strain concentrators; we do not analyse the surface defects (there are many) which negatively influence the mechanic characteristic to a great extent;

(v) the crack does not start and does not pass through the orifices of the ribs because (fact proved experimentally) the holes are limits (obstacles) for the fissures propagation;

(vi) the execution of the model is inappropriate because at the walls joints one has not manufactured adequate junction rays which would have allowed the partial elimination of the formation of corner cavities, taking into account also the fact that having joined walls of different width stressed the tendency of formation of thermal knots, corner cavities, and

Theoretic and Experimental Studies on the Casting of Large Die-Type Parts Made of Lamellar Graphite Grey Pig Irons by Using the Technology of Polystyrene Moulds Casting from Two Sprue Cups

179

axial microcavities (see the simulation, especially the temperature variation during filling);

(vii) the middle area, that is, the zone where the fissure occurs and at the same time the merging area of the two jets (flowing currents) of the liquid pig iron exhibits nonmetallic inclusions oxides, sulphides, and small gaseous inclusions—air bubbles—which may constitute fissure triggers under conditions of very high internal strains generated by a brakes contraction during solidification and cooling;

(viii) there occur areas of primary graphitisation in insufficiently developed eutectic cells, due to the low casting temperature (1220°C), and to a cooling area with accelerated rate;

(ix) the inefficient modification of the pig iron and the exceeding of the delay between modification and casting, over 15 minutes;

(x) due to the fact that the part casting was made with two sprue cups, we encounter two liquid fronts of different temperatures (of which one liquid front has the temperature below the technological casting level), as well as the differences in the chemical composition.

5. Analysis of the Possibility of Defect Prediction Using the Simulation of Casting and Solidification of the ENGJL 200 Pig Iron

All software for simulating alloys casting and solidification has 4 main modules, that is, [2, 5]:

(i) database, where the materials are found with their physical and chemical properties;

(ii) preprocessor, where we introduce the parameters necessary to the simulation, that is, type of alloy, properties of the casting mould, alloy temperature, mould temperature, coefficients of heat transfer, specific coefficients which may determine defects and mechanical properties;

(iii) solver;

(iv) postprocessor, which presents the results of the simulation based on the data introduced by the operator.

It is important to point out that the assessment of the part cast through classical methods and then subjected to simulation results in obtaining certain data which help a lot the user of the software to make the best predictions. As shown in point 2, the second section of the study refers to the simulation of the part casting under conditions identical with those in which the analysed part was cast. The casting was done in polystyrene mould with furanic resin cores. The casting temperature was of 1250°C, then alloy is EN GJL 200 grey pig iron, the chemical analysis shown in Table 1. The casting of the part was done with two sprue cups, technology presented in Figure 13, where in red we see the position of the two runner basins, and of the two mould feeders. The casting with two basins was chosen as the mass of the part together with the casting network is 12,000 kg.

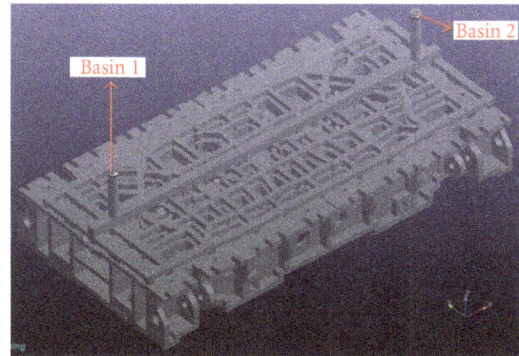

FIGURE 13: Presentation of the casting model with two basins.

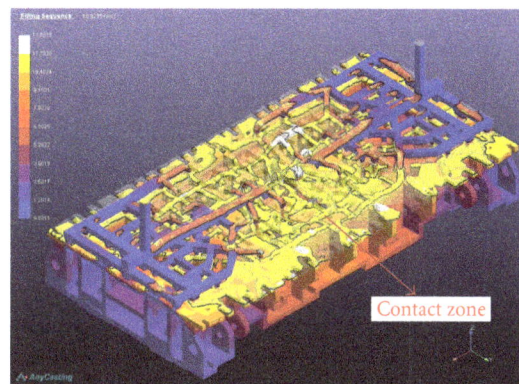

FIGURE 14: Convergence area of the two front.

The casting of the part was made with two basins, we encounter two liquid fronts with different temperatures (of which one liquid front with the temperature below the technological casting level), as well as differences in the chemical composition.

In the case of casting in polystyrene moulds the feeding is made from the upper section through a feeding network resulting in the downward casting towards the base of the casting mould [6]. Consequently the filling of the mould is made from base to the upper part. Figure 14 shows the simulation of the pig iron filling from two sprue cups, using the filling sequence criterion [1, 6]. As remarked, the two liquid fronts converge in the middle area of the part, where the casting defect appears. By analysing the process, according to the colour code displayed in the left side, we find that the most rapid filling takes place in the feeding network at 0.0011-second time, in the area indicated with the red arrow the liquid from basin 1 meets the liquid from basin 2 after around 11,7036 seconds. Moreover, we remark in the same area a turbulence specific to the encounter of two liquid jets. Figure 14 does not show all the filling sequences, we selected those which represent the most accurately the moment of interest.

Another criterion used and presented in Figure 15 is the retained melt volume, measured in cm^3 of volume of liquid metal retained in certain areas and is used in order to visualise the zone where liquid volumes remain, that is,

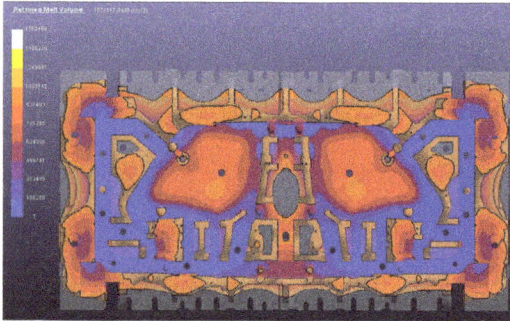

FIGURE 15: The retained melt volume criterion.

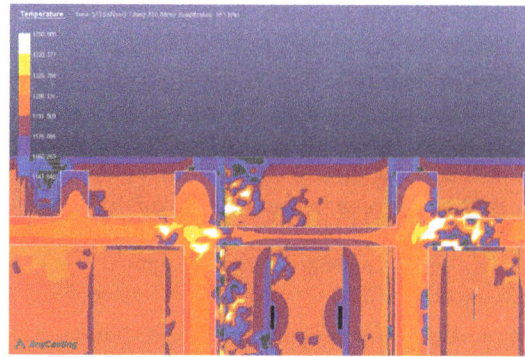

FIGURE 16: Temperature variation during alloy solidification. Section in the defect zone.

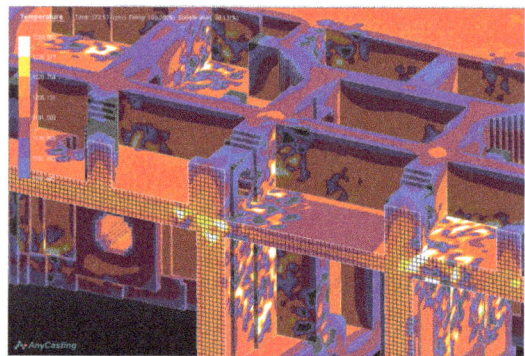

FIGURE 17: Temperature variation during alloy solidification. Meshed section in the defect area.

which zone solidifies more slowly than the others. Another interpretation refers to the fact that it shows the areas that retain the largest quality of liquid metal. In the median area, where the two fronts unite, the volume of liquid metal ranges between 312,499 and 1,093,742 cm^3 at a total volume of 10,791,170,000 cm^3. One remarks the separation of two fronts of liquid that will contract later on from the median area, which cross the thermal area from the solidified phase, where the mechanic strength is reduced and the local wall width also small. This fact leads to the predisposition to cracking of the part as a result of the existence of uncompensated mechanical strains [3, 5].

This situation is favoured both by the metallographic microstructure, made of primary graphite separations, within insufficiently developed lamellas, as well as by the high rate of passage through the solidification range. Moreover, the very median zone, as it has the less ribs and is by 30 mm thinner compared to the neighbouring areas, does not enable a sufficient mechanical consolidation at the passage through the solidification range. This leads us to the assumption of the existence of a prefissured state, even from the solidification period [7].

In this state the internal tensions generated by the inappropriate design of the model, the lack of chemical and thermal homogeneity may pass unobserved at the extraction from the mould, during painting (grounding), and the defect risks to be amplified during later manipulation or during transport.

So in Figure 16 we see in the preprocessor another relevant criterion, that is, the temperature variation during the solidification of the casting mould. The solid of the cast part subjected to simulation was made after measurements performed on the cast part. The solid thus realised was introduced in preprocessor and we applied the condition of real casting parameters. Thus we find design errors, that is, the left and right ribs are 50 mm thick and the middle one is 25 mm thick. Consequently, the middle area cools more rapidly than the side ones, which cool more slowly. Moreover, the distortions generated by the differentiation of solidification in time and constrained by the rigidity of the moulding mixture, generates local, uncompensated tensions [8]. The criterion "temperature variation during the alloy mould filling" show that the filling time of the mould is of 373,537 seconds, that is 6,21 minutes. We may deduce very easily also the filling rate. At this value filling is 100%

and we remark a 30% solidification, which explains that during the part filling the solidification processes also start. By analysing the criterion according to the colours code displayed in the left side, Figures 16 and 17, the solidification starts from the lateral sides of the part towards the middle, the temperature ranging between 1147°C (blue-coloured areas) and 1250°C (white-coloured areas). The presence of 1250°C temperatures especially in the joining areas between ribs points out the presence of heat knots in the part. Starting from this fact, simulation shows that fissures are likely to occur starting from the "+-" or "T-" shaped joints of the thick walls (ribs), with the same wall thickness or with different thicknesses, and even in the middle areas of the thick walls which solidify among the last areas; the thermal knot occurs in joints [9].

Grosser seeds or grains are formed preponderantly in the ribs situated towards the central part of the part where the solidification-cooling rate and the aforementioned joints in the walls (T-shaped heat knot) and where it is lower than in the rest of the part; the higher the cooling rate, the smaller the sizes of grains (but we must be careful, as a very high cooling rate risks to lead to the separation of free cementite).

Figure 17 shows a section and the meshed perspective, which allows a more detailed visualisation of the area of interest. By comparing the colours of certain zones of the part with the colour code in the left side, we can study the solidification and cooling of the part, and we remark that

Theoretic and Experimental Studies on the Casting of Large Die-Type Parts Made of Lamellar Graphite Grey Pig Irons by
Using the Technology of Polystyrene Moulds Casting from Two Sprue Cups

181

the solidification front starts from the extreme sections of the part and especially from the sections in contact with the atmosphere. The area which solidifies the most rapidly is that with the thinnest wall [10].

One proceeded to the nondestructive control of the part, to the identification and location of other hidden casting defects. One measured the inductivity of different areas of the part, and found a predisposition to defects in the median area, where the measured items exhibited higher values.

5.1. Conclusions. The simulation software cannot exhibit aspects related to the nonobservance of the technology of pig iron elaboration. By introducing the parameters of the cast part in the software preprocessor and the display of the results in the postprocessor we get an image very close to reality:

(i) simulation highlights the convergence area of the two liquid fronts, zone which, by measurements, coincides with the position of the defect in the cast part;

(ii) it presents by the analysis of the solid all the design deficiencies, that is, the "+" or "T" joints of the thick walls, zones solidify among the last;

(iii) the presence of $1250°C$ temperatures especially in the joining areas between ribs highlight the presence of heat knots in the part.

6. Final Conclusions

Following the analysis of the simulation results and of conclusions in point 4 and point 5.1 some proposals resulted in view of enhancement the casting technology:

(i) it is necessary to modify the part execution design, by increasing the wall thickness in the defect occurrence area;

(ii) in order to have a high quality of the cast iron, it is necessary to correctly apply the modification technology, and the time between modification and casting must be observed;

(iii) casting the part at a temperature which should take into account the heat losses caused by the transfer of the locked form the furnace in the ladle or sprue cup. The recommended temperature in this situation is of $1350°C$;

(iv) due to the fact that the part casting was performed with two runner basins, two liquid fronts of different temperatures converge; consequently it is required that the temperature of the two liquid fronts be equal, as well as their chemical composition;

(v) a series of anomalies found may be avoided by modifying the moulding-casting technology and the use of compressible moulding mixture (preferably from moulding mixtures bound with classic inorganic binders) if we are interested in the part compactness and metallographic structure;

(vi) technological control of heat knots may be done by building some vent holes at moulding-casting, placed on the wall joints;

(vii) it is necessary to correct the moulding-casting technology, by correcting the construction of the model, with the realisation of junction rays on the model and manufacturing air vents made of the same material of the model, alterations which also lead to the modification of the mould execution;

(viii) it is necessary to build a new polystyrene model;

(ix) it is required to manufacture a new part with the updated model.

By comparing the simulation results with the practical ones and vice-versa, the application of simulations of some cast parts provide the user with a very accurate instrument for the anticipation of defects and their apparition and at the same time leads to the building of a very useful database.

References

[1] SR UNE EN, 1560:2011 Founding—designation system for cast iron—materials symbols and material numbers, 2011.

[2] AnyCasting, DataBase, Advanced Casting Simulation Software, version 3.10, 2009.

[3] M. Skarbinski, *Construcția pieselor turnate și proiectarea formelor (traducere din limba polonă)*, Technical Editions, Bucharest, Romania, 1967.

[4] C. Ștefănescu, C. Cosneanu, and V. Dumitrescu, *Îndrumătorul proiectantului de tehnologii în turnătorii*, vol. 1, Technical Editions, Bucharest, Romania, 1985.

[5] E. Piwowarsky, *Fonte de inalta calitate*, Technical Editions, Bucharest, Romania, 1967.

[6] AnyCasting, Advanced Casting Simulation Software, version 3.10, 2009.

[7] C. Marta, *Aplicatii in AnyCasting*, Efimie Murgu Editions, 2011.

[8] B. Liu, H. Shen, and W. Li, "Progress in numerical simulation of solidification process of shaped casting," *Journal of Materials Science and Technology*, vol. 11, pp. 313–322, 1995.

[9] I. Ciobanu, M. Chisamera, S. I. Munteanu et al., "Researches about the determination of the thermal conductivity coefficient for silica sand moulds used in Romanian foundries," *Key Engineering Materials*, vol. 457, pp. 312–317, 2011.

[10] D. M. Stefanescu, *Science and Engineering of Casting Solidification*, Kluver Academic, Plenum, New York, NY, USA, 2002.

19

Basic Characteristics of IEC Flickermeter Processing

Jarosław Majchrzak and Grzegorz Wiczyński

Poznań University of Technology, Poznań, Poland

Correspondence should be addressed to Grzegorz Wiczyński, gwicz@et.put.poznan.pl

Academic Editor: MuDer Jeng

Flickermeter is a common name for a system that measures the obnoxiousness of flicker caused by voltage fluctuations. The output of flickermeter is a value of short-term flicker severity indicator, P_{st}. This paper presents the results of the numerical simulations that reconstruct the processing of flickermeter in frequency domain. With the use of standard test signals, the characteristics of flickermeter were determined for the case of amplitude modulation of input signal, frequency modulation of input signal, and for input signal with interharmonic component. For the needs of simulative research, elements of standard IEC flickermeter signal chain as well as test signal source and tools for acquisition, archiving, and presentation of the obtained results were modeled. The results were presented with a set of charts, and the specific fragments of the charts were pointed out and commented on. Some examples of the influence of input signal's bandwidth limitation on the flickermeter measurement result were presented for the case of AM and FM modulation. In addition, the diagrams that enable the evaluation of flickermeter's linearity were also presented.

1. Introduction

Flickermeter is a common name for a system that measures the obnoxiousness of flicker caused by voltage fluctuations. According to [1], flickermeter is an instrument designed to measure any quantity representative of flicker. Flicker is an impression of unsteadiness of visual sensation induced by a light stimulus whose luminance or spectral distribution fluctuates with time [2]. According to [3], flickermeter is a flicker measuring apparatus to indicate the correct flicker perception level for all practical voltage fluctuation waveforms. The processing performed by this system's signal chain is complicated [4] to such an extent that it is not easy to obtain output values only analytically. Bibliography concerning this subject contains numerous works that describe the operation and features of flickermeter [5–10]. In most of them, authors reconstruct flickermeter's characteristics for the case of amplitude modulation (AM) [3, 11, 12], input voltage with interharmonic component [12–21], or step changes of input voltage phase [12, 22–25]. The bibliography lacks publication on reconstructing the characteristics of flickermeter processing in a systematic and complex manner. Many discussions and comments contain opinions that vaguely describe the way in which flickermeter actually operates. For example, one of the most common mistakes is the erroneous limitation of frequency range for which the characteristics of IEC flickermeter are constructed. This appears to come from wrong conclusion that modulation with frequencies higher than 40 Hz does not cause obnoxious flicker because the only flicker obnoxious for human is limited to the frequency of 40 Hz. Nevertheless, because of the way incandescent light sources function (which is mapped in an IEC flickermeter block diagram with operation of voltage squaring), flicker could be sensed for frequencies of modulating signal higher than 40 Hz.

Herein, the results of numerical simulations that reconstructed the processing of signal chain of IEC flickermeter will be presented. The research results presented here are concerned with the processing of IEC flickermeter; in other words, the flickermeter is built accordingly to IEC 61000-4-15, and thereby does not describe the flicker phenomenon in general, for instance, for light source other than the one defined in the standard. Therefore any modification of the test signal path is abandoned. The novelty is a comprehensive study of flickermeter for various voltage variation.

Owing to space constraint, this paper contains only selected simulation results that are important from the analytical point of view or have some practical significance [26].

2. Test Signals for Flickermeter Signal Chain

The dynamic properties of measuring devices are determined in frequency domain. The results of the research on frequency domain are amplitude characteristics and, if applicable, phase characteristics. The above-mentioned means of determination of dynamic properties form the classic approach to the subject. Therefore, an attempt has been made to use them to test the flickermeter signal chain while taking into account the occurrence of carrier component in the input voltage. At the same time, an assumption has been made that the tests will be carried out in a steady state of the signal chain, that is, after the transient component has completely faded out. This indicates that, for example, step change of voltage will appear after the signal $u_{IN}(t)$ (with some preset amplitude U_m) has been applied previously for a time needed to fade out transient component completely. One of the benefits of carrying out the analysis of signal chain in time domain, besides enriching the set of reference signals, is the possibility to obtain the values of output and inner signals during the transient state (which is helpful when trying to determine if any saturation occurs) and to determine how long it takes for the transient component to fade out.

An ideal input signal $u_{IN}(t)$ in steady state could be expressed as:

$$u_{IN}(t) = U_m \sin(\omega_c t), \qquad (1)$$

where U_m is the voltage amplitude, f_c is the carrier wave frequency (corresponding to the time period T_c), $\omega_c = 2\pi f_c = 2\pi/T_c$ is the pulsatance of carrier wave, and t is time.

The test in the frequency domain will utilize the following input signals [27]:

(i) AM modulated with not suppressed carrier wave

$$u_{IN}(t) = U_m \cos(\omega_c t)\left[1 + \left(\frac{\Delta U}{U}\right) \cdot u_{mod}(t)\right], \qquad (2)$$

where $u_{mod}(t)$ is the modulating signal that satisfies condition $|u_{mod}(t)|_{max} = 1/2$ and $(\Delta U/U)$ is the modulation depth;

(ii) with single interharmonic component

$$u_{IN}(t) = U_m \cos(\omega_c t) + U_i \cos(\omega_i t), \qquad (3)$$

where U_i is the interharmonic component amplitude and $\omega_i = 2\pi f_i$ is the interharmonic component pulsatance;

(iii) FM modulated [28]

$$u_{IN}(t) = U_m \cos\left[\omega_c t + 2\pi k_{FM}\int x(t)dt\right], \qquad (4)$$

where k_{FM} is the carrier frequency modulation depth scaling coefficient and $x(t)$ is the modulating signal, frequency deviation Δf_{FM}

$$\Delta f_{FM} = k_{FM}|x(t)|_{max}. \qquad (5)$$

3. IEC Flickermeter Simulator Structure

The system is composed of two main parts (Figure 1): flickermeter signal chain modeled according to the standard specification and simulation support block. The task of simulation support block is to generate test signals with the chosen time plots, amplitudes, and frequencies to determine the rms value ($u_{RMS}(t)$) of $u(t)$ signal, to determine the value of P_4^2 signal, as well as to record and visualize selected signals. The structure of the test signals source enables generation of signal $u_{IN}(t)$ as a result of AM modulation, FM modulation, or as a sum of carrier and interharmonic components. Low-pass Butterworth-type filter of 6-order with a cut-off frequency f_{LFP} limits the bandwidth of $u_{IN}(t)$ signal.

The AGC block replaces an input transformer with branches taps and input voltage conditioning circuitry. The parameters of RMS/LPF filter were chosen so as to achieve raising/falling time of $u(t)$ signal equal to 1 min for the step change of rms value. By assuming $f_c = 50$ Hz as a frequency of voltage in power network, filters from Figure 1 could be specified as follows: $f_{1d} = 0.05$ Hz, $f_{1g} = 35$ Hz, $f_2 = 8.8$ Hz and $f_3 \cong 0.53$ Hz (the detailed specification in [3]). IFL is a signal that appears on output 5 in standard flickermeter signal chain schema [3]. The value of P_4^2 signal corresponds to the maximal value of IFL signal. Statistical analysis block calculates the value of short-term flicker severity indicator, P_{st}, on the basis of statistical distribution of IFL signal.

All instantaneous values of P_{st}, P_4^2, and other signals are recorded by the recording and visualization block. The process model of flickermeter and simulation researches were carried out by using the software application Matlab and Matlab-Simulink [29, 30], using the solver ODE45 with variable step (Dormand-Prince).

4. Basic Characteristics of Flickermeter Processing

4.1. Specification of Flickermeter Basic Characteristics. The following groups of characteristics were chosen to describe the processing of flickermeter signal chain:

time plots of internal signals $u_{RMS}(t)$ and IFL during of fading of intial transient component,

grouping of P_{st} output signal values for no-modulation state (signal $u_{IN}(t)$ in accordance with (1)),

dependence of output signal P_{st} and internal signal P_4^2 on frequency f_m, modulating signal $u_{mod}(t)$ type, and modulation depth $(\Delta U/U)$ (amplitude modulation—signal $u_{IN}(t)$ in accordance with (2)),

dependence of the modulation depth $(\Delta U/U)$ on frequency f_m and modulating signal type for the preset values of output signal P_{st} or internal signal P_4^2 (amplitude modulation—signal $u_{IN}(t)$ in accordance with (2)),

dependence of output P_{st} signal on interharmonic component frequency, f_i (signal with interharmonic component—signal $u_{IN}(t)$ in accordance with (3)),

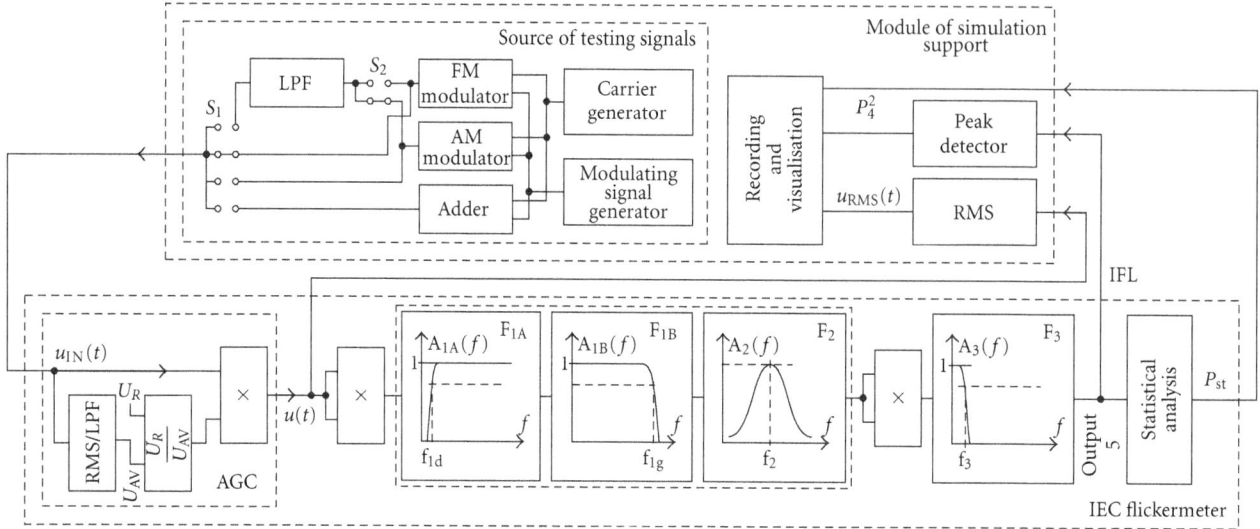

FIGURE 1: Block diagram of simulative test-bed for flickermeter signal chain testing: RMS—True RMS converter, RMS/LPF—True RMS converter with output low-pass filter, U_R/U_{AV}—signal divisor, U_R—reference value, AGC—automatic gain control circuit, IFL—instantaneous flicker level, S_1—selector of input signal $u_{IN}(t)$, S_2—selector of input signal for LPF filter.

dependence of interharmonic component amplitude U_i on f_i frequency for preset values of output P_{st} signal (signal with interharmonic component—signal $u_{IN}(t)$ in accordance with (3)),

dependence of output P_{st} signal on f_m frequency and frequency deviation Δf_{FM} (frequency modulation—signal $u_{IN}(t)$ in accordance with (4)), and

dependence of frequency deviation Δf_{FM} on sinusoidal modulating signal f_m for the preset values of output P_{st} signal (FM modulation—signal $u_{IN}(t)$ in accordance with (4)).

Characteristic groups, 1–3, 5, and 7, were obtained by setting appropriate properties of the input signal $u_{IN}(t)$ while recording the values of the respective signals. The characteristic group denoted as 4 differs from the other groups in the way it was obtained. In that case, an iterative algorithm for determination of modulation depth ($\Delta U/U$) was used, which is presented in Figure 2. In the standard 61000-4-15 [3], the action of the flickermeter for rectangular and sine signal was defined, which modulates the amplitude with a frequency of up to 33 Hz. However, the flicker phenomenon also occurs for higher frequencies. Therefore, this study has presented the characteristics of the full range of frequencies that is likely to cause noticeable flicker. This applies to the modulation characteristics of both AM and FM. As the initial-phase modulating signal did not affect the rate of P_{st}, all simulations were performed for the initial phase equal to zero. It is worth to point out the time-consuming aspect of the simulation process. Observation time was set to 10 min to satisfy the standard requirements. To eliminate the influence of the transient state of signal chain on the simulation result, 3 min of extra time was added before each start of the standard observation period. With the computer used for calculations, determination of the simulations results took significantly more time. It took

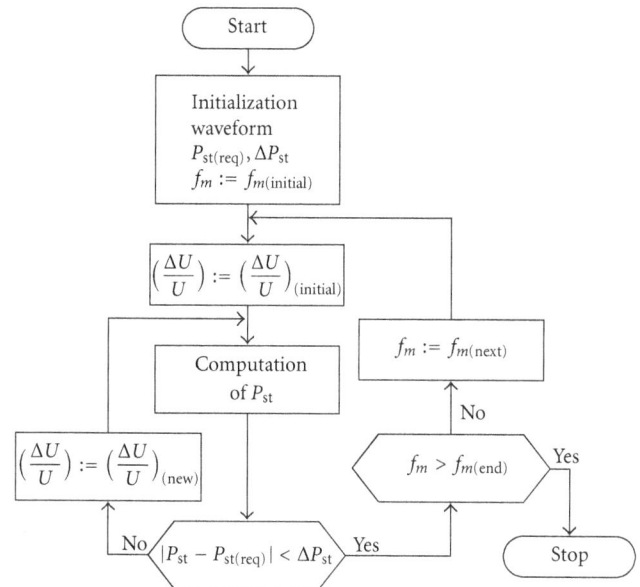

FIGURE 2: Modulation depth ($\Delta U/U$) determination algorithm, which was used during simulations. Waveform—shape of waveform, $P_{st(req)}$—preset value of P_{st}, ΔP_{st}—P_{st} boundary value (calculation of ($\Delta U/U$), for a given f_m frequency value; ends when P_{st} value enters the $P_{st(req)} \pm \Delta P_{st}$ range).

several dozens of hours to obtain multipoint characteristic, especially when iterative algorithm was utilized. Characteristic groups 6 and 8 were build with the use of algorithm presented in Figure 2, while substituting the modulation depth ($\Delta U/U$) with interharmonic component amplitude U_i and frequency deviation Δf_{FM}, respectively.

4.2. P_{st} Measurement Result for Sinusoidal Input Signal $u_{IN}(t)$ (without Modulation). When a steady sinusoidal signal

without modulation (in accordance with (1)) is applied on
flickermeter input, the output P_{st} signal should remain at
zero. However, in case of standard flickermeter signal chain
[3], the P_{st} value for such an input signal is greater than zero.
Table 1 summarizes the values of P_{st} signal obtained with
simulations for different orders of F_{1B} filter.

The test for the absence of voltage variation indicates that
the real flickermeter P_{st} shall not be less than approx. 0.01
(if the meter shows $P_{st} = 0$, it can be inferred that in the
processing the "trick" was used).

4.3. Fading of Initial Transient Component.

By assuming zero
initial conditions, when the signal, in accordance with (1)
is applied, some transient component appears. The fading
time of that component is determined using the features of
flickermeter signal chain. Figure 3 presents the fading out of
the transient component at the two distinctive points of the
signal chain: $u_{RMS}(t)$ signal at the output of AGC block and
IFL signal at the output 5 (see diagrams in Figure 3).

Analysis of Figure 3 leads to the following conclusions:

(i) the maximum value of $u_{RMS}(t)$ signal is about 10^5
times greater than the value at the steady state,

(ii) the maximum value that the IFL signal takes on is
about $4 \cdot 10^{25}$, while the value at the steady state is
zero, and

(iii) the fading out time of $u_{RMS}(t)$ signal transient
component is about 180 s, and the fading out time
of IFL signal transient component is about 120 s.

One of the effects of such a large short-lasting value of
IFL signal may be the occurrence of saturation of flickermeter
signal chain. The occurrence of such state may have an effect
on additional error of P_{st} indicator measurement, which
could be hard to estimate. Such erroneous state occurs when
the fluctuation of voltage is sufficiently strong and repeats
over time. An example of such conditions is the fluctuation
of voltage in power circuit of arc furnace.

4.4. Flickermeter Processing Characteristics for AM Modulation of Input Signal.

The characteristics of flickermeter processing for AM-modulated input signals, described with (2),
can be divided into two groups. The first group contains
characteristics $P_{st} = f(f_m)$ for a preset modulation depth,
$(\Delta U/U) = $ const. The second group contains characteristics
$(\Delta U/U) = f(f_m)$ for $P_{st} = $ const and $P_4^2 = $ const.
Figure 4 presents the graph of $P_{st} = f(f_m)$ dependence
for sinusoidal, triangular, and rectangular modulating signal
$u_{mod}(t)$ with constant $(\Delta U/U)$ value. The modulation depth
was set to the value that guarantees unitary P_{st} indicator value
at frequency $f_m = 8.8$ Hz, that is, for maximum sensing
of flicker. Flickermeter signal chain interrelates the output
P_{st} value with the parameters of input signal: modulation
depth $(\Delta U/U)$, frequency f_m, and the shape of the signal.
To determine $(\Delta U/U) = f(f_m)$ characteristics, the iterative
procedure was used, in which the shape of the modulating
signal was set along with f_m and P_{st} values. Figures 5 and 6
present the sets of $(\Delta U/U) = f(f_m)$ characteristics for three

TABLE 1: Comparison of P_4^2 and P_{st} values for real filters.

Signal	Numerical simulations' results F_{1B} order:			
	6	8	10	12
P_4^2	188 ppm	4.6 ppm	1.8 ppm	1.8 ppm
P_{st}	0.0098	0.0015	0.00086	0.00085

modulating signals: rectangular, sinusoidal, and triangular
with $P_4^2 = 1 = $ const and $P_{st} = 1 = $ const, respectively.

Verification of comparison of the selected characteristics
with points specified in the standard IEC61000-4-15 [3]
was carried out. Normative characteristics of the flickermeter's processing specification are subject only to amplitude
modulation by using signals with frequency of up to 33 Hz.
The credibility of the simulation for higher frequencies has
already been evidenced by the observation of flickering
lights and by comparing with the results of model tests.
Comparison of the determined characteristics leads to the
conclusion that the most "obnoxious" modulation signal
is the rectangular one, and the least "obnoxious" is the
triangular one. Similar to the plot presented in Figure 4,
three local minima are observed to exist, with the distinction
that the minimal value of the modulation depth occurs for
the frequency $f_m = 8.8$ Hz. The other extrema occur for
$f_m \cong 91$ Hz and $f_m \cong 109$ Hz. Relation $(\Delta U/U) = f(f_m)$
for the case of modulation with rectangular signal cab be
distinguished by peculiar non-monotonicity in a frequency
range of 25 Hz–40 Hz.

The two groups of characteristics, $P_{st} = f(f_m, (\Delta U/U) = $
const) and $(\Delta U/U) = f(f_m, P_{st} = $ const) are complemented
with $P_{st} = f((\Delta U/U), f_m = $ const) characteristic, which was
determined for modulation with rectangular signal, as presented in Figure 7. This enables verification of flickermeter
linearity, while assuming the modulation depth as an input
quantity and P_{st} as an output. Accordingly, it is possible to
state that, in general, flickermeter is not a linear system, but
for inputs that correspond to $P_{st} > 0.1$ (which means sensing
of flicker), this system is nearly linear.

Figure 8 presents the plot of $P_{st} = f(f_{LPF}, f_m)$ for AM
modulation with rectangular signal. It gives the information
on how the bandwidth of the input signal $u_{IN}(t)$ reflects
in the flickermeter measurement result. The influence of
bandwidth limitation becomes visible for $f_{LPF} < 100$ Hz.

4.5. Flickermeter Processing Characteristics for Input Signal with Single Interharmonic Component.

An input signal
$u_{IN}(t)$ with single interharmonic component used to determine the characteristics of flickermeter processing is defined
using (3). Figure 9 presents the characteristic $U_i = f(f_i, P_{st} = $
1 const) constructed with the use of the algorithm presented
in Figure 2 while exchanging a modulation depth $(\Delta U/U)$
with amplitude of interharmonic U_i. For the sake of comparison, the characteristic $(\Delta U/U) = f(f_m, P_{st} = $ 1 const)
for AM modulation with sinusoidal signal is also presented.
The main difference between the two is the value of f_i/f_m
frequency, for which the local minimum occurs. For the

(a) $0 \leq t \leq 1\,\text{s}$

(b) $0 \leq t \leq 200\,\text{s}$

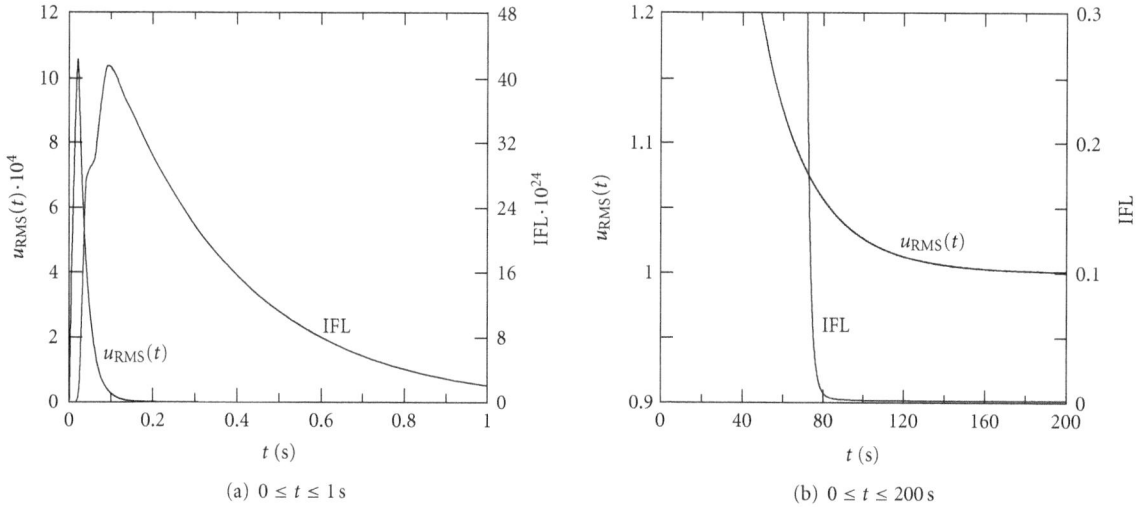

FIGURE 3: Time plots of $u_{\text{RMS}}(t)$ and IFL signals during the fading out of transient component (for zero initial conditions).

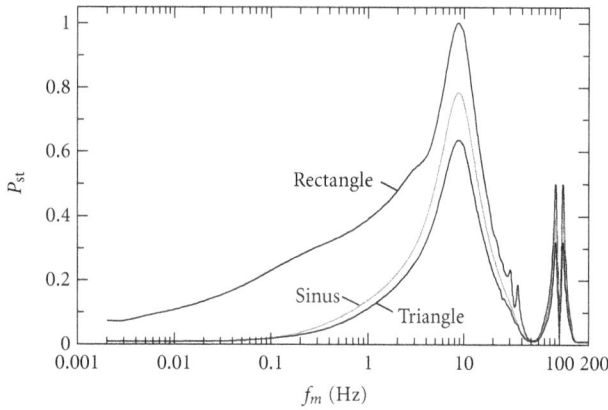

FIGURE 4: $P_{\text{st}} = f(f_m)$ graph for sinusoidal, triangular, and rectangular modulating signal $u_{\text{mod}}(t)$ (signal $u_{\text{IN}}(t)$ described with (2)) with constant modulation depth $(\Delta U/U) = 0.2503\%$.

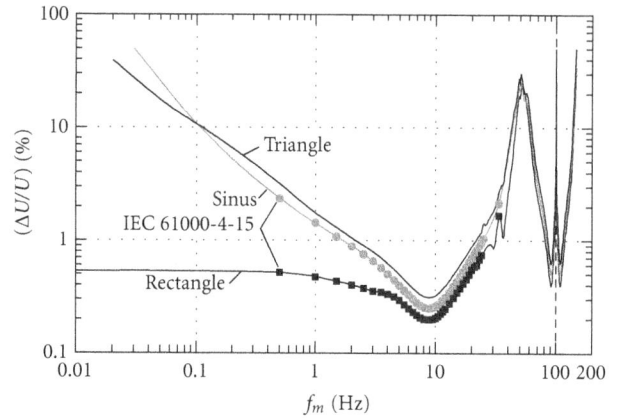

FIGURE 5: Plot of $(\Delta U/U) = f(f_m)$ dependence for rectangular, sinusoidal, and triangular modulating signals with $P_4^2 = 1 = \text{const}$; black dots—Table 2 from [3], gray dots—Table 1 from [3].

input signal with interharmonic component, the maximum flicker sensing occurs at $f_m = 41.2\,\text{Hz}$ and $58.8\,\text{Hz}$. It is worth noting the difference between the signal with single interharmonic component and the AM-modulated signal, defined with (2). Amplitude-modulated signal contains at least two interharmonics: in the case of modulation with sinusoidal signal, it contains two interharmonics, and in the case of modulation with deformed signals, it may contain, theoretically, infinite number of interharmonics.

4.6. Flickermeter Processing Characteristics for FM Modulation of Input Signal. The characteristics of flickermeter processing for input signals obtained as a result of FM modulation were obtained with signal defined as (4). Frequency

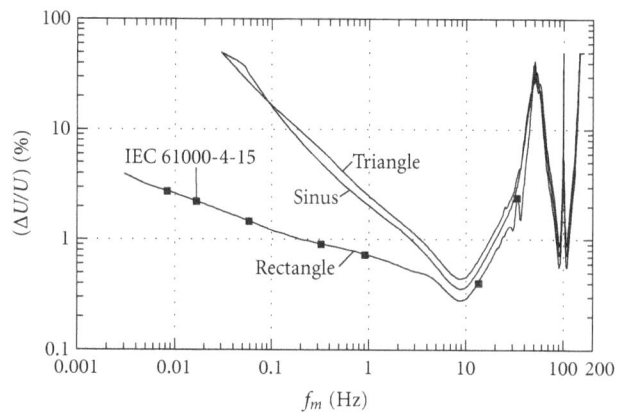

FIGURE 6: Plot of $(\Delta U/U) = f(f_m)$ dependence for rectangular, sinusoidal and triangular modulating signals with $P_{\text{st}} = 1 = \text{const}$; black dots—Table 5 from [3].

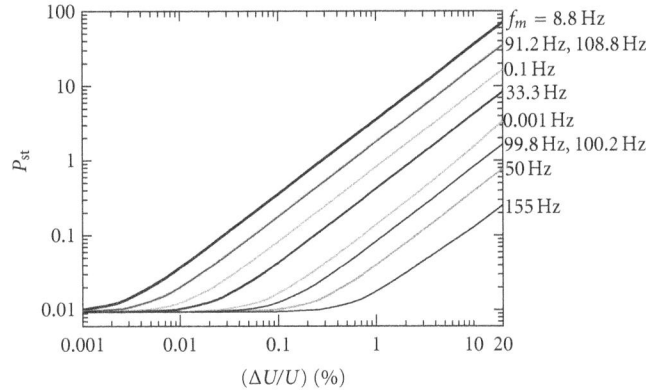

FIGURE 7: $P_{st} = f((\Delta U/U), f_m = \text{var})$ characteristic for AM modulation with rectangular signal.

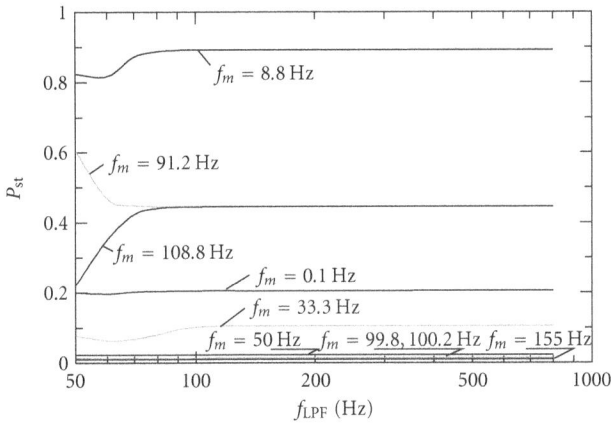

FIGURE 8: Plot of $f_{LPF} < 100\,\text{Hz}$ for AM modulation with rectangular signal.

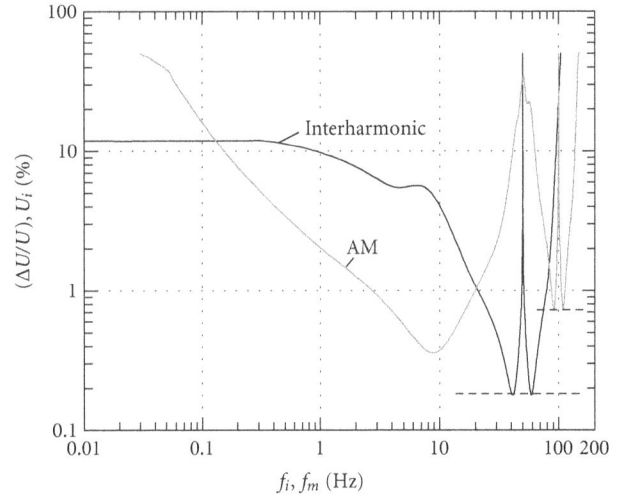

FIGURE 9: Characteristics: $U_i = f(f_i, P_{st} = 1 = \text{const})$ for $u_{IN}(t)$ defined with (3) and $(\Delta U/U) = f(f_m, P_{st} = 1 = \text{const})$ for $u_{IN}(t)$ defined with (2) and for AM modulation with sinusoidal signal.

modulation is a nonlinear operation. This fact complicates the reproduction of flickermeter processing characteristic, because the input signal must be specified in a way that takes into account the working point of the signal chain. Figure 10 presents a $P_{st} = f(f_m, \Delta f_{FM})$ dependency for frequency deviation Δf_{FM} of 0.05 Hz, 0.5 Hz, and 2 Hz, and modulation with rectangular signals. The maximum of $P_{st} = f(f_m, \Delta f_{FM})$ characteristics occur at $f_m = 91.8\,\text{Hz}$. For a case of modulation with rectangular signal, the dependency is highly non-monotonic in a $f_m < 50\,\text{Hz}$ frequency range.

The result of FM modulation is usually a broadband signal. According to [3], "the pass bandwidth of input stage ... should not introduce an extensive suppression at least up to 700 Hz." Figure 11 presents the characteristic $P_{st} = f(f_{LPF}, f_m, \Delta f_{FM} = 1\,\text{Hz})$ obtained for the changing values of f_m frequency. The bandwidth of the input signal $u_{IN}(t)$ was limited with a low-pass LPF filter with a cut-off frequency f_{LPF} adjusted in the range of 50–800 Hz. Surprisingly, to some extent, the resulting characteristic

shows that the limitation of the bandwidth leads to increased value of output P_{st} signal for almost all of the preset values of frequency f_m.

The influence of f_{LPF} bandwidth limitation on P_{st} measurement result depends on f_m frequency value. Figure 12 presents the characteristic $P_{st} = f(f_m, \Delta f_{FM} = 1\,\text{Hz}, f_{LPF})$ for the case of FM with rectangular signal. The increase in the P_{st} value for $f_m < 50\,\text{Hz}$ can be clearly observed.

Figure 13 presents the $P_{st} = f(\Delta f_{FM}, f_m)$ characteristic for modulation with rectangular signal. The evaluation of characteristic linearity is complex for both the cases. For some values of f_m frequency (i.e., 109 Hz, 91 Hz, and 78 Hz), it could be treated as linear, for other values, it is nonlinear, while for the smallest values, it is non-monotonic.

Figure 14 presents the $\Delta f_{FM} = f(f_m, P_{st} = 1)$ characteristic for modulation with sinusoidal and rectangular signals. This characteristic was reconstructed using the algorithm presented in Figure 2, where the modulation depth $(\Delta U/U)$ is replaced by frequency deviation Δf_{FM}.

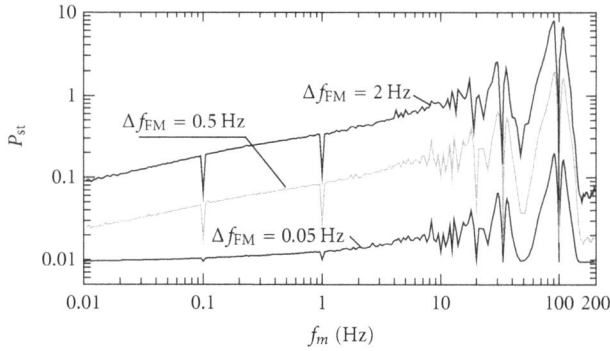

FIGURE 10: $P_{st} = f(f_m, \Delta f_{FM})$ characteristic for FM modulation with rectangular signals.

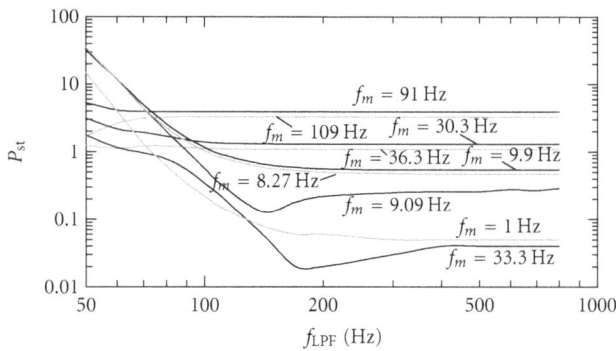

FIGURE 11: $P_{st} = f(f_{LPF}, f_m, \Delta f_{FM} = 1\,Hz)$ characteristics for FM modulation with rectangular signals.

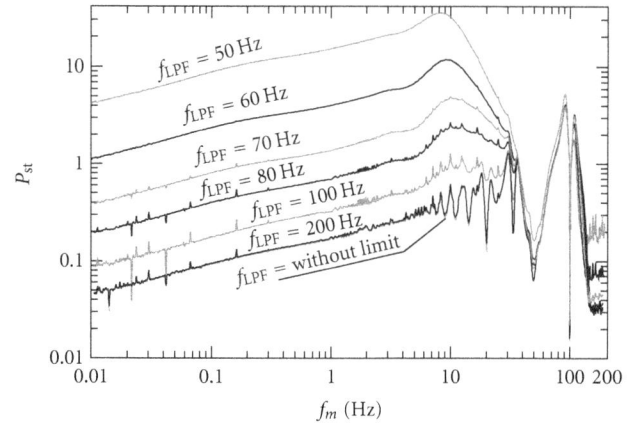

FIGURE 12: $P_{st} = f(f_m, f_{LPF}, \Delta f_{FM} = 1\,Hz)$ characteristic for the case of FM with rectangular signal.

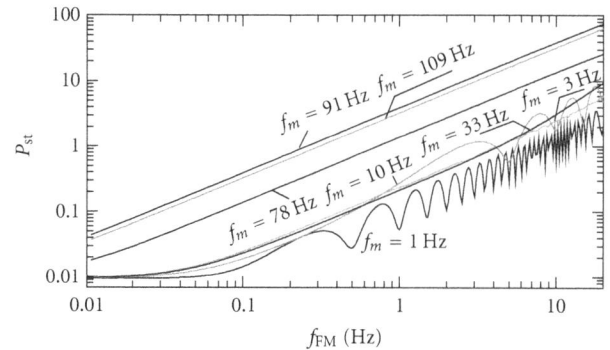

FIGURE 13: $P_{st} = f(\Delta f_{FM}, f_m)$ characteristics for rectangular modulating signals for a changing values of f_m frequency.

5. Discussion of Results

Based on the presented figures, the following conclusions could be derived.

(1) *Fading of Initial Transient Component*

the fading out time of transient component of IFL signal for zero initial conditions (Figure 3) is about 120 s (even though the simulated signal chain includes a block of raising/falling time equal to 60 s, and the fading out time of the transient component of the block equals to 180 s).

(2) *For a Case of No Modulation*

for a case of no modulation, the value of P_{st} indicator is greater than zero (see Table 1) and depends on the order of low-pass F1B filter. For order 6 of this filter, which is recommended in [3], the P_{st} indicator value is about 0.01. This means that the measurement result of a real flickermeter cannot be lower than 0.01.

(3) *For AM Modulation*

(i) if an input signal $u_{IN}(t)$ is a result of AM modulation (2), then the processing characteristic $P_{st} = f(f_m)$

(Figure 4) covers frequencies f_m up to 155 Hz, and three local maxima occur for frequencies $f_m = 8.8, 91.2$ and $108.8\,Hz$. The global maximum does not depend on the shape of the modulating signal and occurs at $f_m = 8.8\,Hz$. These signify that there are three maxima with respect to sensing the obnoxiousness of the flicker for incandescent lamp,

(ii) characteristics $P_{st} = f(f_m)$ (Figure 4), $(\Delta U/U) = f(f_m, P_4^2 = 1)$ (Figure 5), and $(\Delta U/U) = f(f_m, P_{st} = 1)$ (Figure 6) show that for the threshold value of P_{st} indicator ($P_{st} = 1$), the smallest modulation depth occurs for rectangular signal and the greatest modulation depth occurs for triangular signal. Ipso facto, the most obnoxious modulation is the modulation with rectangular signal, followed by the one with sinusoidal signal, and the least obnoxious is the modulation with triangular signal,

(iii) with regard to characteristics $P_{st} = f(f_m)$ (Figure 4), $(\Delta U/U) = f(f_m, P_4^2 = 1)$ (Figure 5), and $(\Delta U/U) = f(f_m, P_{st} = 1)$ (Figure 6), one can observe a non-monotonicity in a f_m frequency range of 28–37 Hz, and hence, this fragment of the characteristics is very useful during the tests of the performance of real flickermeters,

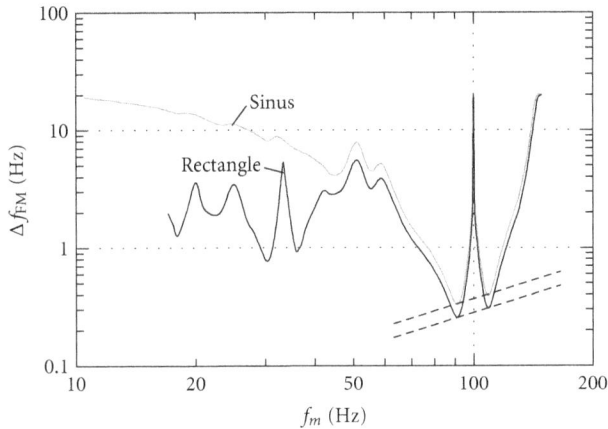

FIGURE 14: $\Delta f_{FM} = f(f_m, P_{st} = 1 = \text{const})$ characteristic for FM modulation with sinusoidal and rectangular signals.

(iv) by taking characteristic $P_{st} = f(\Delta U/U)$ (Figure 5) as a reference criterion when evaluating flickermeter linearity, we can state that the flickermeter signal chain is, in general, nonlinear, but for inputs that correspond to $P_{st} > 0.1$ (which means sensing of flicker), this system is nearly linear,

(v) in a case of AM modulation, the influence of the input signal $u_{IN}(t)$ bandwidth limitation (Figure 8) is visible for $f_{LPF} < 100$ Hz.

(4) *For Input Signal with Interharmonic Component*

comparative combination of $U_i = f(f_i, P_{st} = 1)$ and $(\Delta U/U) = f(f_m, P_{st} = 1)$ characteristics (Figure 9) makes a good basis to conclude on the difference between obnoxiousness of the flicker caused by amplitude modulation and occurrence of single interharmonic in a voltage that supplies incandescent lamp,

(5) *For FM Modulation*

(i) on the basis of $P_{st} = f(f_m, \Delta f_{FM})$ characteristics for FM modulation (Figure 10), it is difficult to estimate the range of f_m frequency in which a changeability of P_{st} indicator value occurs; for modulation with sinusoidal signal, the changeability of P_{st} indicator fades for $f_m > 177$ Hz,

(ii) $P_{st} = f(f_m, \Delta f_{FM})$ characteristic for FM modulation with rectangular signal (Figure 10) is strongly non-monotonic,

(iii) on the basis of $P_{st} = f(\Delta f_{FM}, f_m)$ plot (Figure 13), one can state that the P_{st} indicator value could be greater than that for the frequency deviation Δf_{FM} greater than 0.25 Hz. Thus, FM modulation of the input voltage with $\Delta f_{FM} < 0.25$ Hz should not lead to obnoxious flicker (Figure 14),

(iv) in the case of FM modulation, the measurement result of P_{st} indicator strongly depends on the flickermeter bandwidth; limitation of input signal $u_{IN}(t)$ bandwidth (i.e., decreasing f_{LPF}) surprisingly leads to increased value of P_{st} indicator (Figures 11 and 12),

(v) taking characteristic $P_{st} = f(\Delta f_{FM})$ (Figure 13) as a reference criterion when evaluating flickermeter linearity, we can state that the flickermeter signal chain in the case of FM modulation is, in general, nonlinear.

6. Conclusion

The results give a comprehensive overview of the signal chain and supplement the standard specification for the case of AM modulation of the input signal. The results also complement the specification for the case of FM modulation of the input signal and for the input signal with single interharmonic component. The presented results thoroughly describe the performance of IEC flickermeter in a full frequency range that influences the result of P_{st} indicator measurement (as opposed to other results given in the literature that describe the flickermeter only in limited frequency range). The results of the simulations make it easier to understand the operation of the IEC flickermeter. They describe the influence of the input voltage parameters on P_{st} indicator measurement result. Furthermore, the reaction of the IEC flickermeter to different types of input signals is also demonstrated. The analysis of the presented characteristics helps to determine the requirements with regard to flickermeter signal chain and suggests the potential source of measurement error. Any peculiar fragments of the characteristics define the optimal condition for checking the accuracy of the performance of the IEC flickermeter and, at the same time, help to shorten the time of flickermeter testing. It can be presented in the future.

References

[1] Flickermeter, *International Electrotechnical Vocabulary*, IEC, number 604-01-28.

[2] Flicker, *International Electrotechnical Vocabulary*, IEC, number 161-08-13.

[3] IEC 61000-4-15, *Flickermeter—Functional and Design Specifications*, 2010.

[4] H. De Lange Dzn, "Eye's response at flicker fusion to square-wave modulation of a test field surrounded by a large steady field of equal mean luminance," *Journal of the Optical Society of America*, vol. 51, no. 4, pp. 415–421, 1961.

[5] X. Yang and M. Kratz, "Power system flicker analysis and numeric flicker meter emulation," in *Proceedings of the IEEE Lausanne POWERTECH*, pp. 1534–1539, July 2007.

[6] R. Cai, J. F. G. Cobben, J. M. A. Myrzik, J. H. Blom, and W. L. Kling, "Flicker responses of different lamp types," *IET Generation, Transmission and Distribution*, vol. 3, no. 9, pp. 816–824, 2009.

[7] L. W. White and S. Bhattacharya, "A discrete matlab-simulink flickermeter model for power quality studies," *IEEE Transactions on Instrumentation and Measurement*, vol. 59, no. 3, pp. 527–533, 2010.

[8] P. Clarkson and P. S. Wright, "Sensitivity analysis of flickermeter implementations to waveforms for testing to the requirements of IEC, 61000-4-15," *IET Science, Measurement and Technology*, vol. 4, no. 3, pp. 125–135, 2010.

[9] J. Slezingr and J. Drapela, "Verification of Flickermeters under new edition of IEC, 61000-4-15," in *Proceedings of the IEEE Trondheim PowerTech*, p. 6, June 2011.

[10] I. Sadinezhad and V. G. Agelidis, "Frequency adaptive least-squares-kalman technique for real-time voltage envelope and flicker estimation," *IEEE Transactions on Industrial Electronics*, vol. 59, no. 8, pp. 3330–3341, 2012.

[11] D. Gallo, C. Landi, R. Langella, and A. Testa, "Implementation of a test system for advanced calibration and performance analysis of flickermeters," *IEEE Transactions on Instrumentation and Measurement*, vol. 53, no. 4, pp. 1078–1085, 2004.

[12] E. W. Gunther, "A proposed flicker meter test protocol," in *Proceedings of the Quality and Security of Electric Power Delivery Systems Symposium*, pp. 235–240, October 2003.

[13] L. Peretto, E. Pivello, R. Tinarelli, and A. E. Emanuel, "Theoretical analysis of the physiologic mechanism of luminous variation in eye-brain system," in *Proceedings of the IEEE Instrumentation and Measurement Technology Conference (IMTC '05)*, pp. 128–133, 2005.

[14] M. De Koster, E. De Jaeger, and W. Vancoetsem, "Light flicker caused by interharmonics," in *IEEE Power Engineering Society Transmission and Distribution Committee General Systems Subcommittee, Harmonics Working Group*, 2001.

[15] S. M. Halpin and V. Singhvi, "Limits for interharmonics in the 1-100-Hz range based on lamp flicker considerations," *IEEE Transactions on Power Delivery*, vol. 22, no. 1, pp. 270–276, 2007.

[16] T. Kim, E. J. Powers, W. M. Grady, and A. Arapostathis, "Detection of flicker caused by interharmonics," *IEEE Transactions on Power Delivery*, vol. 58, no. 1, pp. 152–160, 2009.

[17] G. Wiczyński, "Analysis of flickermeter's signal chain for input signal with two sub/interharmonics," *Electrical Review*, vol. 86, pp. 328–335, 2010.

[18] D. Gallo, R. Langella, and A. Testa, "Toward a new flickermeter based on voltage spectral analysis," in *Proceedings of the International Symposium on Industrial Electronics (ISIE '02)*, vol. 2, pp. 573–578, 2002.

[19] D. Gallo, R. Langela, and A. Testa, "Light flicker prediction based on voltage spectral analysis," in *Proceedings of the IEEE Porto Power Tech Conference (PPT '01)*, vol. 1, p. 6, 2001.

[20] W. Mombauer, "Flicker caused by interharmonics," *EtzArchiv*, vol. 12, no. 12, pp. 391–396, 1990.

[21] A. Testa and R. Langella, "Power system subharmonics," in *Proceedings of the IEEE Power Engineering Society General Meeting*, vol. 3, pp. 2237–2242, June 2005.

[22] M. Rogóz, A. Bień, and Z. Hanzelka, "The influence of a phase change in the measured voltage on flickermeter response," in *Proceedings of the 11th International Conference on Harmonics and Quality of Power (ICHQP '04)*, pp. 333–337, September 2004.

[23] J. J. Gutierrez, L. A. Leturiondo, J. Ruiz, A. Lazkano, P. Saiz, and I. Azkarate, "Effect of the sampling rate on the assessment of flicker severity due to phase jumps," *IEEE Transactions on Power Delivery*, vol. 26, no. 4, pp. 2215–2222, 2011.

[24] J. Ruiz, A. Lazkano, J. J. Gutierrez et al., "Influence of the carrier phase on flicker measurement for rectangular voltage fluctuations," *IEEE Transactions on Instrumentation and Measurement*, vol. 61, no. 3, pp. 629–635, 2012.

[25] W. Mombauer, "Flicker caused by phase jumps," *European Transactions on Electrical Power*, vol. 16, no. 6, pp. 545–567, 2006.

[26] G. Wiczyński, "Simple model of flickermeter signal chain for deformed modulating signals," *IEEE Transactions on Power Delivery*, vol. 23, no. 4, pp. 1743–1748, 2008.

[27] S. Haykin, *Communication Systems*, John Wiley & Sons, 1994.

[28] G. Wiczyński, "A model of the flickermeter for frequency modulation of the input voltage," *IEEE Transactions on Instrumentation and Measurement*, vol. 58, no. 7, pp. 2139–2144, 2009.

[29] A. Bertola, G. C. Lazaroiu, M. Roscia, and D. Zaninelli, "A Matlab-simulink flickermeter model for power quality studies," in *Proceedings of the 11th International Conference on Harmonics and Quality of Power (ICHQP '04)*, pp. 734–738, September 2004.

[30] D. Hanselman and B. Littlefield, *Mastering Matlab 6: A Comprehensive Tutorial and Reference*, Prentice Hall, New Jersey, NJ, USA, 2001.

20

Simulation and Modeling Application in Agricultural Mechanization

R. M. Hudzari, M. A. H. A. Ssomad, R. Syazili, and M. Z. M. Fauzan

Department of Agriculture Science, Faculty of Agriculture and Biotechnology, Sultan Zainal Abidin University, 21300 Terengganu, Malaysia

Correspondence should be addressed to R. M. Hudzari, mohdhudzari@unisza.edu.my

Academic Editor: Bauke Vries

This experiment was conducted to determine the equations relating the Hue digital values of the fruits surface of the oil palm with maturity stage of the fruit in plantation. The FFB images were zoomed and captured using Nikon digital camera, and the calculation of Hue was determined using the highest frequency of the value for R, G, and B color components from histogram analysis software. New procedure in monitoring the image pixel value for oil palm fruit color surface in real-time growth maturity was developed. The estimation of day harvesting prediction was calculated based on developed model of relationships for Hue values with mesocarp oil content. The simulation model is regressed and predicts the day of harvesting or a number of days before harvest of FFB. The result from experimenting on mesocarp oil content can be used for real-time oil content determination of MPOB color meter. The graph to determine the day of harvesting the FFB was presented in this research. The oil was found to start developing in mesocarp fruit at 65 days before fruit at ripe maturity stage of 75% oil to dry mesocarp.

1. Introduction

Basically, FFB are classified into four categories, black, hard, ripe: and overripe. However, for initial study, three types of FFB were used, ripe, unripe: and overripe. A camera vision system developed in this research is made up of two critical components. The first component is the hardware component that functions as an image acquisitioned for the system. The second component is the software part which analyzes the image captured by the hardware component. The system made prediction of FFB's maturity by processing the image captured. The main hardware system in this study is a digital camera to capture the image of the sample oil palm fruits and a light meter to detect intensity of the light. Sample pictures were taken in an oil palm plantation at MPOB, Bangi Lama, Selangor, Malaysia. Oil palm FFB maturity prediction in this research was done by determining the Hue values of oil palm at different stages of maturity. The prediction was also made on the relationship between Hue and oil content in the fruit.

The first monitoring period for oil palm fruit and lighting intensity was made starting from December 11, 2008 until December 31, 2008. The second monitoring period was held from August 10, 2009 until October 06, 2009. All experiments were conducted within 8 to 9 weeks monitoring period. The FFB images captured were only after the fruit was completely grown with the fruit color skin changed at from black to reddish color. This is based on the study by Khalid and Abbas [1], who mentioned; at the stage of young fruits ripeness (within 7 to 11 weeks after flower was open-anthesis), the color of fruits skin is black and only changes to reddish black from that duration. They mentioned the fruits, within 15 to 17 weeks after anthesis, had color surface of black plus reddish black while the oil percentage was less than 5%, at 18 to 19 weeks after anthesis, the fruits color was reddish orange with 40 to 48% oil content, at 20 to 22 weeks, the fruit color surface was reddish orange plus orange, and at 22 to 23 weeks after anthesis, the fruit color was mostly orange with more than 50% oil content. The measurement of the oil content was based on percentage of oil with fresh mesocarp ratio. It is the wet base measurement.

This study was carried out from selected immature fruits (black color surface) until to overripe (orange color surface). The factor of lighting intensity under canopy of oil palm plantation was eliminated using Hue color value at specific camera parameters. The radiation from the sunlight was intercepted by the oil palm canopy by intercropping systems for growth resources of solar radiation and caused the intensity to decrease.

2. Methodology

The model was written in Visual Basic 6.0 and it was used to simulate and predict the suitable day to harvest the FFB. Only one progeny (*Dura* X *Pisifera*) was considered for this simulation model. The major input parameters and their mathematical expression were described. The major input variables of this equation are

(i) Hue value of the fruits image,

(ii) mesocarp oil content of the FFB (dry base measurement).

The simulation begun with an experiment which started from the fruit at unripe stage right until it reached overripe stage. Figure 1 shows the simulation of the overall project during the study. The Nikon coolpix 4500 digital camera and Keyence vision system were used to capture fruit image digitally. The camera was set to manual mode to make constant image output. The shutter speed of camera parameter was set to 1/8, and the exposure was set to maximum. This parameter value was experimentally suitable to capture the FFB image for the whole day shift. The image was taken from under the canopy of oil palm tree in real plantation condition. After the capturing image session completed, 3 fruitlets were collected for the chemical test. These fruitlets picked must be from the outside of monitoring area. These steps were repeated until the FFB became overripened. More than 50 images were captured during the photography session.

During the running of the GUI program, the user chose the picture to be analyzed using VB 6.0. The user needs to feed the FFB image into the software and run the function to obtain histogram value of *RGB* in monitored area of image. From maximum of *R*, *G*, and *B* value, the value of Hue would be determined and compared with the Hue value from Keyence vision system. When the user wanted to analyze another FFB images, they have to repeat the same procedure as described earlier. Finally, the graph of oil content of mesocarp versus pixel value was plotted automatically in this program.

2.1. Calculation for Hue Value (HVM). The images captured were fed into the software, and these images were analysed using RGB histogram module in order to get maximum RGB value of all maturity stages of FFB image. Figure 2 shows the flow chart for calculation the Hue digital value. The color value of the images obtained from *RGB* conversion value and would then verify by comparing the value obtained from Keyence vision system. In this experiment, only the Hue (*H*) digital value was measured for determine the color surface of FFB while for saturation (*S*) and intensity (*I*) values, they are not used for color measurement. Same technique also

was adopted by Abdullah et al. [2], which used single value of Hue for distinguishing different maturity stages of FFB, and the values for saturation and intensity were ignored. The histogram method which basically just a graphing of the frequency of each intensity of red, green, blue, or luminance in an image is used.

2.2. Calculation for Oil Extraction Ratio (OERM). Figure 3 shows the steps involved in calculating oil content in mesocarp ratio. The whole dry weight model was used to calculate oil-to-dry mesocarp ratio.

The whole dry weight calculation is as follows:

$$WDW = SST - PST, \tag{1}$$

where

$$SST = Sub - sample\ Weight\ (Wt.)$$
$$+ Thimble\ Wt.\ (weight\ of\ filter\ paper), \tag{2}$$
$$PST = Post\ Soxhlet + Thimble\ Wt.$$

The oil-to-dry Mesocarp was described as below:

$$\underset{DM}{O} = \left[\frac{WDW}{(SST - TW)}\right] \times 100, \tag{3}$$

where WDW is the whole dry wt; TW is the thimble wt.

So, the average of oil content in mesocarp was calculated using the following equation:

$$
\text{Average oil content in meso (\%)}
$$
$$
= \frac{[ODM_1 + ODM_2 + ODM_3]}{3}, \tag{4}
$$

Where ODM_1 = oil to dry Mesocarp sub-sample 1; ODM_2 = oil to dry Mesocarp sub-sample 2; ODM_3 = oil to dry Mesocarp sub-sample 3.

2.3. Calculation Day of Harvesting For FFB (DHM). Figure 4 shows the method to determine the days of harvesting. The relation of Mesocarp oil content with the Hue value of the FFB can be expressed by the following equation:

$$
\text{oil content (\% mesocarp)}
$$
$$
= \text{constant parameter} * \text{Hue value}(0 \sim 255). \tag{5}
$$

The correlation regression, R^2 shows the significant level on parameter relationship. The best regression correlation between Hue values with the mesocarp oil content is always 1.

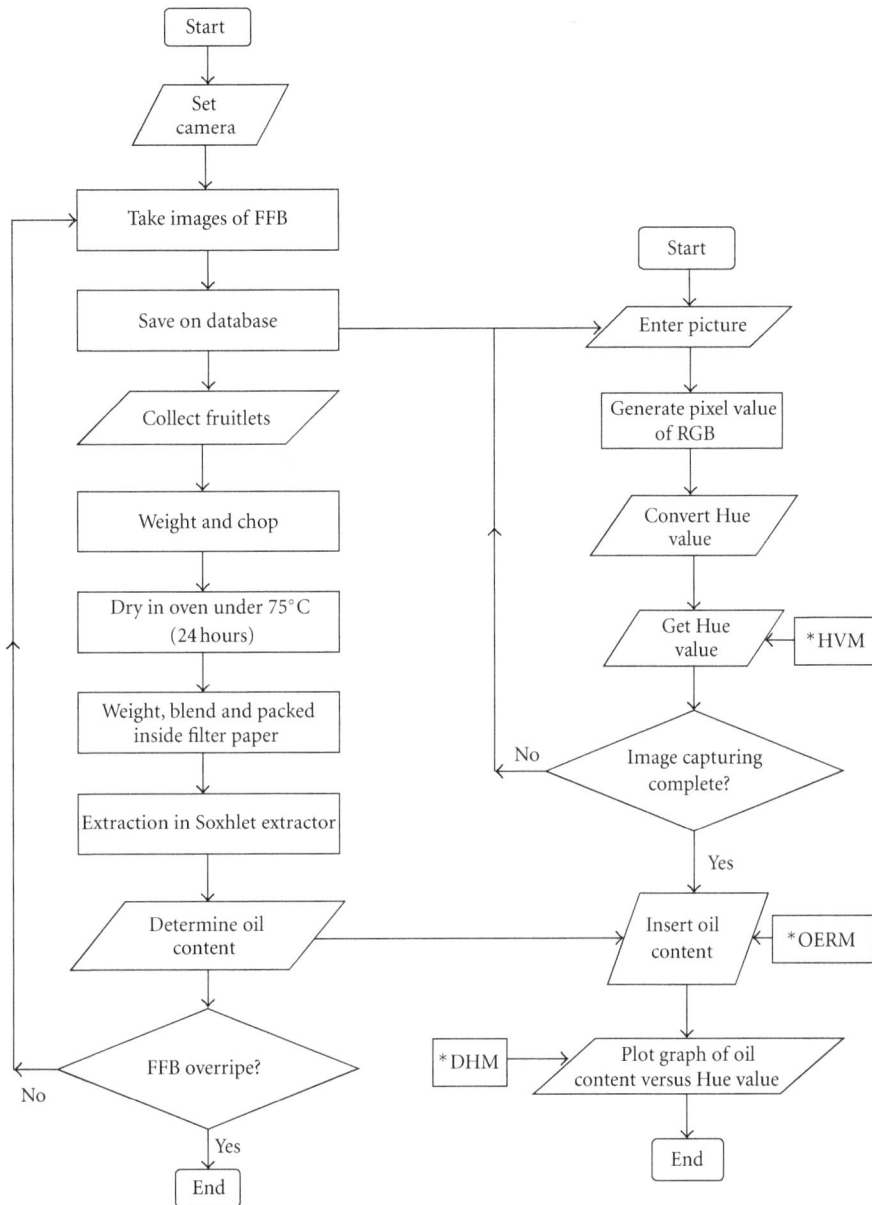

*HVM—Hue value calculation
*DHM—date harvesting calculation
*OERM—oil extraction calculation

FIGURE 1: Flow chart of the involved work.

Figure 5 shows the flow charts for calculating harvesting day of FFB (DHM). The calculation and for the estimation day for harvesting the FFB model is described as below:

$$\frac{75\% - 0}{K - 0} = \frac{0 - 65}{N - 65},$$

$$N = \left[\frac{(K - 0)(-65)}{(75\% - 0)} \right] + 65, \qquad (6)$$

$$N = \frac{(-65K)}{75\%} + 65,$$

where 75% is the maximum mesocarp oil content obtained from ripe FFB as mentioned by Harun and Noor [3], K is mesocarp oil content calculated in %, and 65 is the number of days for harvesting the FFB and also is the initial day of development with 0% mesocarp oil content FFB. The maximum of mesocarp oil content, 75%, and L were obtained by the chemical analysis of Soxhlet extractor in the laboratory. The zero day was meant for the actual harvesting day for FFB at 75% oil content and N is the day calculated.

Assuming that the value of K is 35% mesocarp oil content. From Figure 5 and flow chart, the number of days before harvesting will be 35 days.

```
            ┌──────────┐
            │  Start   │
            └────┬─────┘
                 ↓
      ┌────────────────────┐
      │ Generate histogram for │
      │   analysis image    │
      └──────────┬─────────┘
                 ↓
      ┌────────────────────┐
      │ Define the average pixel │
      │  values of R, G, and B │
      └──────────┬─────────┘
                 ↓
      ┌────────────────────┐
      │  Calculate Hue value │
      │ using Gonzalez formula │
      └──────────┬─────────┘
                 ↓
            ┌──────────┐
            │   End    │
            └──────────┘
```

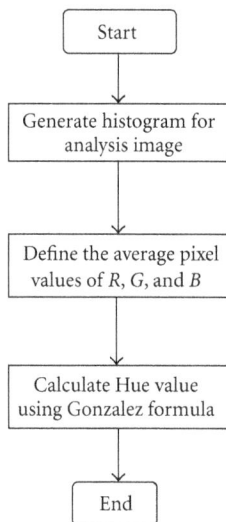

Figure 2: Flow chart for calculating the Hue digital value.

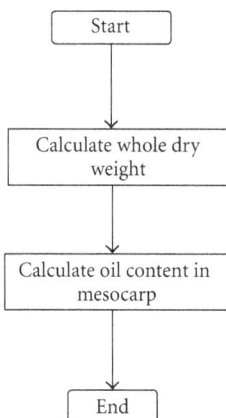

```
            ┌──────────┐
            │  Start   │
            └────┬─────┘
                 ↓
      ┌────────────────────┐
      │  Calculate whole dry │
      │       weight        │
      └──────────┬─────────┘
                 ↓
      ┌────────────────────┐
      │ Calculate oil content in │
      │      mesocarp       │
      └──────────┬─────────┘
                 ↓
            ┌──────────┐
            │   End    │
            └──────────┘
```

Figure 3: Flow chart for calculating oil content.

3. Result and Discussion

The MPOB colorimeter was used to determine the ripeness of oil palm fruit based on mesocarp surface color. This equipment was used in this study to validate the ripeness of the oil palm fruits after determining the mesocarp oil content using Soxhlet extraction process. Figure 6 shows the relationship of the Hue digital value with oil palm fruit maturity level. From the Table 1, the Hue value of 158 to 179 indicates unripe FFB, 185 to 212 indicates underripe FFB and 224 to 255 indicates ripe FFB.

3.1. Estimation for Harvesting the 20-Year-Old FFB. The suitable days for harvesting were calculated based on equation. The model for maturity simulation was based on linear interpolation method in determining the maximum oil content which was assigned as ripe FFB and the date for harvesting. Harun and Noor [3] mentioned that the 75% of oil to dry mesocarp indicated the ripe FFB. The experiments dates were manipulated for linear interpolation so that the day for harvesting the FFB with at 75% mesocarp oil content, meant zero day for harvesting is determined. The ripeness

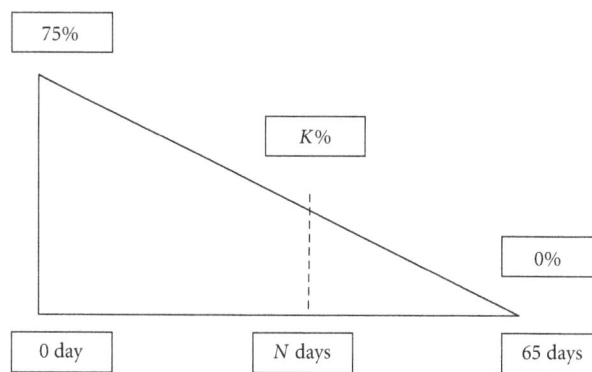

Figure 4: The method to determine the harvesting days.

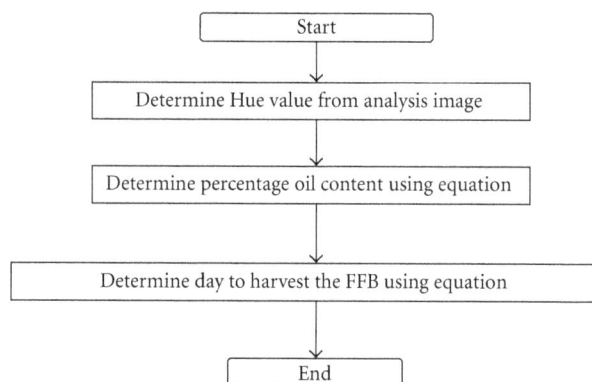

```
            ┌────────────────────┐
            │       Start        │
            └─────────┬──────────┘
                      ↓
      ┌──────────────────────────────────┐
      │ Determine Hue value from analysis image │
      └────────────────┬─────────────────┘
                       ↓
      ┌──────────────────────────────────┐
      │ Determine percentage oil content using equation │
      └────────────────┬─────────────────┘
                       ↓
      ┌──────────────────────────────────┐
      │ Determine day to harvest the FFB using equation │
      └────────────────┬─────────────────┘
                       ↓
            ┌────────────────────┐
            │        End         │
            └────────────────────┘
```

Figure 5: Flow chart for calculating the harvesting day of FFB.

Table 1: The correlation of moisture content and weeks of fruits after anthesis (adapted from [1]).

Maturity stage	Moisture content
Under ripe, less than 18th weeks after anthesis	40%–80%
Nearly ripe, 18th to 20th weeks	35%–40%
Ripe, 20th to 22nd weeks	33%–35%
Fully ripe, 23rd onwards	Less than 33%

stage of FFB was confirmed on visual image evaluation. Figure 7 shows the estimation model of mesocarp oil content with harvesting day prediction. High correlation of R^2 found for relationship of mesocarp oil content versus date for harvesting with equation of $Y = -1.3295X + 71.503$ and regression squared, $R^2 = 0.81$ was high and acceptable. The linear regression method was used to generate an equation for interpolating the days required for harvesting the FFB. From the graph, the oil starts to develop the fruit before 54 days to harvest the FFB for 20 years old oil palm fruits.

Figure 8 shows the model to predict the harvesting days for FFB 20-year-old oil palm tree. The graph relationship of mesocarp oil content with Hue digital value was combined with the graph of mesocarp oil content with days to harvest FFB (indicated as red line). The mesocarp oil content was found developed at 54 days before harvesting the FFB. The zero day meant the day for harvest the FFB at maximum mesocarp oil content of 75% (indicated as blue dash line).

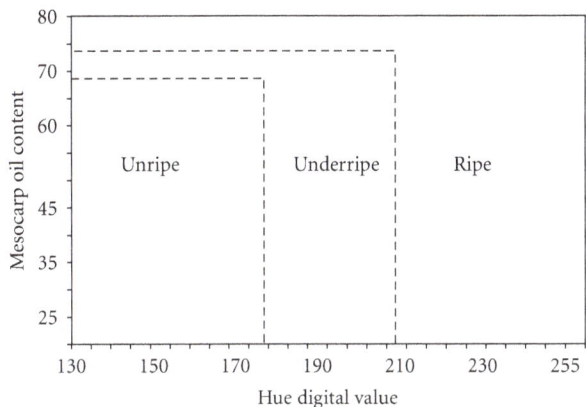

FIGURE 6: The relationship of the Hue digital value in relation to maturity level of oil palm FFB.

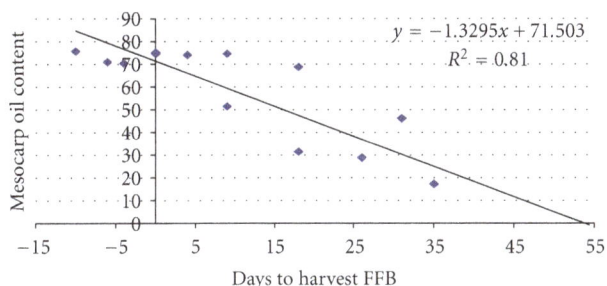

FIGURE 7: Estimation model of mesocarp oil content for harvesting date prediction.

Figure 8 is used as a standard graph to predict the harvesting days for FFB of 20 years old.

3.2. Estimation for Harvesting the 16-Year-Old FFB.

The data included for harvesting the 16-year-old FFB also was published as mentioned by Razali et al., [4]. Otherwise there is a need to find a relationship between the age of FFB cycle time on all data experiments; for earlier time of harvestings was at 5 years old, 16 years old stand as at middle while 25 years stand as at the end of time cycle for FFB. Figure 9 shows the estimation model of mesocarp oil content for harvesting date prediction. High correlation of R^2 found for relationship of mesocarp oil content versus date for harvesting with equation of $Y = -1.125X + 76.386$ and regression squared, $R^2 = 0.904$ was high and acceptable. The oil was developed in the fruit at 66 days before harvesting the FFB for 16-year-old oil palm fruits.

Figure 10 shows the model to predict the harvesting days for FFB 16 years old oil palm tree. The graph relationship of mesocarp oil content with Hue digital value was combined with the graph of mesocarp oil content with days to harvest FFB (indicated as red line). The mesocarp oil content was found developed at 66 days before harvesting the FFB. Along 66 to 0 days, the minor unit in the graph was divided into 35, and every unit had 1.89 days. The zero days meant the day for harvest the FFB at maximum mesocarp oil content of 75%, (indicated as blue dash line). Let us say if the captured

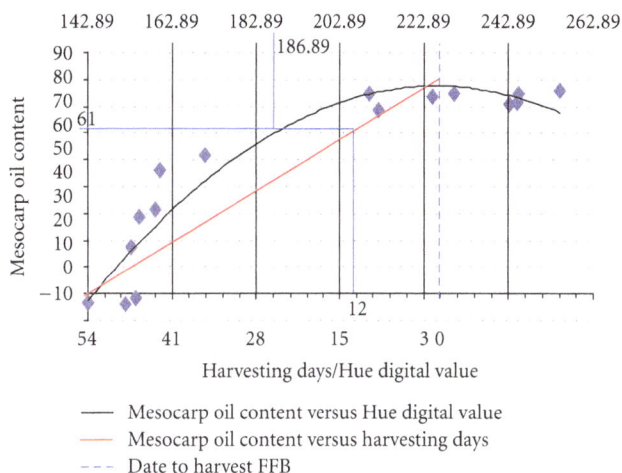

FIGURE 8: The model of the harvesting days for FFB 20-year-old oil palm tree.

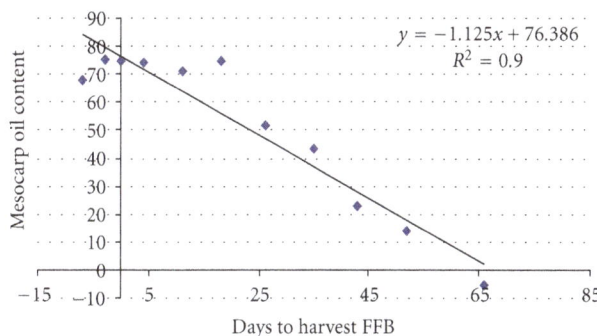

FIGURE 9: Estimation model of Mesocarp oil content for harvesting date prediction.

image of FFB had Hue of 180, the oil content was 57%, and the fruit will require 17 days or 408 hours for harvesting the FFB (indicated as blue line).

3.3. Estimation for Harvesting the 5 Year-Old FFB.

Figure 11 shows the estimation model of mesocarp oil content for harvesting date prediction. High correlation of R^2 found for relationship of mesocarp oil content versus date for harvesting with equation of $Y = -0.1547X + 74.279$ and regression squared, $R^2 = 0.80$ was high and acceptable. From the graph, the oil content was 65% developed in the fruit at 60 days before harvesting the FFB for 5 year-old oil palm fruits.

The regression squared of R^2 for 5 years old tree FFB was found low compared with 16 and 20 years old. That was due to experiment date did not start at 0% mesocarp oil content. In actual experiment, we do not know the content of the oil in the fruit FFB. The determination of fruits mesocarp oil content was only measured using the basis procedure of standard bunch analysis. The Soxhlet extractor machine, oven, and so forth were used during experiment as earlier discussed. Figure 12 shows the model predicting the harvesting days for FFB 5 years old oil palm tree. The graph

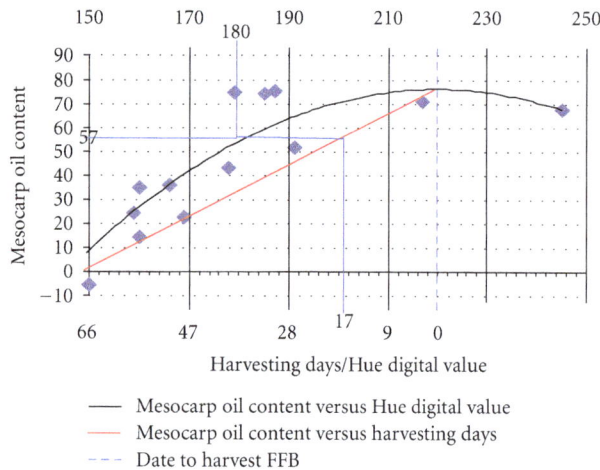

FIGURE 10: The model of the harvesting days for FFB 16 years old oil palm tree.

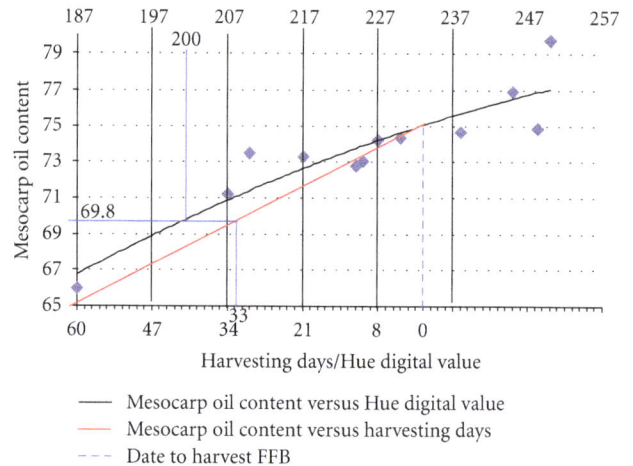

FIGURE 12: The model of the harvesting days for FFB 5 years old oil palm tree.

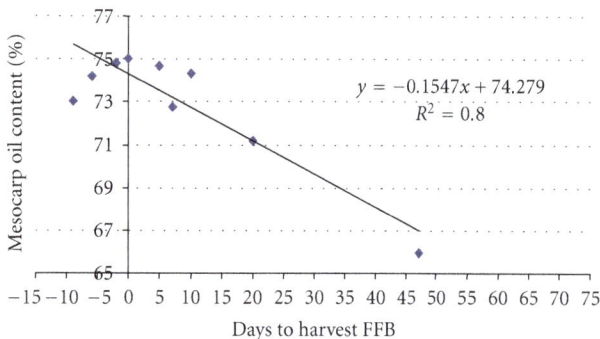

FIGURE 11: Estimation model of mesocarp oil content for harvesting date prediction.

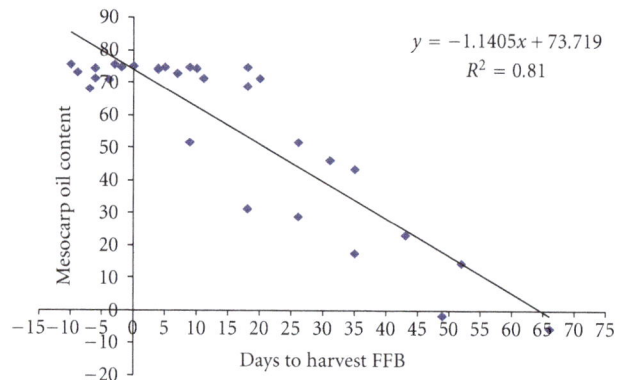

FIGURE 13: Overall estimation model of mesocarp oil content and harvesting days prediction.

relationship of mesocarp oil content with Hue digital value was combined with the graph of mesocarp oil content with days to harvest FFB (indicated as red line). The mesocarp oil content was found 65% developed at 60 days before harvesting the FFB. Along 60 to 0 days, the minor unit in the graph was divided into 46 and every unit had 1.30 days. Zero day meant the day for harvest the FFB at maximum mesocarp oil content of 75% (indicated as blue dash line). Let us say if the captured image of FFB had Hue of 200, the oil content was 69.8%, and the fruit will require 33 days or 792 hours for harvesting the FFB (indicated as blue line).

3.4. Overall Estimation for Harvesting the FFB. Figure 13 shows the overall estimation model of mesocarp oil content for harvesting days prediction of FFB of 20, 16, and 5 years old oil palm tree. High correlation of R^2 found for relationship of mesocarp oil content versus date for harvesting with equation of $Y = -1.1405X + 73.719$ and regression squared, $R^2 = 0.81$ was acceptable. From the graph in Figure 12, the oil was found to start developing in mesocarp fruit at 65 days before fruit at ripe maturity stage with indicated 75% oil to dry mesocarp.

The relationship between mesocarp oil content with Hue value of the FFB can be expressed by the following equation:

$$Y = -0.0116X^2 + 5.2376X - 514.88, \qquad (7)$$

where Y is mesocarp oil content in % and X is Hue value. The mesocarp oil content was determined by Hue digital value using equation of (7). The Nikon digital camera used to capture the FFB image before uploading into analysis software that was developed using programming source code of Visual Basic. Figure 13 shows the estimation model of mesocarp oil content and harvesting days prediction for oil palm FFB. The harvesting days was determined based on the 75% mesocarp oil content which indicated as a ripe stage for FFB [3]. Linear interpolation technique is used to fix the date for the oil content of mesocarp reaching at 75% as earlier discussed. The harvesting days of FFB was determined by percentage of mesocarp oil content using linear equation as shown in Figure 14. The development of oil in mesocarp fruit starts at 65 days before fruit at ripe maturity stage. Along 65 to 0 days, the minor unit in the graph was divided into 66 and every unit had 2.10 days for harvesting the FFB. Zero days meant the day for harvesting the FFB with maximum

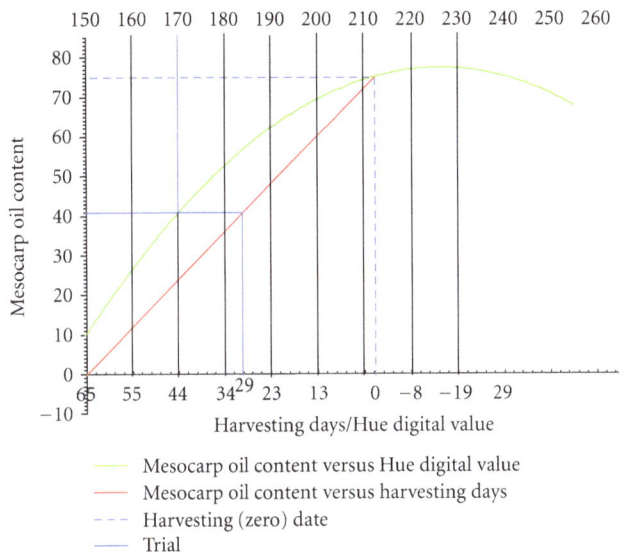

FIGURE 14: The graph for determining the day of harvesting the FFB.

mesocarp oil content of 75% (indicated as blue dash line). Let us say if the captured image of FFB had Hue of 170, the oil content was 41% and the fruit will require 29 days for harvesting the FFB (indicated as blue line).

3.5. Validation of Day Estimation Simulation for Harvesting the FFB. A fruit bunch normally takes around 20 to 22 weeks to ripen after anthesis [5–7]. In this experiment, the time taken for FFB to ripe and to be harvested was 22 weeks after the flower fully open for pollination process (anthesis). Similar finding was also mentioned by Khalid and Abbas [1] where they had developed a microstrip sensor in order to determine the harvesting day for oil palm fruits. They mentioned mesocarp moisture content can indicate the age of the fruits, in terms of number of weeks, after anthesis. Table 1 shows the correlation of moisture content and the age of the fruits after anthesis with respect to number of weeks before harvesting time. It means that from Hue value of the FFB image, the number of days required for optimum harvesting is approximately 14 days or within 2 weeks. Khalid and Abbas [1] were claimed that his method is better on assessing the fruit maturity by its accuracy and time efficiency compared with traditional fruit picker. Otherwise the element of contact measurement method was overcome by this research study. On actual experiment, the FFB was considered matured when there were loose fruits found laying on the ground, that was after 65 days of constant monitoring starting from unripe stage (also confirmed after running the mesocarp oil extraction). The suitable days for harvesting were calculated based on (11) which indicates maximum 75% oil to dry Mesocarp of [3, 8].

That was an agreement, the FFB normally took around 20 to 22 weeks to ripen after anthesis [5–7]. Thus, the developed simulation model to predict harvesting day for the FFB was similarly established by Khalid and Abbas [1]. They used the harvesting period in number of weeks which still in a range for harvesting day as in this research. The prediction

model for harvesting based on Hue value used in this project was more advanced compared to the used of microstrip sensor [9, 10]. Razali et al. [9] used the RGB (red, green, blue) color space scheme for predicting the maturity stage of outdoor FFB's image and found that only the red color space component having correlation with maturity index while the green and blue color space components were ignored. These make the Hue color component which stands for the whole color of FFB surface is found better compared with that the only red components effected in RGB color space. This method of work can be used in real-time prediction in an actual oil palm plantation, and this method also used nondestructive device application. The camera can shoot the image even from the ground level rather than to pick the sample fruit on the tree in order to run the microstrip sensor application [4].

4. Conclusion

The estimation of day harvesting prediction was calculated based on developed model of relationships for Hue values with mesocarp oil content. When the standard camera captures the FFB image, the image will be analysed using Hue digital values which correlate to the oil content of the fruit mesocarp. The simulation model is regressed and predict the day of harvesting or a number of days before harvest of FFB. The result from experimenting on mesocarp oil content can be used for real-time oil content determination of MPOB color meter. The graph to determine the day of harvesting the FFB was contributed in this research. The oil was found to start developing in mesocarp fruit at 65 days before fruit at ripe maturity stage of 75% oil-to-dry mesocarp.

Acknowledgments

The authors are thankful to all staff members especially from Department of Agriculture and Biological Engineering, Universiti Putra Malaysia, Serdang, Selangor, Malaysia which provided all materials and tools for conducting this research.

References

[1] K. Khalid and Z. Abbas, "Microstrip sensor for determination of harvesting time for oil palm fruits (tenera. Elaeis guineensis)," *Journal of Microwave Power and Electromagnetic Energy*, vol. 27, no. 1, pp. 3–10, 1992.

[2] M. Z. Abdullah, C. G. Lim, and B. M. N. M. Azemi, "Stepwise discriminant analysis for colour grading of oil palm using machine vision system," *Food and Bioproducts Processing*, vol. 79, no. 4, pp. 223–231, 2001.

[3] M. H. Harun and M. R. M. Noor, "Fruit set and oil palm bunch components," *Journal of Oil Palm Research*, vol. 14, no. 2, pp. 24–33, 2002.

[4] M. H. Razali, W. I. W. Ismail, A. R. Ramli, M. N. Sulaiman, and M. H. Harun, "Development of image based modeling for determination of oil content and days estimation for harvesting of fresh fruit bunches," *International Journal of Food Engineering*, vol. 5, no. 2, pp. 1633–11637, 2009.

[5] K. T. Ng and A. Southworth, *Optimum Time of Harvesting Oil Palm Fruits*, Incorporated Society of Planters, Kuala Lumpur, Malaysia, 1973.

[6] I. M. Siregar, "Assessment of ripeness and crop quality control," in *Proceeding of the Malaysian Inter Agricultural Oil Palm Conference*, pp. 740–754, Kuala Lumpur, Malaysia, 1976.

[7] A. A. Azis, "A simple floatation technique to gauge ripeness of oil palm fruits and their maximum oil content," in *Proceeding of the International Palm Oil Development Conference (PORIM '90)*, pp. 87–91, Kuala Lumpur, Malaysia, 1990.

[8] W. I. W. Ishak and R. M. Hudzari, "Image based modeling for oil palm fruit maturity prediction," *Journal of Food, Agriculture and Environment*, vol. 8, no. 2, pp. 469–476, 2010.

[9] M. H. Razali, W. I. wan Ismail, A. R. Ramli, and M. N. Sulaiman, "Modeling of oil palm fruit maturity for the development of an outdoor vision system," *International Journal of Food Engineering*, vol. 4, no. 3, pp. 1396–1396, 2008.

[10] M. R. Hudzari, W. I. W. Ismail, A. R. Ramli, M. N. Sulaiman, and M. H. B. Harun, "Prediction model for estimating optimum harvesting time of oil palm fresh fruit bunches," *Journal of Food, Agriculture & Environment*, vol. 9, no. 3, 4, pp. 570–575, 2011.

21

Application of Nontraditional Optimization Techniques for Airfoil Shape Optimization

R. Mukesh,[1] K. Lingadurai,[1] and U. Selvakumar[2]

[1] *Department of Mechanical Engineering, Anna University, Tamil Nadu, Dindigul 624622, India*
[2] *Department of Information Technology, IBBT, Ghent University, 9050 Ghent, Belgium*

Correspondence should be addressed to R. Mukesh, pr.mukeshphd@gmail.com

Academic Editor: Antonio Munjiza

The method of optimization algorithms is one of the most important parameters which will strongly influence the fidelity of the solution during an aerodynamic shape optimization problem. Nowadays, various optimization methods, such as genetic algorithm (GA), simulated annealing (SA), and particle swarm optimization (PSO), are more widely employed to solve the aerodynamic shape optimization problems. In addition to the optimization method, the geometry parameterization becomes an important factor to be considered during the aerodynamic shape optimization process. The objective of this work is to introduce the knowledge of describing general airfoil geometry using twelve parameters by representing its shape as a polynomial function and coupling this approach with flow solution and optimization algorithms. An aerodynamic shape optimization problem is formulated for NACA 0012 airfoil and solved using the methods of simulated annealing and genetic algorithm for 5.0 deg angle of attack. The results show that the simulated annealing optimization scheme is more effective in finding the optimum solution among the various possible solutions. It is also found that the SA shows more exploitation characteristics as compared to the GA which is considered to be more effective explorer.

1. Introduction

The computational resources and time required to solve a given problem have always been a problem for engineers for a long time though a sufficient amount of growth is achieved in the computational power in the last thirty years. This becomes more complicated to deal with when the given problem is an optimization problem which requires huge amount of computational simulations. These kinds of problems have been one of the important problems to be addressed in the context of design optimization for quite some years. When the number of result(s) influencing variables are large in a given optimization problem, the required computational time per simulation increases automatically. This will severely influence the required computational resources to solve the given design optimization problem. Due to this reason, a need arises to describe a general geometry with minimum number of design variables. This leads to a search activity of finding some of the best parameterization methods. Nowadays, various parameterization methods are employed: partial differential equation approach (time consuming and not suitable for multidisciplinary design optimization), discrete points approach (the number of design variables becomes large), and polynomial approach (the number of design parameters depends on the degree of the polynomial chosen and suitable for multidisciplinary design optimization) are the three basic approaches to describe the geometry of a general airfoil [1–3]. Previous research works in design optimization suggest that the parameterization schemes highly influence the final optimum design which is obtained as a result of the optimization [4]. In this work, the parametric section (PARSEC) parameterization scheme is employed. The panel method is used to compute the flow field around the airfoil geometry during the design optimization process. Both SA and GA are employed to carry out the design optimization problem. This is not the first time that the mentioned optimization schemes (GA and SA) have been applied for the airfoil shape optimization. Here, in the current work, the capability of the strategies is investigated while they are applied to the airfoil kind of surfaces. Three

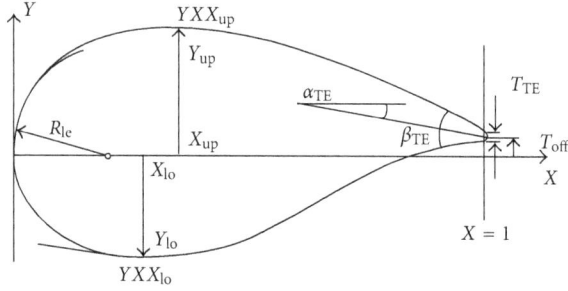

FIGURE 1: Control variables for PARSEC.

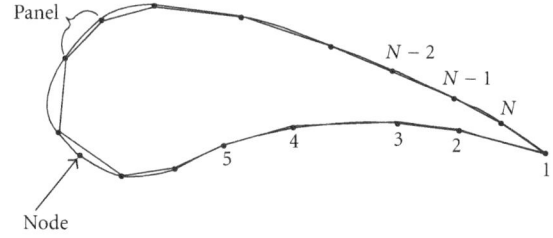

FIGURE 2: Nodes and Panels.

MATLAB codes are developed to implement PARSEC, panel, and SA approaches. A freely available FORTRAN code is picked for the GA. The results and issues faced during the whole design process are discussed in the following sections.

2. PARSEC

In PARSEC parametrisation scheme, an unknown linear combination of suitable base functions is used to describe the airfoil geometry [5]. This approach is considered to be more suitable for design optimization problems, since the geometric constraints on the airfoil shape can be described by some simple linear constraints. Twelve design variables are chosen to have direct control over the shape of the airfoil. The twelve design variables are upper leading edge radius (R_{leu}), lower leading edge radius (R_{lel}), upper crest point (Y_{up}), lower crest point (Y_{lo}), position of upper crest (X_{up}), position of lower crest (X_{lo}), upper crest curvature (YXX_{up}), lower crest curvature (YXX_{lo}), and trailing edge offset (T_{off}), trailing edge thickness (T_{TE}), trailing edge direction angle (α_{TE}), and trailing edge wedge angle (β_{TE}), as shown in Figure 1. The leading edge radius parameters provide more control at the leading edge of the airfoil geometry. The mathematical relations for the PARSEC approach are given as follows:

$$
\begin{aligned}
y_u &= \sum_{i=1}^{6} a_i x^{i-(1/2)}, \\
y_l &= \sum_{i=1}^{6} b_i x^{i-(1/2)},
\end{aligned}
\tag{1}
$$

where y_u is the upper y coordinate, y_l is the lower y coordinate, and a_i, b_i are the unknown coefficients to be solved from the specified values of the twelve design variables. The previous polynomial equations are solved using a set of geometrical conditions.

3. Panel Technique

The panel method is used to solve the potential equations without being computationally expensive. It provides more reasonably accurate results. These two properties make the panel method to be more suitable for design optimization problems where the number of simulations is incredibly large. Since the current problem deals with the incompressible subsonic flow region, this approach is employed in this

work. The solution procedure for panel technique consists of discretizing the surface of the airfoil into straight line segments or panels and assuming the following conditions: (a) the source strength is constant over each panel but has a different value for each panel and (b) the vortex strength is constant and equal over each panel [6, 7]. The compressibility and the viscosity of air in the flow field are neglected. But it is required to satisfy the condition that the net viscosity of the flow should be such that the flow leaving the trailing edge is smooth. The curl of the velocity field is assumed to be zero. Hence,

$$
\phi = \phi_\infty + \phi_\delta + \phi_v,
\tag{2}
$$

where ϕ, which is expressed as a summation of the free stream potential, source potential, and vortex potential, is the total potential function. Except the free stream potential, the other potentials have potentially locally varying strengths. Figure 2 depicts the notations of the panel approach.

As the number of panels increases, the accuracy of the solution increases. Indeed, the computational time will increase as the number of panels increases. $N + 1$ node points define N panels. The tangential velocity (V_{ti}) at the centre of each panel is estimated by imposing a flow tangency condition at each panel. The coefficient of pressure (C_p) at each panel is calculated using the following relation:

$$
C_p(x_i, y_i) = 1 - \left[\frac{V_{\text{ti}}^2}{V_\infty^2} \right].
\tag{3}
$$

4. Simulated Annealing

Simulated Annealing [8, 9] is one kind of non-traditional based optimization algorithm for searching global optimum. It is a point-by-point method. It resembles the cooling process of molten metals through annealing, and the formation of the crystal depends upon the cooling rate. The process of slow cooling is called as annealing. The cooling phenomenon is simulated by controlling a temperature-like parameter, and it can be done by introducing the concept of Boltzmann probability distribution. In addition to that, Metropolis suggested one idea to implement the Boltzmann probability function in simulated systems for better optimization.

The main steps of simulated annealing are given as follows.

  (a) Choose an initial point and a high temperature T.

  (b) A second point is created at random in the vicinity of the initial point.

(c) The difference between these two points is calculated.

(d) If the second point has a larger function value, the point is accepted.

(e) In the next generation, another point is created at random in the neighbourhood of the current point and the Metropolis algorithm [10, 11] is used to accept or reject the point.

(f) The algorithm is terminated when an optimized value is obtained.

The initial value (lower bound values) and the number of iterations (1×10^5) are the two important parameters of the simulated annealing. So, we have to choose these two parameters according to our optimization problem to be solved. The temperature is a controlling parameter which is used to find out the functional value (coefficient of lift) from the given points (lower and upper bound values), and this process will be continued until the optimized value (maximum vertical aerodynamic force) is obtained.

5. Genetic Algorithm

Genetic algorithms (GA), in contrast to gradient optimization approaches, offer an alternative approach with several attractive features. The basic idea associated with the GA is to search for optimal solutions using an analogy to the theory of evolution. During solution advance (or "evolution" using GA terminology), each chromosome is ranked according to its fitness vector—one fitness value for each objective. The higher ranking chromosomes are selected to continue to the next generation while the probability of the selection of lower ranking chromosomes is less. In every generation, a new set of artificial creatures (strings) is created using bits and pieces of the fittest of the old; an occasional new part is tried for good measure. While randomized, genetic algorithms are not simple random walk. They efficiently exploit historical information to speculate on new search points with expected improved performance. The newly selected chromosomes in the next generation are manipulated using various operators (combination, crossover, or mutation) to create the final set of chromosomes for the new generation. These chromosomes are then evaluated for fitness, and the process continues iterating from generation to generation—until a suitable level of convergence is obtained or until a specified number of generations have been completed. GA optimization requires no gradients; it does not need the sensitivity of derivatives. It theoretically works well in nonsmooth design spaces containing several or perhaps many local extrema. It is also an attractive method for multiobjective design optimization applications offering the ability to compute the so-called "pareto-optimal sets" instead of the limited single design point traditionally provided by other methods. The basic genetic algorithm comprises four important steps. They are initialisation, selection, crossover, and mutation [12, 13].

In GA the PARSEC parameterization scheme variables act as the optimization parameters (design variables) which will influence the coefficient of lift (objective function). A pool of optimization parameters will be generated by the

TABLE 1: Optimization objectives and constraints.

Angle of attack	5.0 deg
Flow constraint	Subsonic and incompressible
Geometric constraint	Max thickness must be less than 10% chord length T_{TE} and T_{off} the airfoil is zero
Aerodynamic constraint	Lift not less than the original one
Objective	Maximize coefficient of lift

GA within the defined range of values of the optimization parameters to start the optimization process. Then, based on two GA operators, crossover and mutation, the best optimization parameters at each generation which will increase the coefficient of lift will be selected. This process will be continued until the whole design space is completely explored.

6. Optimization of NACA 0012 Airfoil

The aerodynamic shape optimization process is carried out with an intention of increasing the vertical aerodynamic force subject to aerodynamic and structural constraints. The structural constraints are implemented by fixing the values of trailing edge thickness and trailing edge offset parameters during the optimization in both of the optimization schemes. These constraints are placed in order to avoid the optimizer to get converged at inefficient locations and to avoid getting unrealistic aerodynamic shapes. Since the panel method is only applicable for low speed flows, a flow constraint is placed to keep the assumptions valid throughout the whole optimization process. The flow constraint is implemented by fixing the angle of attack at 5.0 deg. For each design parameter, lower and upper bound values are defined. Each generation produced by the SA and genetic algorithms has the best set of twelve PARSEC parameters. The corresponding airfoil profile is generated using PARSEC parametrisation. Then, the panel method is used to compute the flow around the airfoil at 5.0 deg angle of attack. From the pressure distribution, the lift coefficient is calculated using the trapezoidal rule. This new coefficient of lift is compared to the original one. The SA and genetic algorithms in the end will lead to the best set of PARSEC parameters which will maximise the objective function within the search space. The design conditions, optimization objectives and constraints, which are used during the optimization process using SA and GA, are tabulated in Table 1.

7. Result and Analysis

The initial PARSEC parameters have been given approximately by specifying their lower and upper bound values. There is no need for specifying this accurately. The geometry of the airfoil is expressed by the best twelve PARSEC parameters resulting from the SA algorithm which exhibits a considerable increase in the coefficient of lift as compared to the best solution found by the genetic algorithm. There is a

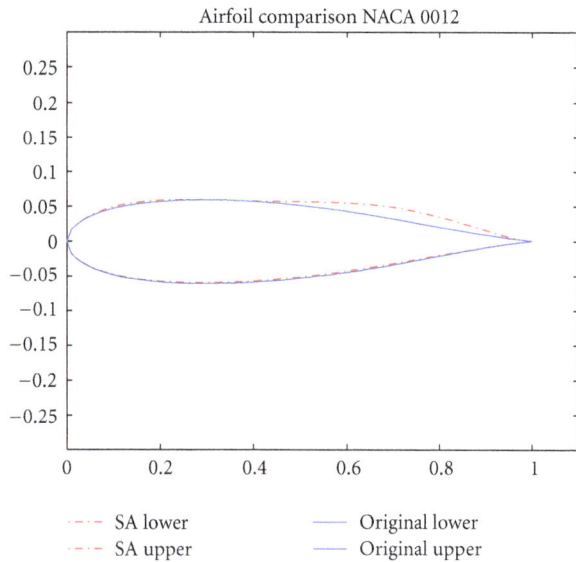

FIGURE 3: Original NACA 0012 airfoil versus optimized airfoil using SA.

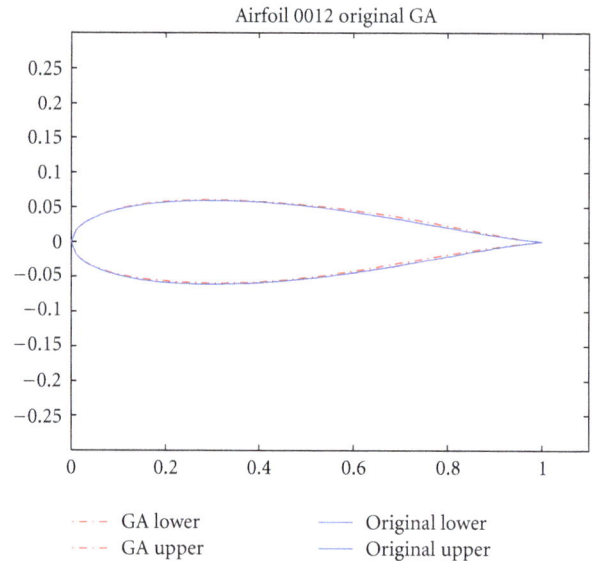

FIGURE 5: Comparison of original NACA 0012 airfoil and optimized airfoil using GA.

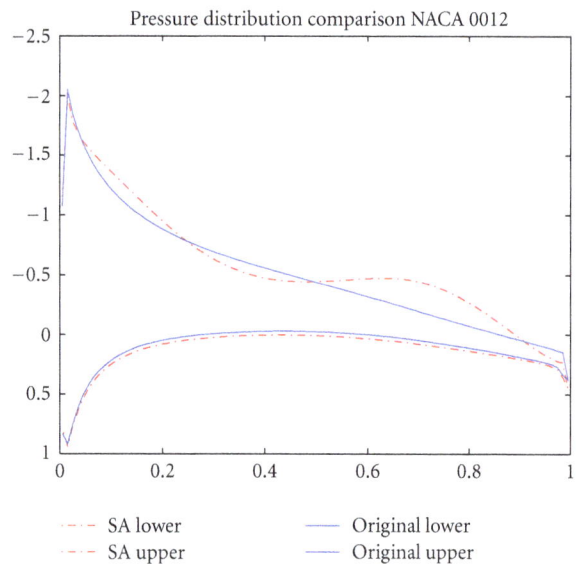

FIGURE 4: Comparison of pressure distribution over the surface of original NACA 0012 airfoil and optimized airfoil using SA.

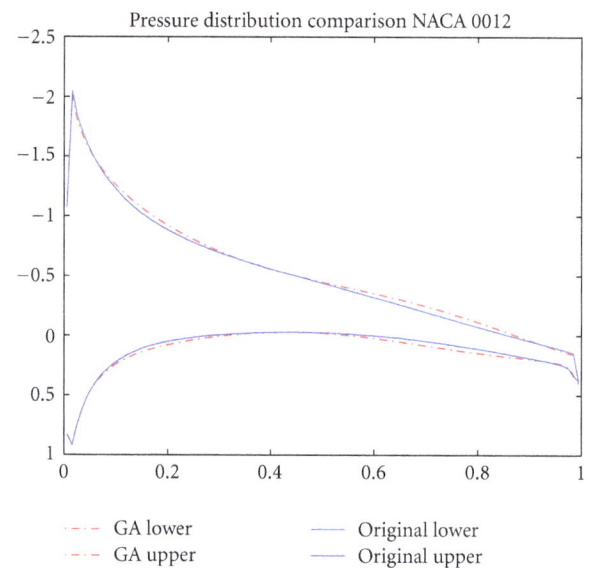

FIGURE 6: Comparison of pressure distribution over the surface of original NACA 0012 airfoil and optimized airfoil using GA.

history for the SA to be good for problems involving highly nonlinear functions where the function has large number of peaks and valleys. It is again witnessed from the obtained results that the SA has not got stuck with the local optima or extrema. The comparison between the original NACA 0012 airfoil geometry and the optimized airfoil geometry using SA is indicated in Figure 3. The comparison of pressure distribution over the surface of the original NACA 0012 airfoil and the optimized airfoil using SA is shown in Figure 4. It can be seen from these figures that the actual airfoil geometry is modified in such a way that the airflow is highly accelerated in the upper surface of the optimized airfoil as compared to the actual airfoil. From this, it can be

clearly understood that the increase in the lift coefficient is caused by the pressure variation in the upper surface of the optimized airfoil. Figure 5 shows the comparison between the original NACA 0012 airfoil geometry and the optimized airfoil geometry found by GA. The comparison of pressure distribution over the surface of the original NACA 0012 airfoil and the optimized airfoil found by GA is given in Figure 6.

The comparison of geometry and its corresponding pressure distribution between the optimum designs which are found by both SA and GA is depicted in Figures 7 and 8 respectively. It can be clearly seen that the variation of the geometry found by the GA is quite less compared to the SA,

TABLE 2: Optimized PARSEC parameters.

Parameter	Original value	Optimized value using SA	Optimized value using GA
(R_{leu}) upper leading edge radius	0.0155	0.0140	0.0145
(R_{lel}) lower leading edge radius	0.0155	0.0152	0.0160
(X_{up}) position of upper crest	0.296632	0.2500	0.2900
(Y_{up}) upper crest point	0.060015	0.0605	0.0610
(YXX_{up}) upper crest curvature	−0.4515	−0.4600	−0.4480
(X_{lo}) position of lower crest	0.296632	0.2900	0.3100
(Y_{lo}) lower crest point	−0.06055	−0.0590	−0.0590
(YXX_{lo}) lower crest curvature	0.453	0.4588	0.4599
(T_{TE}) trailing edge thickness	0	0	0
(T_{off}) trailing edge offset	0.001260	0.0011	0.0012
(α_{TE}) trailing edge direction angle	0	0	0
(β_{TE}) trailing edge wedge angle	7.36	7.300	7.2484

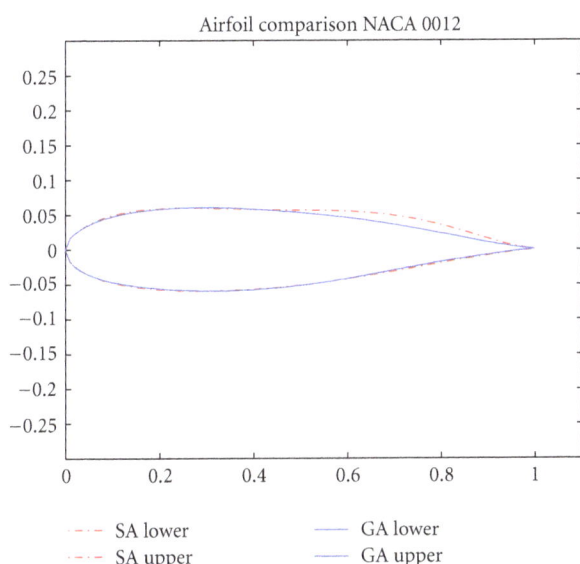

FIGURE 7: Comparison of optimized airfoil using both GA and SA.

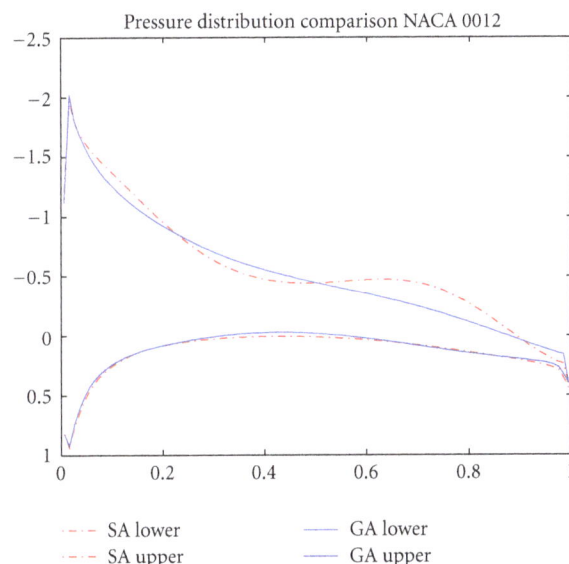

FIGURE 8: Comparison of pressure distribution over the surface of optimized airfoil using both GA and SA.

though the same design space is given to them to be explored. It can also be noticed that the geometry found by SA has more negative pressure at the upper surface which is one of most important requirements for an efficient aerodynamic design. The optimized values of PARSEC parameters which are found by both GA and SA and their corresponding coefficient of lift values are tabulated and compared with the actual values in Tables 2 and 3, respectively. It can be clearly seen that airfoil geometry which is found by SA has more coefficient of lift as compared to the airfoil geometry which is found by GA.

8. Conclusion

A problem of optimizing the actual NACA 0012 airfoil geometry for the previously discussed flow and geometrical conditions are formulated and solved using two optimization schemes, simulated annealing and genetic algorithm. The optimized airfoil geometries have an improved coefficient of

TABLE 3: Original versus optimized coefficient of lift.

Angle of attack	$Cl_{original}$	$Cl_{optimized\ using\ SA}$	$Cl_{optimized\ using\ GA}$
5.0	0.55	0.69429	0.62571

lift of 0.6942 (SA) and 0.6257 (GA) as compared to the actual NACA0012 airfoil geometry which has 0.55 at 5.0 deg angle of attack. The PARSEC parametrisation scheme is used to express the shape of the airfoil. The result shows that the PARSEC parameters show proper control over the aerodynamic performance of the airfoil by effectively controlling the aerodynamic shape of the airfoil. The PARSEC approach eases the way of understanding the impact of individual geometrical parameters on the aerodynamic properties of the airfoil. It is once again witnessed that the panel method gives reasonably accurate results without being computationally expensive. It is concluded from the results that the SA algorithm is so effective in finding the best solution among many possible solutions within a search space as compared to the GA optimization scheme in the current formulated problem. During the optimization process, plenty of airfoil data is obtained. It can be effectively used for the airfoil design by making use of these data for constructing mathematical models. The constructed mathematical models can be suitably applied to new design studies of innovative configurations.

References

[1] P. Castonguay and S. K. Nadarajah, "Effect of shape parameterization on aerodynamic shape optimization," in *Proceedings of the 45th Aerospace Sciences Meeting and Exhibit (AIAA '07)*, pp. 561–580, January 2007.

[2] R. Balu and V. Ashok, Airfoil shape optimization using paras-3D software and genetic algorithm VSSC/ARD/TR/095/2006, Vikram Sarabhai Space Centre, Kerala, India, 2006.

[3] G. S. Avinash and S. A. Lal, *Inverse Design of Airfoil Using Vortex Element Method*, Department of Mechanical Engineering, College of Engineering, Thiruvananthapuram, Kerala, India, 2010.

[4] R. Balu and U. Selvakumar, "Optimum hierarchical Bezier parameterization of arbitrary curves and surfaces," in *Proceedings of the 11th Annual CFD Symposium*, pp. 46–48, Indian Institute of Science, Bangalore, India, August 2009.

[5] H. Sobieczky, *Parametric Airfoils and Wings*, vol. 68 of *Notes on Numerical Fluid Mechanics*, Vieweg, 1998.

[6] J. L. Hess, "Panel methods in computational fluid dynamics," *Annual Review of Fluid Mechanics*, vol. 22, no. 1, pp. 255–274, 1990.

[7] J. Katz and A. Plotkin, *Low-Speed Aerodynamics from Wing Theory to Panel Methods*, McGraw-Hill, New York, NY, USA, 1991.

[8] B. Behzadi, C. Ghotbi, and A. Galindo, "Application of the simplex simulated annealing technique to nonlinear parameter optimization for the SAFT-VR equation of state," *Chemical Engineering Science*, vol. 60, no. 23, pp. 6607–6621, 2005.

[9] Margarida, F. Cardoso, R. L. Salcedo, and S. F. De Azevedo, "The simplex-simulated annealing approach to continuous non-linear optimization," *Computers and Chemical Engineering*, vol. 20, no. 9, pp. 1065–1080, 1996.

[10] N. Metropolis, A. W. Rosenbluth, M. N. Rosenbluth, A. H. Teller, and E. Teller, "Equation of state calculations by fast computing machines," *The Journal of Chemical Physics*, vol. 21, no. 6, pp. 1087–1092, 1953.

[11] S. Kirkpatrick, C. D. Gelatt, and M. P. Vecchi, "Optimization by simulated annealing," *Science*, vol. 220, no. 4598, pp. 671–680, 1983.

[12] D. E. Goldberg, *Genetic Algorithms in Search, Optimization and Machine Learning*, Addison-Wesley, Reading, Mass, USA, 1989.

[13] R. Mukesh and U. Selvakumar, "Aerodynamic Shape Optimization using Computer Mapping of Natural Evolution Process," in *Proceedings of the International Conference on Mechanical and Aerospace Engineering at University of Electronics Science and Technology of China*, vol. 5, pp. 367–371, April 2010.

Permissions

The contributors of this book come from diverse backgrounds, making this book a truly international effort. This book will bring forth new frontiers with its revolutionizing research information and detailed analysis of the nascent developments around the world.

We would like to thank all the contributing authors for lending their expertise to make the book truly unique. They have played a crucial role in the development of this book. Without their invaluable contributions this book wouldn't have been possible. They have made vital efforts to compile up to date information on the varied aspects of this subject to make this book a valuable addition to the collection of many professionals and students.

This book was conceptualized with the vision of imparting up-to-date information and advanced data in this field. To ensure the same, a matchless editorial board was set up. Every individual on the board went through rigorous rounds of assessment to prove their worth. After which they invested a large part of their time researching and compiling the most relevant data for our readers. Conferences and sessions were held from time to time between the editorial board and the contributing authors to present the data in the most comprehensible form. The editorial team has worked tirelessly to provide valuable and valid information to help people across the globe.

Every chapter published in this book has been scrutinized by our experts. Their significance has been extensively debated. The topics covered herein carry significant findings which will fuel the growth of the discipline. They may even be implemented as practical applications or may be referred to as a beginning point for another development. Chapters in this book were first published by Hindawi Publishing Corporation; hereby published with permission under the Creative Commons Attribution License or equivalent.

The editorial board has been involved in producing this book since its inception. They have spent rigorous hours researching and exploring the diverse topics which have resulted in the successful publishing of this book. They have passed on their knowledge of decades through this book. To expedite this challenging task, the publisher supported the team at every step. A small team of assistant editors was also appointed to further simplify the editing procedure and attain best results for the readers.

Our editorial team has been hand-picked from every corner of the world. Their multi-ethnicity adds dynamic inputs to the discussions which result in innovative outcomes. These outcomes are then further discussed with the researchers and contributors who give their valuable feedback and opinion regarding the same. The feedback is then collaborated with the researches and they are edited in a comprehensive manner to aid the understanding of the subject.

Apart from the editorial board, the designing team has also invested a significant amount of their time in understanding the subject and creating the most relevant covers. They scrutinized every image to scout for the most suitable representation of the subject and create an appropriate cover for the book.

The publishing team has been involved in this book since its early stages. They were actively engaged in every process, be it collecting the data, connecting with the contributors or procuring relevant information. The team has been an ardent support to the editorial, designing and production team. Their endless efforts to recruit the best for this project, has resulted in the accomplishment of this book. They are a veteran in the field of academics and their pool of knowledge is as vast as their experience in printing. Their expertise and guidance has proved useful at every step. Their uncompromising quality standards have made this book an exceptional effort. Their encouragement from time to time has been an inspiration for everyone.

The publisher and the editorial board hope that this book will prove to be a valuable piece of knowledge for researchers, students, practitioners and scholars across the globe.

List of Contributors

Parham Azimi, Mohammad Reza Ghanbari and Hasan Mohammadi
Department of Industrial and Mechanical Engineenng, Qazvin Branch, Islamic Azad University, Barajin, Daneshgah St., Nokhbegan Blvd., P.O. Box 34185141, Qazvin, Iran

Quamrul H. Mazumder
Mechanical Engineering, University of Michigan-Flint, Flint, MI 48502, USA

Mohammad Pirani, Hassan Basirat Tabrizi and Ali Farshad
Department of Mechanical Engineering, Amirkabir University of Technology, P.O. BOX 15875-4413, Tehran 159163411, Iran

Tahar Hassaine Daouadji and Abdelaziz Hadj Henni
Departement of Civil Engineering, Ibn Khaldoun University of Tiaret, BP 78 Zaaroura, 14000 Tiaret, Algeria
Laboratoire des Mat´eriaux et Hydrologie, Universit´e de Sidi Bel Abbes, BP 89 Cit´e Ben M'hidi, 22000 Sidi Bel Abbes, Algeria

Abdelouahed Tounsi and Adda Bedia El Abbes
Laboratoire des Mat´eriaux et Hydrologie, Universit´e de Sidi Bel Abbes, BP 89 Cit´e Ben M'hidi, 22000 Sidi Bel Abbes, Algeria

Hamidreza Allahbakhsh and Ali Dadrasi
Mechanical Department, Islamic Azad University, Shahrood Branch, Shahrood, Iran

Jitendra Mohan
Department of Electronics and Communications, Jaypee Institute of Information Technology, Noida 201304, India

Sudhanshu Maheshwari
Department of Electronics Engineering, Z. H. College of Engineering and Technology, Aligarh Muslim University, Aligarh 202002, India

Parham Azimi
Faculty of Industrial and Mechanical Engineering, Islamic Azad University of Qazvin, Daneshgah St., Nokhbegan Blvd., P.O. Box 34185141, Qazvin, Iran

Hamid Reza Charmchi
Sales Department, Iran Khodro Industrial Group, Tehran, Iran

Wei Wu
Flight Control and Navigation Group, Rockwell Collins, Warrenton, VA 20187, USA

Nurilla Avazov and Matthias Pätzold
Faculty of Engineering and Science, University of Agder, P.O. Box 509, 4898 Grimstad, Norway

Wen Zhang and Baolun Yuan
College of Opto-Electronic Science and Technology, National University of Defense Technology, Changsha 410073, China

Mounir Ghogho
School of Electronic and Electrical Engineering, University of Leeds, Leeds LS2 9JT, UK
International University of Rabat, Rabat 11 100, Morocco

Sirod Sirisup
Large-Scale Simulation Research Laboratory, National Electronics and Computer Technology Center, Prathum Thani 12120, Thailand

Montri Maleewong
Department of Mathematics, Faculty of Science, Kasetsart University, Bangkok 10900, Thailand
Centre of Excellence in Mathematics, CHE, Si Ayutthaya Road, Bangkok 10400, Thailand

G. Solís-Perales
Departamento de Electr´onica, CUCEI Universidad de Guadalajara, Avenida Revolucion No. 1500, 44430 Guadalajara, JAL, Mexico

R. Peón-Escalante
Facultad de Ingenier´ıa, Universidad Aut´onoma de Yucat´an, Avenida Industrias no Contaminantes, Apdo. Postal 150 Cordemex, M´erida, Yucat´an, Mexico

Tomáš Balogh and Martin Medvecký
Institute of Telecommunications, Faculty of Electrical Engineering and Information Technology, Slovak University of Technology, Ilkovicova 3, 812 19 Bratislava, Slovakia

Pénélope Leyland and Angelo Casagrande
EPFL STI GR-SCI-IAG, Station 9, 1015 Lausanne, Switzerland

Yannick Savoy
APCO Technologies, Chemin de Champex 10, CH-1860 Aigle, Switzerland

S. Didouh, M. Abri and F. T. Bendimerad
Telecommunications Laboratory, Faculty of Technology, Abou-Bekr Belkaid University, 13000 Tlemcen, Algeria

Anders G. Andersson, Elianne M. Lindmark, Patrik Andreasson and T. Staffan Lundstr¨om
Division of Fluid Mechanics, Lule°a University of Technology, 971 87 Lule°a, Sweden

Dan-Erik Lindberg, Kjell Leonardsson and Hans Lundqvist
Department of Wildlife, Fish and Environmental Studies, Swedish University of Agricultural Sciences, 901 83 Ume°a, Sweden

Mohammad Asif Zaman and Md. Abdul Matin
Department of Electrical and Electronic Engineering, Bangladesh University of Engineering and Technology, Dhaka 1000, Bangladesh

Constantin Marta, Ioan Ruja, Cinca Ionel Lupinca and Monica Rosu
Department of Engineering, University "Eftimie Murgu" of Resita, Piat¸a Traian Vuia, nr.1-4, 320085 Resita, Jud., Caras-Severin, Romania

Jarosław Majchrzak and Grzegorz Wiczyński
Pozna´n University of Technology, Pozna´n, Poland

R. M. Hudzari, M. A. H. A. Ssomad, R. Syazili and M. Z. M. Fauzan
Department of Agriculture Science, Faculty of Agriculture and Biotechnology, Sultan Zainal Abidin University, 21300 Terengganu, Malaysia

R. Mukesh and K. Lingadurai
Department of Mechanical Engineering, Anna University, Tamil Nadu, Dindigul 624622, India

U. Selvakumar
Department of Information Technology, IBBT, Ghent University, 9050 Ghent, Belgium

www.ingramcontent.com/pod-product-compliance
Lightning Source LLC
Chambersburg PA
CBHW080659200326
41458CB00013B/4919

* 9 7 8 1 6 3 2 4 0 2 1 2 7 *